AF388904

Advances in Internal Combustion Engines and Motor Vehicles

Advances in Internal Combustion Engines and Motor Vehicles

Editors

Dziubak Tadeusz
Karczewski Mirosław
Wróblewski Piotr

Basel • Beijing • Wuhan • Barcelona • Belgrade • Novi Sad • Cluj • Manchester

Editors
Dziubak Tadeusz
Faculty of Mechanical
Engineering
Military University of
Technology
Warsaw, Poland

Karczewski Mirosław
Faculty of Mechanical
Engineering
Military University of
Technology
Warsaw, Poland

Wróblewski Piotr
Faculty of Engineering
University of Technology and
Economics H. Chodkowska
Warsaw, Poland

Editorial Office
MDPI
St. Alban-Anlage 66
4052 Basel, Switzerland

This is a reprint of articles from the Special Issue published online in the open access journal *Energies* (ISSN 1996-1073) (available at: https://www.mdpi.com/journal/energies/special_issues/Internal_Combustion_Engines_Motor_Vehicles).

For citation purposes, cite each article independently as indicated on the article page online and as indicated below:

Lastname, A.A.; Lastname, B.B. Article Title. *Journal Name* **Year**, *Volume Number*, Page Range.

ISBN 978-3-0365-8646-5 (Hbk)
ISBN 978-3-0365-8647-2 (PDF)
doi.org/10.3390/books978-3-0365-8647-2

Cover image courtesy of Dziubak Tadeusz, Karczewski Mirosław and Wróblewski Piotr

Contents

Preface

This Special Issue, entitled "Advances in Internal Combustion Engines and Motor Vehicles", presents issues related to improving the design and optimizing the operation of modern internal combustion engines, particularly air and fuel supply systems, piston–ring–cylinder friction pairs, and research into their operation with regard to specific special applications, e.g., in military vehicles.

The aim of this Special Issue was to present the possibilities of reducing the emission of harmful exhaust components and greenhouse gases using compression–ignition engines in operation today, and to assess the influence of flow resistance in the air supply system on changing the emission of exhaust components and useful parameters of heavy-duty vehicle internal combustion engines. In addition, the knowledge in the area of the influence of the use of alternative fuels in bi-fueled compression-ignition (CI) engines and of air-filter pressure drop on engine performance and exhaust emissions was significantly expanded.

The papers show that the internal combustion engine, as a source of propulsion for motor vehicles, work machinery and military vehicles, still plays a significant role in land transport. The combustion processes of hydrocarbon fuels taking place in the engine, in addition to the primary function of converting chemical energy into mechanical energy, will have to be carried out in a way in which climate and environmental protection is the most important aspect. Increasingly stringent regulations on motor vehicle emissions, together with the depletion of fossil fuels, emphasize the importance of finding and developing renewable and alternative fuels. In recent years, potential sources of alternative fuels have been considered, namely liquefied natural gas, compressed natural gas, biogas, biodiesel, vegetable oils and e-fuels, which are replacing fossil fuels due to their advantages such as availability, biodegradability and renewability.

It was pointed out that technological solutions that improve the overall energy efficiency of the powertrain are an important factor in reducing CO_2 and other engine exhaust emissions. These include, for example, hybrid concepts—allowing benefits to be derived from the synergy of internal combustion and electric engines—and the idea of 'downsizing', in which the efficiency of internal combustion engines is increased by means of instrumentation. In addition to the technological solutions used today to minimize fuel consumption and CO_2 emissions, designers are looking for other methods for both petrol and diesel engines, for example, controlling the combustion process using in-cylinder pressure sensors, controlling the filling process by changing valve lift and camshaft phases, improving powertrain efficiency by recovering energy from exhaust heat, or reducing the pressure drop in the intake system by using composite filter materials.

This Special Issue can be of invaluable help to those involved in the design and operation of fuel and air supply systems for engines with CI, and especially to employees of scientific institutes and students in the area of alternative fuel use and engine research.

Dziubak Tadeusz, Karczewski Mirosław, and Wróblewski Piotr
Editors

Review

A Study on the Effect of Inlet Air Pollution on the Engine Component Wear and Operation

Tadeusz Dziubak * and Sebastian Dominik Dziubak

Faculty of Mechanical Engineering, Military University of Technology, 2 gen, Sylwestra Kaliskiego St., 00-908 Warsaw, Poland; sebastian.dziubak@wat.edu.pl
* Correspondence: tadeusz.dziubak@wat.edu.pl; Tel.: +48-261-837-121

Abstract: This paper systematically reviews the research progress in the field of the influence of air pollutants in the engine inlet on the accelerated wear of the elements of the association: piston, piston rings, cylinder liner (P-PR-CL), and plain bearing (journal–panel). It was shown at the outset that the primary component of air pollution is road dust. Its main components are dust grains of hard minerals (SiO_2, Al_2O_3), which penetrate the oil film area between two frictionally mating surfaces causing their abrasive wear. Therefore, the effect of three dust parameters (grain size and hardness, and dust concentration in air) on the accelerated wear of the friction pair: piston, piston rings, cylinder liner(P-PR-CL), and plain bearing (journal–pan) is presented extensively. It was noted that the wear values of the same component were obtained by different researchers using different testing techniques and evaluated by different indices. It has been shown that the greatest wear of two frictionally cooperating surfaces is caused by dust grains with sizes equal to the thickness of the oil film at a given moment, which in typical combustion engine associations assumes varied and variable values in the range of 0–50 μm. The oil film thickness between the upper ring and the cylinder liner varies and depends on the crankshaft rotation angle, engine speed and load, and oil viscosity, and takes values less than 10 μm. It was shown that the maximum wear of the cylinder liner, resulting from the cooperation with the piston rings, occurs in the top dead centre (TDC) area and results from unfavorable (high temperature, low piston speed) operating conditions of these elements. From the extensive literature data cited, it follows that abrasive wear is caused by dust grains of specific dimensions, most often 5–20 μm, the greater the wear the greater the hardness of the grains and the sulfur content of the fuel. At the same time, it was shown that the main bearing, crankshaft bearing, and oil ring experienced maximum wear by a different range of particle size, respectively: 20–40, 5–10, and 20–80 μm. It was shown that the mass of dust that enters the engine cylinders and thus the wear of the components is determined by the concentration of dust, the value of which is definitely reduced by the air filter. However, it was pointed out that the low initial filtration efficiency and the presence of large dust grains in the purified air in the initial period of the filter operation (after replacement of the filter element with a new one) may have an impact on the accelerated wear of mainly (P-PR-CL) association. The next stage of the paper presents the effects of excessive wear of the cylinder liner and piston rings of the engine, resulting from actual vehicle operation and bench tests on the decrease in compression pressure and engine power, increase in the intensity of exhaust gas blow-by into the oil sump and increase in oil consumption and exhaust gas toxicity. This paper addresses the current problem of the effect of engine inlet air contaminants on the performance of the air flow meter, which is an essential sensor of the modern internal combustion engine. The phenomenon of deposition of contaminants (mineral dust, salt, carbon deposit, and moisture) on the measuring element (wire or layer anemometer) of the air flow meter has been analyzed. The empirical results presented show that the mineral dust layer on the measuring element of the air flow meter causes a 17.9% reduction in output voltage, and the dust and oil layer causes a 46.7% reduction in output voltage. This affects the decrease in engine power and exhaust toxicity.

Keywords: mechanical engineering; internal combustion engines; air pollution; engine component wear; power loss

Citation: Dziubak, T.; Dziubak, S.D. A Study on the Effect of Inlet Air Pollution on the Engine Component Wear and Operation. *Energies* **2022**, *15*, 1182. https://doi.org/10.3390/en15031182

Academic Editor: Flavio Caresana

Received: 13 January 2022
Accepted: 2 February 2022
Published: 5 February 2022

Publisher's Note: MDPI stays neutral with regard to jurisdictional claims in published maps and institutional affiliations.

1. Introduction

In recent decades, dust pollution has been widely recognized as a serious environmental and engineering problem affecting the reliability of mechanisms and machines from nano to macro scale [1]. No industrial sector is immune to this problem, mainly affecting equipment that requires clean environments for proper operation and as intended. Examples of engineering applications where cleanliness requirements are high are mainly all lubricated machine components in relative motion (bearings, gears, axles, motor pistons, dynamic seals, etc.), miniature devices such as microelectromechanical systems, valves, mechanisms, and machines in the food, medical and pharmaceutical, aerospace, and electronics industries.

Contaminants found in machine associations come from a variety of sources. These can be residues from manufacturing and assembly processes or created by abrasive wear on machine components. Contaminants can be introduced from the environment or during equipment maintenance.

The problem of solid contaminants has arisen due to increased loads and tighter tolerances, as well as the miniaturization of equipment. Loads and tolerances mean that the lubricating film in contacts such as the ball-race in a rolling bearing, the pivot-cup association of a plain bearing, and the piston–ring–cylinder system has generally been reduced, making the mating surfaces more susceptible to increasingly smaller particles. Miniaturization means that small particles now have a greater impact because the ratio of particle size to the size of the parts involved has increased.

Modern engineering practice and research has shown the deleterious effects of particles as small as 0.1–100 μm in contact with typical oil films on the order of a few nanometers to a few micrometers [1]. Each time a particle is compressed into narrow gaps, it inevitably creates local surface stresses that are elastic (reversible) at best and plastic (permanent) at worst. Any surface damage depends on the size, hardness, and fracture toughness of the particle, the hardness of the surface, the coefficient of friction at the "particle-surface" interface, the relative velocity of the surface, and the type of contact, i.e., rolling, sliding, spinning, or a combination thereof. Thus, various failure modes such as denting, abrasion, thermal damage, peeling, chipping, and even scuffing have been recognized.

The problem of the durability of internal combustion engines is closely related to the purity of the air drawn into the engine from the environment. Internal combustion engines of motor vehicles are particularly affected by particulate pollutants that are found in the air drawn from the atmosphere which is the primary operating medium of the engine. The engine needs at least 14.5 kg of air to burn 1 kg of fuel. When operating at rated conditions, passenger car engines draw about 150–400 m^3/h of air and truck engines in the range of 900–1400 m^3/h. A T-72 tracked vehicle engine needs more than 3500 m^3/h of ambient air to operate.

There is a large variety of solid and gaseous pollutants in the ambient air which are drawn in with the air by internal combustion engines. Along with fuel and oil, pollutants also enter the engine cylinders, but their quantity is much smaller. The concentration and type of pollutants in the air depends on many factors, such as the type and condition of the ground, the presence of vegetation, the type of industrial plants, the activities of the automobile, forest and landfill fires, and volcanic eruptions.

The sources of pollutants emitted into the atmosphere (dusts and gases) are many and varied. In general, they can be divided into natural and anthropogenic (artificial). Pollutants emitted to the atmosphere (dust and gases) can be both primary, emitted directly to the atmosphere, and secondary, formed in the atmosphere as a result of chemical reactions.

Natural pollutants result from the activities of nature and include:

- Volcanic eruptions, which produce volcanic ash and the gases SO_2, CO_2, H_2S, among others;
- Forest, savannah, and steppe fires (CO_2, CO, and dust emissions);
- Dump fires;
- Marine aerosols, and material of plant and animal origin;
- Marshes that give off gases, for example: CH_4, CO_2, NH_3, H_2S;

- Mineral dust carried from the ground by motor vehicle traffic, activity of working and agricultural machines;
- Erosive soils and rocks, mineral dust carried from the ground by the wind–sandstorms;
- Biological contamination: microorganisms (mites, fungi, bacteria), plant pollens, parasites;
- Green areas from which pollen comes.

Artificial pollutants result from human activities and include:

- Energy industry—burning of fuels in heating plants and power plants;
- Mining industry, metallurgical industry, chemical industry, construction industry (production of cement) emitting to the atmosphere large amounts of dusts and harmful gases (NH_3, H_2S, HCl) and heavy metals;
- Municipal and household sector (households), collection and disposal of waste and sewage, e.g., landfills, sewage treatment plants;
- Motorization—mainly road transport, but also water and air transport being the source of atmospheric emissions of gaseous pollutants originating from the process of combustion of engine fuels: CH, CO, NO_x, SO_3, SO_2, PM2.5, and PM10 particulates and chemical compounds, as well as dusts, from abrasion of tires and road surfaces, brake pads, and discs and clutch discs.

In the overall stream of pollutants that are introduced into the atmosphere surrounding the earth, 85% are pollutants of natural origin. These are dusts and harmful gases.

The users of vehicles are interested in these pollutants that, when they get inside the engines together with the air, have a destructive impact on engine parts and assemblies, causing their accelerated wear, which results in the reduction of effective parameters of engine operation and in their reduced durability and reliability. These are mineral dusts found in the air mainly during the operation of vehicles on unpaved roads and off-roads. Mineral dust is one of the dominant constituents of atmospheric aerosols and accounts for ~60% of the dry aerosol mass globally [2]. This applies mainly to special-purpose vehicles, work machines, and military vehicles equipped with high-powered internal combustion engines and high air demand.

Passenger cars and trucks travel mainly on paved roads, whose surfaces are the place of deposition of particles from many different sources. First and foremost is mineral dust that has been carried by the wind from soils surrounding the road, generated from field work or road construction activities, and from sandy and desert areas that cover about one-third of the earth's surface [3]. Large amounts of dust are generated by open pit coal mine operations [4–6].

The road surface may contain salt used for de-icing roads, plant fragments, pollen, animal hair and mold, and other biological materials carried from nearby locations [7,8]. The author of this paper [9] conducted an analysis of the contaminants retained on the filter. The contaminant particles varied greatly in size, shape, and chemical composition. The contaminants with the largest sizes were biological material visible to the naked eye. These included small insects such as beetles, wasps, bees, and flies, as well as their legs, feelers, and wings. Plant remains, represented by fragments of wood, seeds, and leaves, were also observed, though less frequently. Inorganic particles and synthetic fibers were also found. Analysis of dust particles retained on the filter showed that the most abundant elements found in large irregular particles were C, O, Si, Al, Ca, Mg, Na, K, Ti, Fe, and Si [9].

A significant amount of particulate pollution is generated by anthropogenic activities such as agriculture, industrial production. Pollutants from anthropogenic sources are also particles that have been released into the atmosphere as a result of abrasion processes in the friction pairs of brakes and clutches of cars [10–13] and as a result of abrasion of the tire tread against the road surface, which then fall by gravity [14–18]. It has been shown in [17] that under actual operating conditions, the mass of a car and the intensity of braking have a significant effect on the emission of PM2.5 and PM10 particles from brake and tire wear. The highest concentrations of PM2.5 (520–4280 $\mu g/m^3$) and PM10 (950–8420 $\mu g/m^3$) particles from brake wear were observed for the vehicle with the highest mass, while the lowest peak concentrations of PM2.5 (250–2440 $\mu g/m^3$) and PM10 (430–3890 $\mu g/m^3$) were observed for the vehicle with the lowest mass. Similarly, the highest PM2.5 (340–4750 $\mu g/m^3$) and PM10 (810–8290 $\mu g/m^3$) concentrations were from the tires of the highest mass vehicles. As expected, vehicles with lower mass emitted much lower PM2.5 (340–4750 $\mu g/m^3$) and PM10 (810–8290 $\mu g/m^3$) concentrations from tire wear.

The internal combustion engines of automobiles are the source of emissions of a significant amount of exhaust gases, which include non-toxic (N, O, H_2O, CO_2, H) and toxic (CO, NO_x, CH, SO_2, RCHO, 3,4-benzopyrene) components [19,20]. Internal combustion engines also emit PM (particulate matter), which are formed by complex chemical and physical processes occurring during fuel combustion in the engine cylinder and exhaust [19–22]. Particulate matter (PM) is soot which is a by-product of incomplete and incomplete combustion of fuel, on the surface of which there are: unburned hydrocarbons from fuel and lubricating oil, water vapor, sulfur and nitrogen compounds, and ash, as well as products of frictional wear of metal elements from frictionally cooperating engine joints [23].

Road surfaces are primarily a deposition site for mineral dust that has been carried by the wind from soils surrounding the road, generated by field work, or from road construction activities. Falling atmospheric dust has a complex mineral–chemical composition. The diameter of these particles is greater than 10 μm and less than 100 μm [24]. Its basic constituents are silica (SiO_2) and alumina (Al_2O_3), whose hardness evaluated on the Mohs 10-point scale has a value of 7 and 9, respectively. The proportion of these two components of mineral dust reaches 60–95. The residue (4–19%) consists of oxides of various metals: F_2O_3, CaO, MgO, and organic components [25,26]. The road surface also contains particles that have been released into the atmosphere as a result of abrasion processes in the friction pairs of brakes and clutches and as a result of the abrasion of the tire tread against the road surface, which then fall by gravity. Roads contain pollutants that are expelled from engines with exhaust gases, such as soot and wear products of frictionally mating engine components. The road surface may contain de-icing salt, plant fragments, pollen, animal hair and mold, and other biological materials carried from nearby locations. Contaminants on the road surface, due to vehicle traffic or by wind, are lifted up as dust, which is commonly referred to as road dust [27–29].

In the case of motor vehicles, airborne dust is hazardous and when it enters the engine, it can degrade engine performance. Silica and corundum are the cause of accelerated wear of cooperating and key engine components, such as piston rings and cylinder liners, crankshaft journals and bushings, valves, and guides. The minimum service life of these parts determines the life of the engine.

Impurities which are sucked with air into engine cylinders have different chemical and granulometric composition and moreover their content in air (dustiness) is variable and depends on many factors: terrain conditions, type of road, traffic conditions (traffic volume, traffic speed, traffic of other vehicles, share of studded tires and exhaust emissions), meteorological conditions (types of precipitation and their frequency, wind direction, and strength), road surface condition (dry, wet, covered with ice, sprinkled with salt or sand), vehicle design and its running gear (wheeled, tracked), and location and design of the air intake to the engine.

Dust concentration values in the air vary within wide limits, from 0.01 mg/m^3 near rural buildings to about 20 g/m^3 when a column of tracked vehicles is moving over desert ground [30]. During dust storms, the dust concentration takes values in the range of 0.1–3 g/m^3 [31]. According to the author of the paper [32], the dust concentration in the air can take values in the range of 0.001–10 g/m^3. The maximum values of dust concentration in the air within a few meters of the sand road on which off-road vehicles traveled was variable in a wide range of 0.05–10 g/m^3 [33]. The dust concentration on highways takes small values but in a wide range of 0.0004–0.1 g/m^3, while during the movement of a column of vehicles on a sandy terrain the dust concentration takes much larger values 0.03–8 g/m^3 [34].

Dust concentrations on streets depend on traffic volume, but typically range from 0.09 to 0.13 mg/m^3. Much higher dust concentrations, reaching up to 0.65 mg/m^3, have been measured while passing through a tunnel, where heavy traffic, high speed, and confined space created favorable conditions for high dust concentrations [35].

The author of the paper [36] states that the concentration of dust in the air behind a moving column of trucks, armored personnel carriers, and tracked vehicles depends on the type of ground, vehicle speed, wind speed, and the type of chassis. At a distance of several meters behind a column of tanks moving at 30 and 10 km/h on dry ground, the dust concentration reached a maximum value of 1.17 and 0.48 g/m^3, respectively. The authors of the paper [37] report that the dust concentration at the periphery of the armor of a tracked vehicle operated over sandy terrain increases with increasing driving speed and for 18 km/h takes values in the range of 2.1–3.8 g/m^3. At the air filter inlet, the concentration has much lower values and is 0.8–1 g/m^3.

Numerous studies show that the dust concentration at the inlet to the vehicle combustion engine intake system usually does not exceed 2.5 g/m^3 [32,33]. A dust concentration in the air of 0.6–0.7 g/m^3 is the reason for reduced visibility, and at a concentration of about 1.5 g/m^3, visibility decreases to zero [38].

The concentration of dust inside a car with open windows was about 2.5 mg/m^3 [39] when it was moving behind another car on a gravel road.

The author of the paper [31] divides the road conditions on which most cars and light trucks travel into two groups:

- Normal driving conditions (streets, highways, residential, and rural roads) with dust concentrations below 1 mg/m^3;
- Severe driving conditions (gravel roads, desert areas) with dust concentrations higher than 1 mg/m^3.

Dust, which is sucked with air into engine cylinders, only partly takes part in destructive influence on engine elements. The authors of [40] state that about 30% of the pollutants getting with the air into the engine super-piston space may get out in basically the same form with the exhaust gases from cylinders to the exhaust system. This increases the emission of particulate matter (PM) from the engine

The dust particles contained in the exhaust gas cause erosive wear to the exhaust valves and turbine blades of the supercharger as a result of the high velocity of the exhaust gas flowing through narrow valve gaps. Some of the dust particles whose melting point is much lower than the average temperature in the combustion chamber (1600K) melt [41]. Only 10–20% of the dust that enters the engine through the intake system settles on the cylinder liner walls. This part of the dust forms together with oil a kind of abrasive paste, which in contact with the surfaces of the piston–piston ring–cylinder (P-PR-CL) association causes abrasive wear. As the piston moves in the bottom dead center (BDC) direction, the rings scrape the oil from the cylinder face together with the contamination into the oil sump. The contaminants contained in the oil are distributed via the oil system to those areas of the engine that are lubricated by oil, e.g., crankshaft journal–crankcase, camshaft journal–crankcase, valve guide-valve stem, causing accelerated wear. The primary role of the piston ring pack is to maintain an effective gas seal between the combustion chamber and the crankcase. Piston rings fulfil this task by forming a labyrinth seal by fitting tightly into the grooves in the piston and the cylinder wall. The tribological effect of piston rings causes their wear. Within the framework of works on the development of internal combustion engines, it was found that the course of wear of the friction pair "ring–cylinder liner" and the distribution of the oil film is largely influenced by the flow of exhaust gases and their degree of contamination in the labyrinth seals of piston rings [42]. It has been indicated that this process is exacerbated when the maximum operating pressure in the combustion chamber is high and insignificant total cylinder volumes are used [43]. The wear of the cylinder liner and piston ring pack has a significant effect on the loss of tightness of this association and consequently on the performance of internal combustion engines due to the loss of compressed medium and loss of power, increase in fuel and oil consumption [44–46]. There is also an increase in exhaust gas blow-by into the crankcase, which causes an increase in the temperature of the mating parts and the lubricant, which reduces its viscosity. There is then a high susceptibility to oil film discontinuity between the kinematic pairs of the main engine mechanism [47]. The second function of piston rings is to transfer heat from the piston to the cylinder walls and from there to the engine coolant. Another function of the piston ring package is to reduce the amount of oil that enters the combustion chamber from the crankcase. This flow path is a primary contributor to engine oil consumption and leads to increased emissions of toxic exhaust components as a result of oil mixing and reacting with hot exhaust gases.

The friction pair consisting of the cylinder liner and piston rings (P-PR-CL) are the basic components of the friction linkage of the engine piston–crank system, which, due to the extremely harsh operating conditions resulting from the action of high temperature (oil viscosity drop) and high pressure and high forces for a long time, are particularly susceptible to wear and have a significant impact on engine efficiency and service life [48–50]. According to the authors of works [48,51–57], depending on the type of engine and its operating conditions, 20–50% of friction losses in internal combustion engines occur in frictional contact between the piston ring assembly and the cylinder liner. Friction results in the loss of the major part (48%) of the energy produced in the engine, of which the friction losses of the piston shell, piston rings, and bearings comprise 66%, and the valve train, crankshaft, transmission, and gears account for about 34% [58]. A small reduction in friction can result in significant savings in fuel consumption, given the large number of vehicle systems operating daily in the world. There are an estimated 1.2 billion light-duty vehicles and 380 million heavy-duty vehicles in the world, and these numbers continue to grow [59].

The energy demand of the transportation sector is enormous, with daily liquid fuel requirements exceeding about 11 billion liters. Stringent emission standards have forced the automotive sector to significantly improve the tribological behavior of engine systems. Advances in the tribology of engine systems may be a more economical alternative than developing new technologies to achieve better engine performance. Analysis of energy loss in an internal combustion engine can show that about 15% of the total fuel energy is lost due to mechanical friction. A 1.5% reduction in fuel consumption can be achieved with a 10% reduction in mechanical losses. In addition to reducing fuel consumption, tribological improvements can also reduce oil consumption, reduce emissions, improve durability, and increase engine power [60].

The intensity of wear of engine elements depends on many factors, including the operating conditions of the engine (rotational speed, load), which is connected with creating the appropriate thickness of the lubrication wedge between the mating surfaces of engine elements. The decisive influence on the wear of friction mating surfaces has the parameters of dust sucked in with the air, whose values change within wide limits and depend on the conditions in which the engine is used. The amount of wear on the "cylinder-ring" friction couple is also influenced by the type and quality of fuel and the parameters of its injection into the cylinder. This is mainly manifested in the context of deposits formation on engine elements and exhaust emission. To this end, work is being done to optimize the performance and reduce exhaust emissions of engines tested on unconventional fuels [61–67].

One of the major causes of increased engine wear, especially with most engine manufacturers opting for exhaust gas recirculation (EGR) technology to reduce nitrogen oxide (NO_x) emissions [68–70] is soot contamination of engine oil [71–73]. Due to the increasing requirements to meet exhaust emissions, diesel engines and gasoline direct injection (GDI) engines are being equipped with catalytic particulate filters. This is an effective solution by which soot can be captured and burned [74,75].

The current tendency in the construction of internal combustion engines to increase specific power results in the fact that the elements of the P-PR-CL assembly of an internal combustion engine are subjected to increasingly higher mechanical and thermal loads, which promotes their faster wear. The experience from exploitation shows that the continuous increase of engine power does not remain without influence on wear, durability and reliability of the most loaded elements (cylinder liner and piston rings), which determine the durability of the whole engine.

The purpose of this paper is an extensive and structured analysis of the influence of road dust, sucked together with air through the engine intake system, on the intensity of wear of engine elements with special attention paid to the P-PR-CL piston–ring–cylinder combination. A detailed analysis of the influence of the fuel type and injection system on the formation of deposits and their impact on the wear of engine elements was intentionally omitted in the analysis. The analysis of the soot effect on engine wear was also not considered. The authors are aware of the importance of these factors in the context of engine durability. It was considered that the paper would become too voluminous and the most important problem, i.e., the influence of mineral dust on engine wear, would not be properly exposed. It was pointed out that despite the use of inlet air filtration systems, mineral dust grains with dimensions less than 2–5 µm enter the engine and are the cause of accelerated wear. Larger dust grains can only enter the engine cylinders if the filtration system fails. Modern tribological systems of internal combustion engines, due to the increasing thermal and mechanical loads on the engines, are characterized by increasingly smaller clearances and oil thickness between the friction pair surfaces. It is currently believed that mineral dust grains above 1 µm cause accelerated wear.

The paper also analyzes the phenomenon of the effect of road dust and other air pollutants (salt, oil, moisture) that settle on the measuring element of the air flow meter causing a change in its characteristics. As a result, the information transmitted to the on-board computer and the mass of air flowing is incorrect, which interferes with engine operation causing a decrease in power. Data were quoted showing the influence of contaminants contained in the engine oil, where they enter from the air in contact with the cylinder liner, on the accelerated wear of elements of these associations, which are lubricated by oil. The paper draws attention to the effects that result from accelerated wear of engine components. The influence of excessive wear of PR-CL junction on the decrease of compression pressure, engine power and on the blow-by of cargo and exhaust fumes to the crankcase is presented. Innovative technologies were presented, e.g., texturing the surfaces of piston rings and cylinder liners, applying layers that reduce friction and prevent their increased wear. Attention was drawn to the necessity of using air filters of high efficiency and accuracy, especially in the case of vehicles operating in conditions of high dust concentration in the air. The number of available literature items concerning this problem is extensive; therefore, the authors used only selected items. In the available literature, there is no study which comprehensively covers the problem of the influence of the dimensions of dust grains, their hardness, and dust concentration on the wear of the PR-CL friction pair of an internal combustion engine. The results of wear of the elements obtained during laboratory tests, bench tests on an engine dynamometer, and during real engine operation were analyzed.

2. The Effect of Dust in the Intake Air on Engine Component Wear

The basic task of the air supply system is to supply the engine with air of the appropriate quantity and quality (purity), for which the air filter is responsible. The basic filter material of motor vehicle operating fluids are fibrous materials, mainly filter paper, which is characterized by high mass efficiency of over 99.9% and high accuracy of dust grains above 2–5 µm, but low dust absorption of about 200–250 g/m^2 [76–78].

Therefore, the engine inlet air (downstream of the filter) contains dust grains with sizes below 5 µm. Dust grains with larger sizes can enter with the air into the engine cylinders only in case of failure (leakage of connections) of the inlet system or loss of filtering properties of the paper cartridge.

The effects of dust particles contained in engine intake air and oil on engine components are varied and consist of (Figure 1):

- Abrasive wear of P-PR-CL coupling elements;
- Erosive wear of compressor and turbine;
- Abrasive wear of junction elements to which lubricating oil is supplied;

- Settling of dust on the measuring element of the air flow meter;
- Settling of molten dust particles on the catalytic surface.

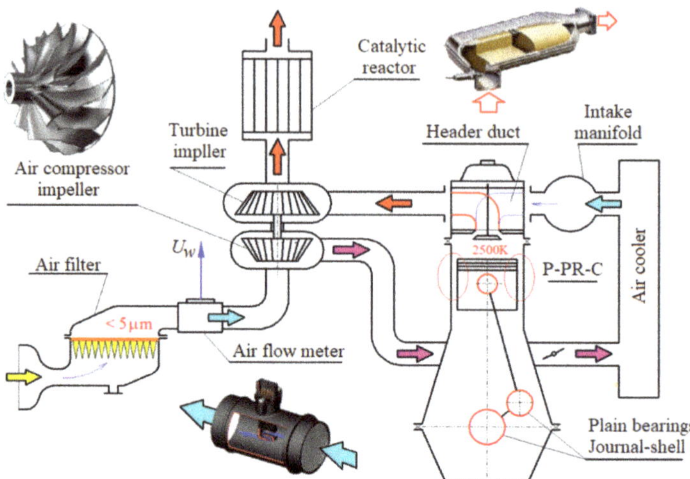

Figure 1. Functional diagram of the intake system of a motor vehicle combustion engine.

In machine operation, tribological wear processes are divided into the following processes: abrasive, adhesive, oxidative, fatigue, corrosive, and cavitation. From the point of view of filtration of exploitation fluids, the important processes are those that are caused by impurities in the fluids. This is wear caused by the abrasive action of loose solid particles of impurities. Abrasive wear is a phenomenon of destruction of the surface layer of cooperating bodies (moving in relation to each other) during friction as a result of the impact of roughness projections of one element on the surface layer of the other element. This movement is accompanied by dry or mixed friction. This type of wear consists in the fact that hard bumps or loose abrasive grains are embedded (pressed) in the surface layer of mating elements and during relative movement cause its destruction. During the operation of engines, the abrasive grains enter the mating fluids, mainly with the air. Abrasive wear is a mechanical process. The character of the action of abrasive particles on the surface layer material depends mainly on the type of their relative motion and on the character and value of forces acting in the area of contact between the particle and the machine component.

In internal combustion engine systems, there are many friction pairs that require a lubricant for proper operation, which minimizes friction and reduces their wear value. The piston–ring–cylinder (P-PR-CL) friction pair is the primary assembly that determines the wear and life of an internal combustion engine. Other frictional pairs include the main and crankshaft and camshaft plain bearings (journal–crankcase), valve stem-guide, camshaft cam-valve galleries, piston pin–crankcase, and valve seat–valve seats (Figure 2). The tribological behavior of the (P-PR-CL) assembly has long been recognized as an important factor affecting the performance of internal combustion engines in terms of fuel consumption, power loss, oil consumption, emissions of harmful exhaust components, and working fluid blow-by into the crankcase.

The piston, which is the mobile combustion chamber closure, serves to convert the chemical energy of the fuel into useful kinetic energy. The primary role of the piston rings, which are pressed against the cylinder surface by elastic force, is to seal the gas flow between the combustion chamber and the crankcase. The piston rings, which are located in the piston grooves, move along the cylinder liner in a reciprocating motion with a high, but also variable sliding speed. In addition, the piston rings move axially and radially in the piston groove, rotate in the ring groove relative to the sleeve axis and deform, causing inclination relative to the cylinder liner. This can cause changes in the oil film parameters The pressure of the rings on the surface of the sleeve is assisted by the gas pressure acting on the rear part of the ring. When the total of the forces acting on the ring along the cylinder axis changes its direction, the ring breaks away from the lower piston groove surface and moves axially to the upper piston groove surface. This results in gas being blown into the crankcase. The cyclic piston ring movements cause oil to flow over the piston crown, which is burned off [79].

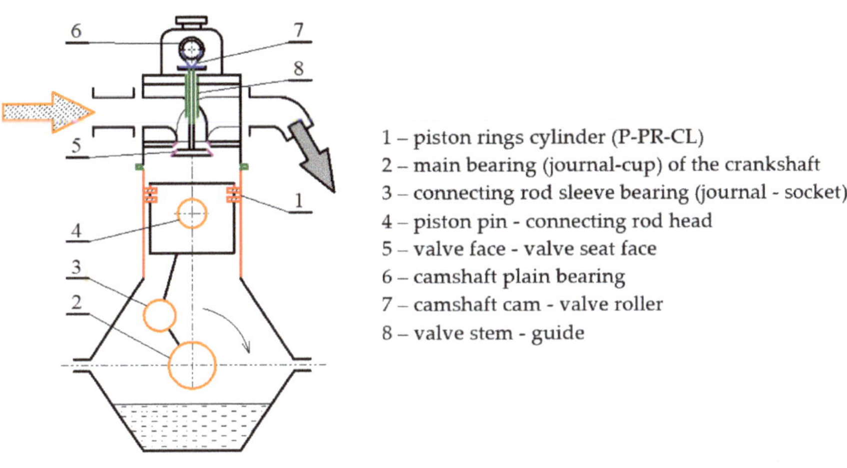

Figure 2. Friction couplings in internal combustion engine systems.

The air flowing into the combustion chamber usually makes a helical motion towards the piston crown, which is caused by a suitably shaped intake duct. The inertia force acts on the dust grains in the airstream and directs the heavier grains to the cylinder surface where they are fixed in the oil film.

In the abrasive wear of the engine component surfaces, only that part of the dust grains that penetrated the engine's superstructure with the air and settled in the oil film on the cylinder liner wall are involved. The contamination grains suspended in the oil film, which are mainly of mineral origin, get between the two mating surfaces of the piston–piston ring–cylinder (P-PR-CL) or plain bearing (journal–shell) joint and form an oil film (Figure 3). If the minimum oil film thickness h_{min} between the mating surfaces is greater than the particle size d_p, there is no contact between the dust grains and the mating surfaces and they do not cause damage to the surfaces.

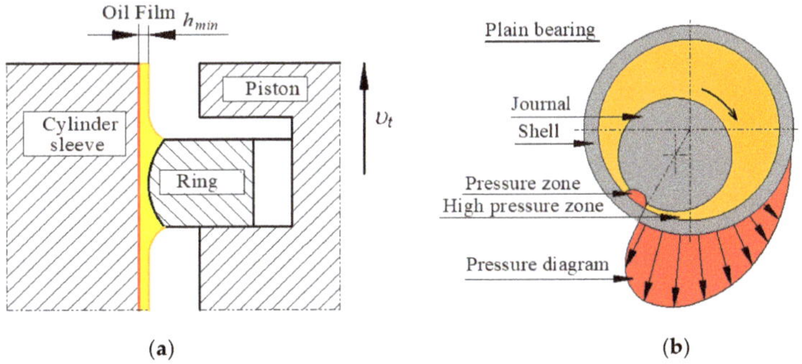

Figure 3. Oil film formation in the following combinations: (**a**) piston–piston rings–cylinder (P-PR-CL), (**b**) plain bearing: journal–shell.

In an internal combustion engine, which operates with variable speed and load, the oil film thickness h_{min} is not a constant value and depends directly proportional to the oil viscosity depending on the temperature, relative speed in the lubricated surfaces and inversely proportional to the loading force N (engine load) according to the relation [80]:

$$h_{min} = C\frac{\eta \cdot v_w}{N} \tag{1}$$

where C—a coefficient that depends on the dimensions of the association elements.

Abrasive wear occurs when, as a result of an increase in the applied force N (increase in load, increase in temperature and decrease in viscosity), two mating surfaces come into close proximity at a distance $d_p = h_{min}$ (Figure 4). Dust grains are generally irregular lumps with sharp multilateral cutting edges. When a hard particle (mineral dust grain) has gained contact with two surfaces and the action of the force N does not give way, the particle usually penetrates one of the surfaces. The dust particle so fixed acts as a cutting tool causing micro-volume cutting and deformation of the layer of the opposite surface. The dust grains can cause scratching, furrowing, and break off metal shavings that are embedded in the material of the components that make up the engine's component assembly. In addition, engine components are subject to different loads, which changes the clearance values and oil operating conditions. Most of the dynamic clearances are in the range of 0–20 μm.

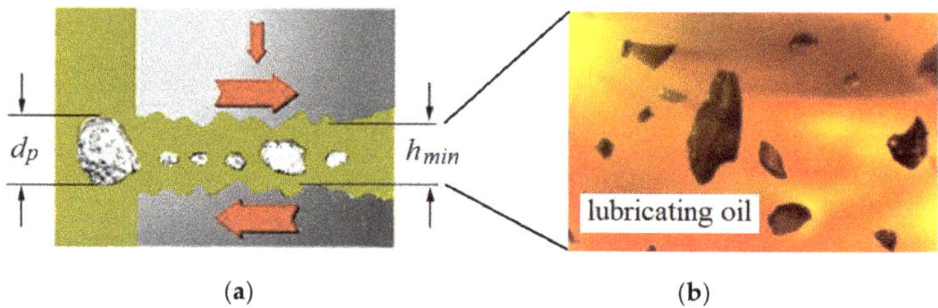

Figure 4. Effect of particulate matter on tribological pair: (**a**) impurities suspended in oil [81], (**b**) condition of maximum wear of the association. Figure made by the authors based on data from [82].

The prevailing view in the literature is that wear is caused by dust grains with size d_p equal to the minimum h_{min} thickness of the oil film between the mating surfaces at a given moment. For any other value of the h_{min}/d_p quotient, the association wear decreases [82,83]. Accurate calculation of this thickness is difficult, both in associations with reciprocating and rotating motion, because:

- The surface outline is not exactly geometric;
- The temperature and viscosity of the oil film changes;
- Unit pressures are constantly changing due to the cyclic nature of operation and varying engine load;
- The clearances in the linkages increase as the wear of the engine components increases—with the increase of its service mileage.

Between the cylinder liner and piston rings, the oil film thickness is determined mainly by the piston velocity, which varies sinusoidally (Figure 5), and oil viscosity, which depends on the temperature of mating elements. The highest value of piston (piston ring) velocity, the most critical zone where intense wear occurs, is in the top dead center (TDC) area due to a combination of various adverse factors. The change in the direction of piston movement (alternate acceleration and deceleration) in the extreme positions of the cylinder liner causes its velocity to be the lowest in this area, and zero at BDC and TDC. This leads to a reduction in oil film thickness or its complete disappearance. P-PR-CL components are subjected to the highest temperature, resulting from direct contact with the hot exhaust gas, which causes a decrease in viscosity and lubricating wedge thickness. Consequently, there may be periods of even direct metallic contact between the piston ring and the bushing, resulting in increased friction coefficient and wear [84].

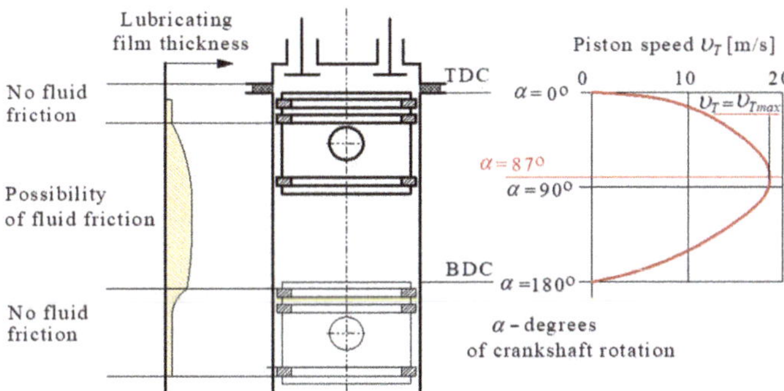

Figure 5. Effect of piston speed on lubrication wedge formation and lubrication film thickness.

The change in oil film thickness between the upper ring and cylinder liner as a function of crankshaft rotation angle for three (800, 3000, 7000 rpm) The rotational speeds of the engine operating at full load are shown in Figure 6. The oil film thickness increases its value as the engine speed increases—the reciprocating motion speed increases [85]. The oil film thickness reached its maximum value in the middle of the stroke, where the piston speed is the highest, which induces hydrodynamic lubrication. For 800; 3000 and 7000 rpm, the oil film thickness at mid-stroke (90 and 450° crank angle) varied from 0.7–1.5, 2.4–2.6, and 3.4–4 μm, respectively. At BDC and TDC, the oil film thickness reaches minimum values. These are probably the points where the piston ring comes into metallic contact with the cylinder liner (boundary lubrication). Under these conditions theoretically any particle of arbitrarily small size can be the cause of wear.

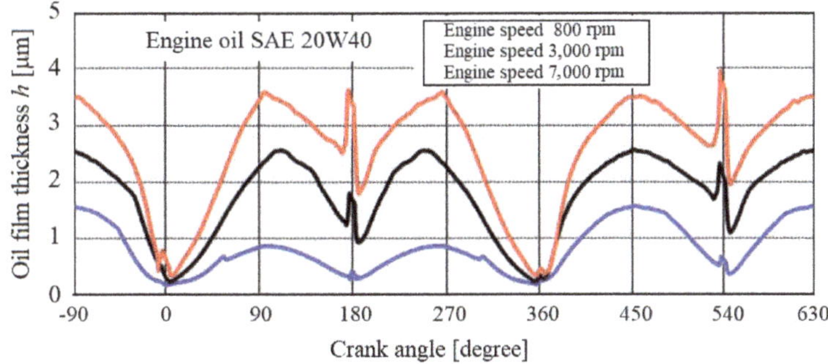

Figure 6. Effect of engine speed on oil film thickness between top ring and cylinder liner at SAE 20W40 engine oil. Figure made by the authors based on data from [85].

From the theoretical and experimental studies conducted by the authors of the paper [86], it can be seen that the minimum oil film thickness h_{min} changes its value depending on the crankshaft rotation angle (piston position in the cylinder), which is influenced by the rotational speed, engine load and oil viscosity. During the compression stroke (3000 rpm, 40 Nm load), the minimum thickness of the oil film on the sealing ring was $h_{min} = 9$ μm, and during the operating stroke—$h_{min} = 5$ μm, which is 50% less and results from the pressure of exhaust gases on the rear part of the ring.

As a result of changes in the thickness of the oil film between the mating surfaces, the dust grains found there may be crushed and fragmented. The resulting smaller grains can penetrate between the two mating surfaces where the oil film thickness takes on small values. In typical combustion engine mating, the oil film thickness reported by the authors of the paper [87] assumes varying values (Figure 7).

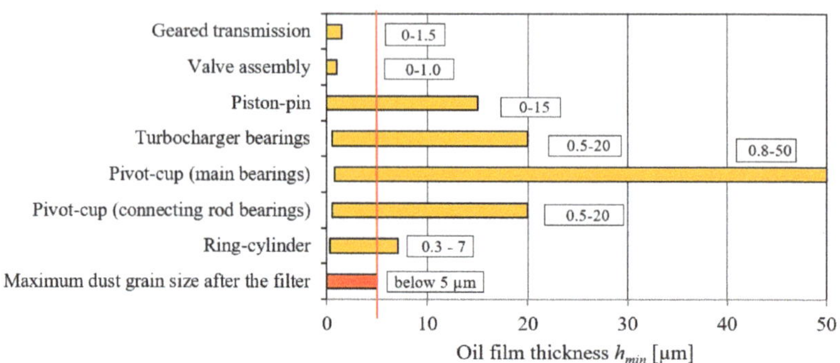

Figure 7. Oil film thicknesses in typical combustion engine associations. Figure made by the authors using data from [87].

It is believed that the abrasive aggressiveness of dust decreases when the dust grain sizes are less than 5 μm. Such a value is considered to be the upper acceptable size of dust grains that can be passed through air, fuel, and oil filters. Therefore, work is being undertaken to increase the filtration accuracy of dust grains in this range, for example by using nanofiber filter baffles [88–92].

Dust entering with the air into the engine cylinders most intensively affects the first piston ring, the piston and the upper part of the cylinder.

The influence of the parameters of the dust sucked in together with the air to the engine on the value of wear is considered in relation to the wear of these very engine elements. The value of their wear is determined mainly during experimental research on real engines. Due to great difficulties in carrying out such tests, the available literature contains few and partial results, mainly concerning the influence of three dust parameters (grain size, dust concentration in the air, and grain hardness) on the accelerated wear of the first piston ring, piston, top part of the cylinder, and main and connecting-rod journals.

The effect of particle size on engine wear is a complex process. Particles whose diameter is smaller than the thickness of the oil film should not damage the surface. However, they weaken the oil film and can cause an increase in oil density.

The magnitude of dust-induced wear of engine components depends on the properties of the dust drawn in with the air, and mainly on [93]:

- The size of the dust grains;
- Dust concentration in the air;
- Granulometric and chemical (hardness) composition of dust;
- Shape of dust grains.

At the same time, the effect of dust is closely related to:

- Clearance values between mating parts;
- Engine design parameters (compression ratio, piston stroke);
- Presence and type of lubricating oil;
- Engine operating parameters (load, rotational speed);
- Mechanical properties of the materials from which the components are made);
- Frequency of air filter maintenance.

The author of the paper [94] shows that there is a close relationship between the dimensions of SiO_2 grains in the air sucked into the engine and the amount of wear of the cylinder liner and piston. All dust grains above 1 m cause increased wear, but the greatest abrasive effect occurs at grain sizes of 7–18 μm—Figure 8a.

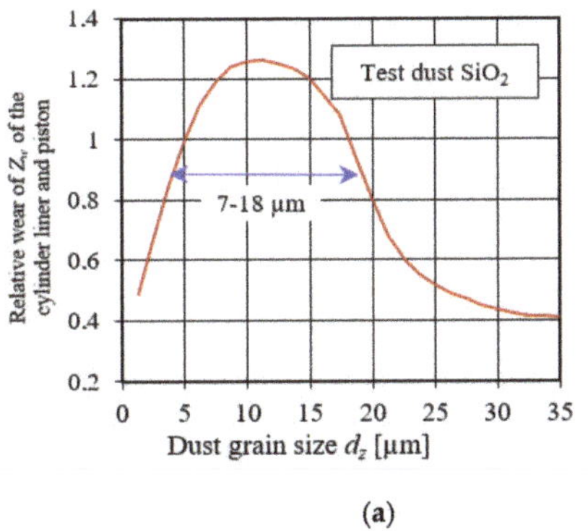
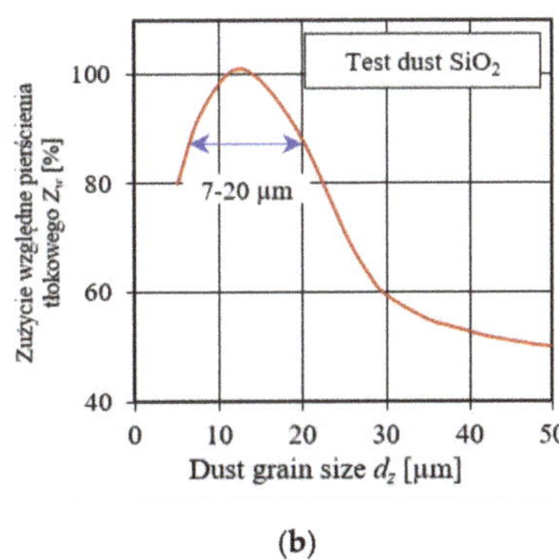

Figure 8. Effect of SiO$_2$ dust grain size on relative wear of: (**a**) cylinder liner and piston [94], (**b**) piston ring. Figure made by the authors using data from [95].

A similar nature of wear, in this case of the upper piston ring of a Diesel engine, is shown in the graph in Figure 8b. The greatest abrasive effect of this ring was found for grains in the 7–20 μm range [95]. The same wear value is caused by dust grains of small (less than 5 μm) as well as large sizes—above 20 μm.

The graph shown in Figure 9 confirms, the relationship between particle size and component wear. Five classified fractions of road dust were used in the study. The maximum effect of piston ring wear was obtained for particles of 21.5 μm diameter. Above and below this size, ring wear is significantly reduced.

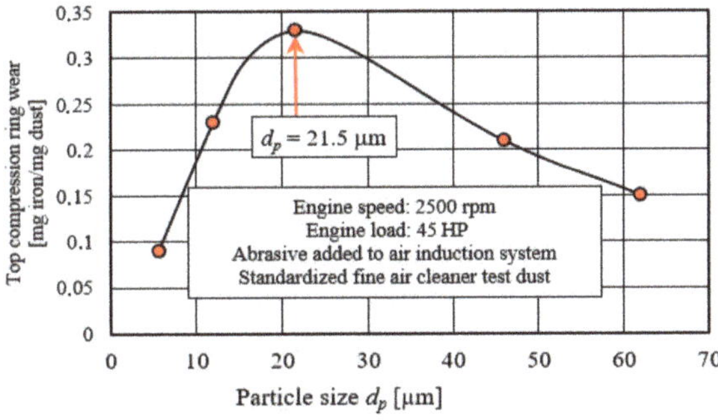

Figure 9. Influence of the road dust grain size on the relative wear of the upper piston ring. Figure made by the authors using data from [96].

This phenomenon is explained by the fact that dust grains of large size do not penetrate between two mating surfaces at the first moment and do not cause wear. However, due to the variation of operating conditions and engine load, the oil film thickness is constantly changing. In such a situation, large dust grains can penetrate between two moving parts and when they come into contact with two surfaces they crumble into smaller grains and only then do they cause wear.

This phenomenon is confirmed by the graphs in Figure 10. The greatest wear of the cylinder face of the ZS engine was caused by quartz (silica) dust grains with sizes of 20–40 μm. Dust grains with dimensions in the range d_p = 0–4 μm cause the same value of wear as grains with dimensions d_p = 40–60 μm [97]. Regardless of the size of the dust grains, the maximum wear occurred in the upper part of the cylinder liner, which is typical for P-PR-CL association, and this is due to the fact that at this location the upper piston ring reaches the TDC position. It is known from the kinematics of the piston–crank system that near TDC and BDC the piston ring travel velocity along the cylinder liner surface has the smallest value, and at TDC and BDC it is zero. The lubrication wedge disappears, and then even the smallest dust grains come into contact with the mating surfaces, causing their abrasive wear.

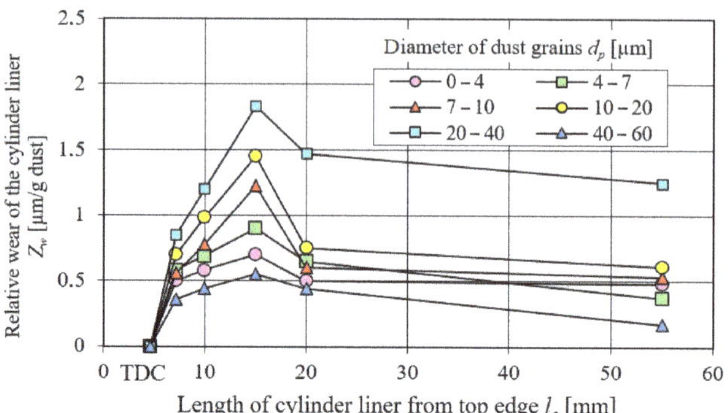

Figure 10. Cylinder face wear for different test dust fractions. Figure made by the authors using data from [97].

According to the author of the paper [98], the highest value of piston ring wear (241 mg/h) was observed for the dust fraction 5–10 μm, and the lowest wear (5 mg/h) was caused (as expected) by dust grains with sizes of 0–5 μm. The wear value with dust grains of size 10–20 μm and standard "fine" dust 0–80 μm is 50% less than the wear for the dust fraction 5–10 μm (Figure 11). A dust concentration of 1.0 grain/1000 cubic feet of air (28.317 m^3) was used.

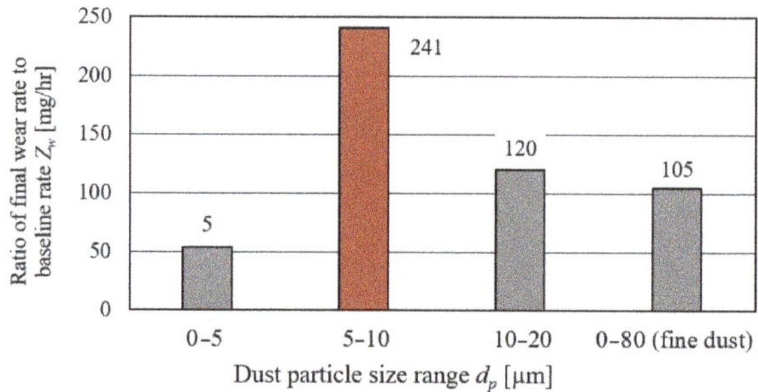

Figure 11. Influence of dust grain size on piston ring wear. Figure made by the authors using data from [98].

From the data presented in Figure 12, it can be seen that the highest values of the wear rate of the main and connecting-rod journal surfaces of a Diesel engine are caused by dust grains with sizes of 5–10 μm, while dust grains with sizes of 0–5 μm are the cause of increased wear of the piston rings of this engine [85]. Even contaminants containing large particles up to 80 μm did not cause as much damage as contaminants with particles concentrated in the size range of 0 to 10 μm.

Studies by the authors of the work [100] have shown that dust with a grain size of 0-5 μm causes more engine wear than grains in the 0–40 μm range. Similar trends were observed by the authors of the work [85], who showed that about four times more wear of diesel engine rings is caused by dust grains with grain sizes of 0–5 and 5–10 μm than by particles of 10–20 μm. According to another study [101], engine wear is an exponential function of particle diameter in the 2–20 μm range, and wear caused by 4–10 μm particles is about three times greater than for 2–4 µm particles. The obvious inconsistency in the size of particles causing maximum engine wear may be due to the use of different engines with different dynamic clearances between moving parts and different measurement techniques.

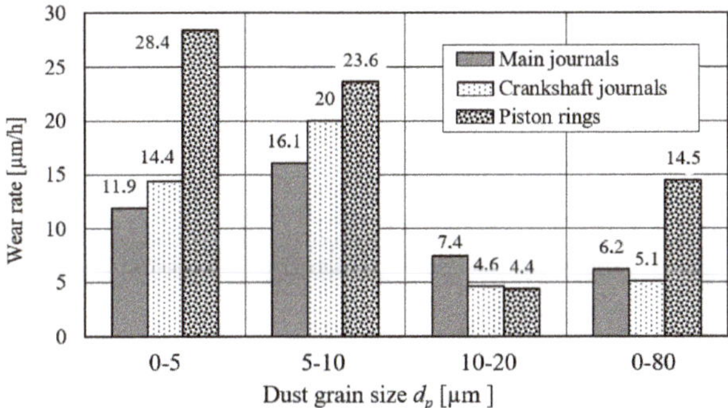

Figure 12. Effect of dust grain size on wear rate of Diesel engine components. Figure made by the authors using data from [85,99].

Tests performed by the authors of the paper [102] showed that the main bearing, crankshaft bearing and oil ring experienced maximum wear by a different range of particle sizes. During the tests, a sample of 2 g of AC Coarse test dust with particle size ranges of 0–5, 5–10, 10–20, 20–40, 40–80, and 80–200 µm was introduced into the engine lubrication system (without filters). For the oil ring, the maximum wear was caused by particle sizes of 5–10 µm, for the main bearing it was particles in the 20–40 µm range, and for the connecting rod bearing the most dangerous particles were 20–80 µm (Figure 13). This sensitivity to different particle sizes can be explained by the fact that each of these associations has a different clearance between the mating surfaces.

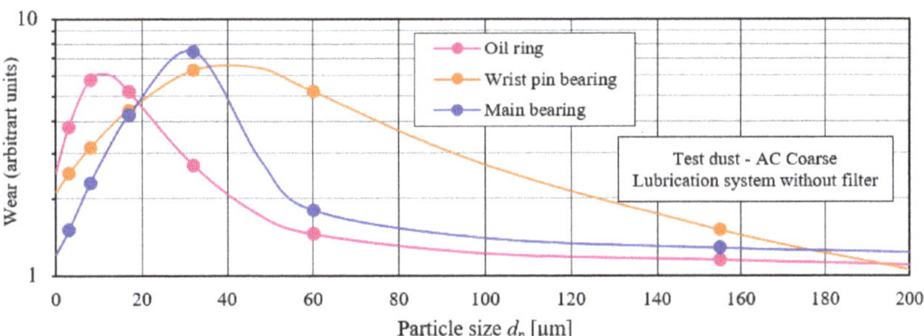

Figure 13. The particle wear sensitivity functions for the main bearing, wrist pin bearing, and oil ring. Figure made by the authors based on data from [102].

It can be seen from the graphs in Figure 14 that as the particle radius R increases, the wear rate of the engine cylinder liner and piston ring increases rapidly, but only up to a certain critical size, in this case about 7 µm, and then the wear rate decreases and is almost constant. This may be related to the fact that the clearance between the piston ring and the cylinder liner is small enough to limit the entry of larger size particles and thus eliminate their effect on wear. Moreover, Figure 14 shows that the wear of the piston ring is almost ten times greater than that of the cylinder liner, even though both components are affected by the same value of dust concentration in the air [103].

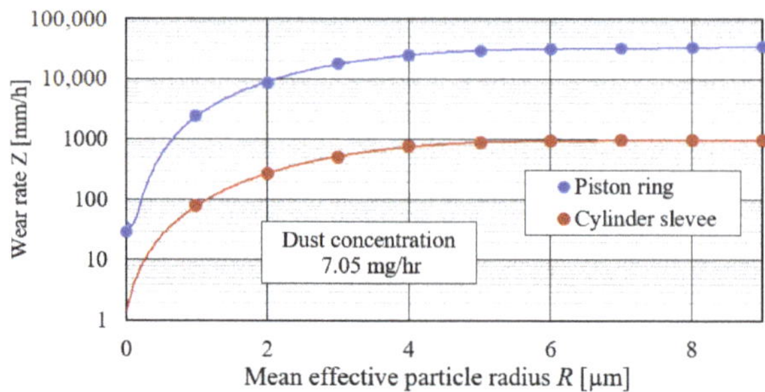

Figure 14. Effect of particle size on wear rate of cylinder liner and piston ring. Figure made by the authors using data from [103].

From the data in Figure 15, it can be seen that the engine load has a significant effect on the abrasive wear of the upper seal ring. Changing the engine load from 0–20 HP causes a slight increase in ring wear. An increase in load above 22.5 HP results in a more than fourfold increase in wear intensity, which may be due to a decrease in oil film thickness between the friction surfaces and the occurrence of grain contact with the friction surfaces [96].

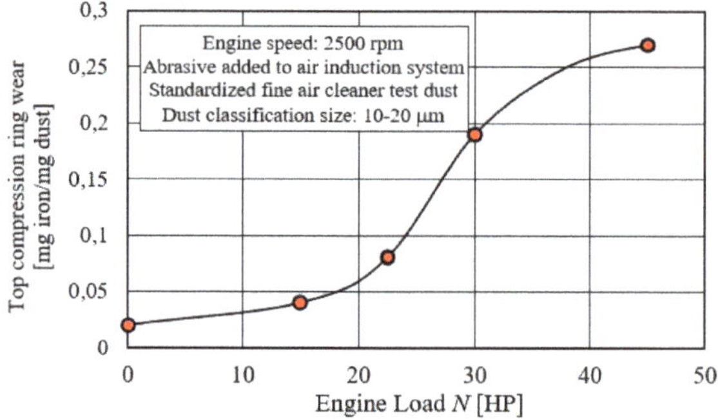

Figure 15. Effect of engine load on top seal ring wear due to 10–20 μm of dust. Figure made by the authors using data from [96].

Figure 16 shows, for different operating times of the friction pair, the effect of dust particle size on the bearing wear rate [82]. Tests were carried out without filter and with the addition of 0.4 g of road dust in the indicated particle size ranges. For the first half hour after dust addition, the fraction in the 10–20 μm range caused bearing wear of k_w = 124.8 mg/h. This is significantly greater than the bearing wear caused by the other fractions, and also greater wear caused by the same mass of test dust in the 0–80 μm range. During the next 30 min after dust addition, the rate of bearing wear by all fractions is almost identical and takes values around k_w = 14 mg/h. These values are similar to the values of bearing wear by the 0–5 μm fraction for the first half hour after dust addition. After the third hour of friction pair operation, a slight wear rate was recorded, but only for test dust in the 0–80 μm range. The authors believe that large dust particles are crushed into smaller ones by the action of the oil pump and between the surfaces of engine components.

The paper [104] presents the results of a study on the effect of three granulations of quartz dust: 2.5–7, 8–20, and 16–40 μm contained in oil on the wear of engine plain bearings. The highest wear of plain bearings is caused by quartz dust particles contained in oil with size up to 20 μm.

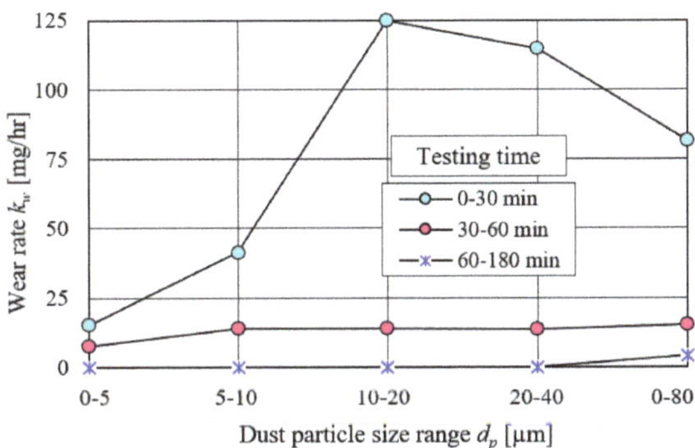

Figure 16. Effect of particle size of different dust fractions on wear rate for different bearing running time. Figure made by the authors based on data from. Figure made by the authors using data from [82].

It can be seen from the graphs presented in Figure 17 that during engine operation at 50% load, as the size of dust grains increases in the range of 2.5–22.5 µm, the wear of cylinder liners increases almost linearly regardless of the type of dust [86]. Dust grain sizes of 2.5–7.5 µm and engine operation at 100% load cause 3.5 times more bushing wear than during engine operation at 50% load.

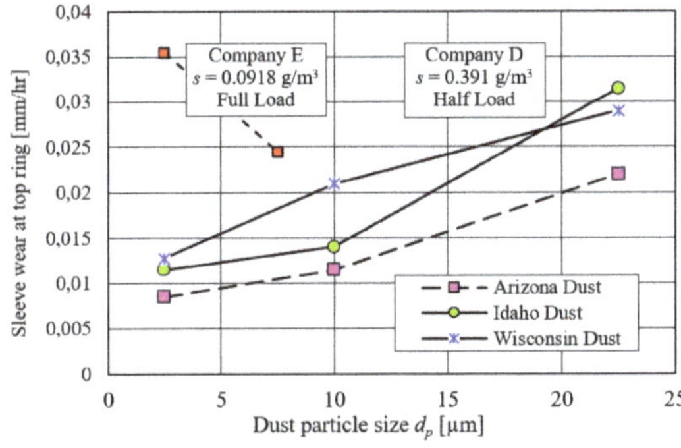

Figure 17. Effect of grain sizes of different dusts on the intensity of cylinder liner wear at the height of the upper piston ring and. Figure made by the authors using data from [105].

The abrasive wear of engine component surfaces is directly related to the hardness of the particles. The hard dust grains are silica SiO_2 and alumina Al_2O_3, whose hardness, judged by the Mohs 10-point scale (diamond has a value of 10 and talc has a value of 1), takes on a value of 7 and 9, respectively. The hardness of both dust components far exceeds that of most engine construction materials. The hardness of hardened carbon steel is about 6.5 on the Mohs scale.

The effect of the hardness of the dust grains on the wear of the upper piston ring is shown in Figure 18. In the case of diamond dust, the wear is much higher than in the case of road dust of equivalent grain size, and unlike road dust, whose primary component is SiO_2, it steadily increases with particle size over the entire range studied.

Lower hardness dust causes accelerated wear when the abrasive comes into contact with the surface, then the dust grains are smoothed and ground. Wear is then limited to the first piston ring. On the other hand, dust of high hardness does not undergo the process of grinding and comminution, but causes wear of all piston rings and then, after getting into oil, it takes part in the process of accelerated wear of other friction pairs (main journal–

crankcase, connecting rod–crankcase, camshaft journal–crankcase, valve stem-guide, valve cam–cam, and turbocharger bearings), which are reached by engine oil.

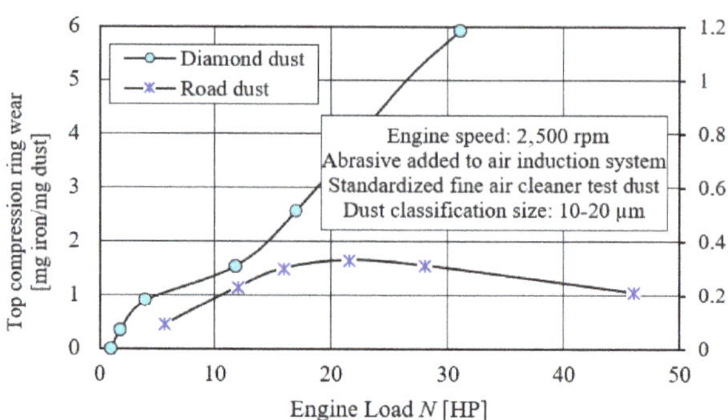

Figure 18. Effect of dust grain size with different hardness on upper piston ring wear. Figure made by the authors using data from [96].

The effect of the size of dust grains of different hardness (corundum and quartz) on the wear of a chrome plated piston ring and cylinder liner is shown in Figure 19. The course of the ring and liner wear rate for both minerals is practically the same. It can be seen from Figure 19 that Al_2O_3 particles with a diameter in the range of 0–10 μm caused several times more wear than SiO_2 particles of the same size and hardness (6–7) on the Mohs scale [105]. The ring wear caused by SiO_2 is four times less than Al_2O_3. For particles up to 5 mm in size, the wear intensity is higher than for particles of larger size. This may be due to the fact that at the given friction conditions a certain part of particles larger than 5 μm does not enter the tribological areas of rings and sleeves and does not participate in abrasive wear.

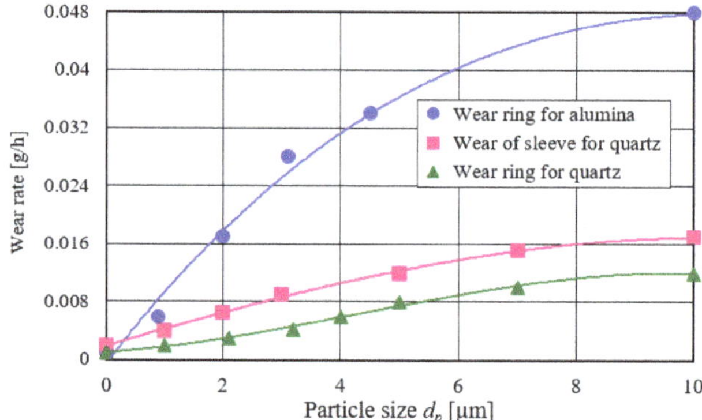

Figure 19. Effect of dust grain size of different hardness on wear of chrome plated piston ring and cylinder liner. Figure made by the authors using data from [105].

The concentration of dust in the air is another factor intensifying the wear of engine components. The life of an engine is characterized by varying values of air dustiness. For an air filter with a certain filtration efficiency, a higher dust concentration in the air results in a greater mass of dust entering the engine cylinders. Several studies have shown that engine wear increases significantly as the concentration of dust taken in with the air by the engine increases [94,96,103,106–108].

The author of the paper [106] presented data on the concentration of dust in the air depending on the conditions of vehicle use (Table 1) and then related it to the wear rate of engine components—Figure 20.

Table 1. Dust concentration values as a function of vehicle operating conditions Data for the table were taken from the paper [106].

Condition	Dust Concentration s (mg/m^3)
Average Ambient	0.010–0.139
95th Percentile Ambient	0.089
99th Percentile Ambient	0.112
Paved Roads	0.139–57
Dirt Roads	0.139–6113
Dust Storms	0.1–176
Worst Dust Storm	3000
0 Visibility	883
20 × 0 Visibility	17,657

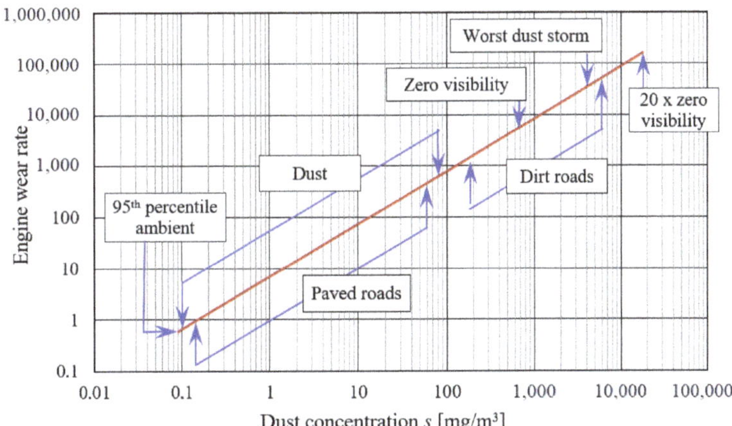

Figure 20. Effect of dust concentration versus vehicle operating conditions on engine wear rate. Figure made by the authors using data from [106].

The established linear relationship between engine wear and dust concentration shows that when the dust concentration in the environment increases by more than 5 orders of magnitude, the intensity of engine wear also increases by the same amount. An engine operating on a dirt road or in a gravel pit in zero visibility, wears out about 10,000 times faster than an identical engine operating in a rural environment (Figure 20).

Similarly, the author of the paper [96] showed that in the range of dust concentration in the inlet air from 0.0042 to 0.287 g/1000 m^3 of air, the wear rate of the upper piston ring is a linear relationship (Figure 21).

During testing, standard engine operating conditions were maintained so as to minimize piston ring wear from causes other than abrasion. The test dust used was Arizona fine road dust also called "Standardized fine air cleaner test dust" with typical chemical analysis shown in Table 2 and granulometric composition Table 3 [96].

According to the author of the paper [107], the engine wear, depending on the concentration of dust in the air increases exponentially. At the dust concentration s = 20–30 mg/m^3 the wear becomes 1000 times higher than at the concentration s = 1–2 mg/m^3, which is 10 times lower.

Figure 22 shows the exponential relationship between the Csepel Diesel engine life and the airborne dust concentration (mass of SiO$_2$ dust drawn in with the air) of 10–15 μm [94].

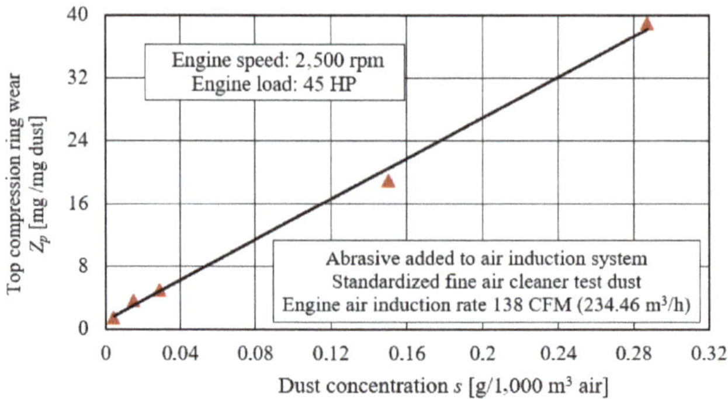

Figure 21. Effect of dust concentration on upper piston ring wear. Figure made by the authors using data from [96].

Table 2. Chemical analysis of fine grade Arizona Road dust used in abrasive wear tests [96].

Components of Dust	Percent by Weight [%]
SiO_2	67–69
Fe_2O_3	3–5
Al_2O_3	15–17
CaO	2–4
MgO	0.5–1.5
Total Alkalis	3–5
Ignition loss	2–3

Table 3. Particle size distribution of dust used in abrasive tests [96].

Dust Particle Sizes d_p [μm]	Mass Fraction of Grains in the Dust F_m [%]
$0 \div 5$	39 ± 2
$5 \div 10$	18 ± 3
$10 \div 20$	16 ± 3
$20 \div 40$	18 ± 3
$40 \div 80$	9 ± 3

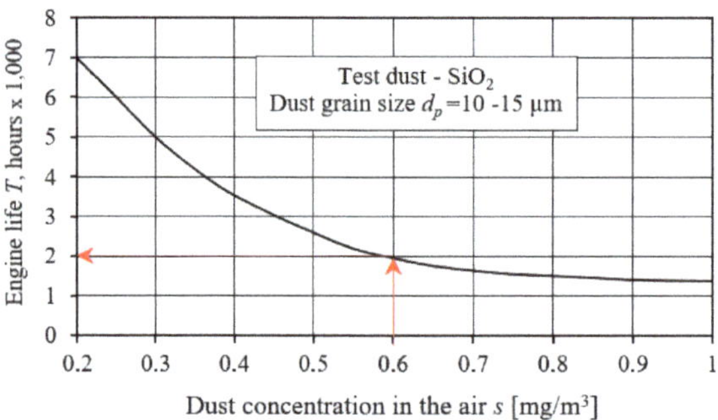

Figure 22. Dependence of engine life on dust concentration in air. Figure made by the authors using data from [94].

From the graph in Figure 22, it can be seen that if the engine draws in ambient air with a dust concentration of 0.2 mg/m^3, the engine life (understood here as the number of kilometers driven by the vehicle or hours of engine operation up to a certain wear limit) is 7000 h. If the mass of aspirated dust increases three times, the life of the engine decreases to approximately 2000 h. An increase in the dust concentration in the air from 0.2 to 1 mg/m^3 results in a fivefold increase in the mass of dust supplied to the engine cylinders and a sixfold decrease in engine life.

The relationship between Diesel engine life and the amount of dust in the air (dust concentration) and air filter permeability is shown in Figure 23 [94]. The increase in the coefficient ε of air filter permeability (decrease in filtration efficiency) decreases the engine life the more the dust concentration in the air has a higher value.

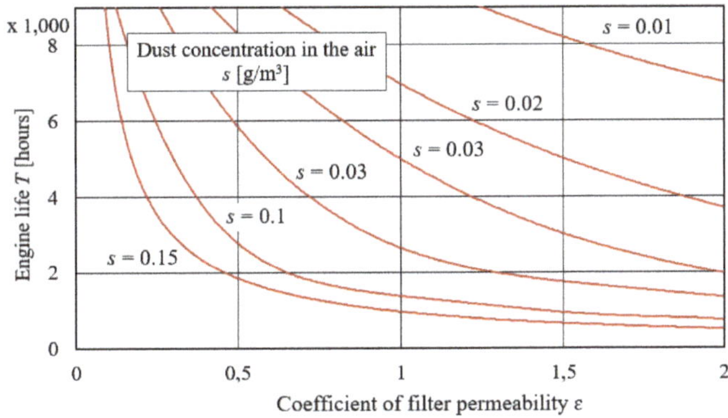

Figure 23. Dependence of engine life on coefficient of filter permeability ε for different dust concentrations in the air. Figure made by the authors using data from [94].

According to the authors of the work [108], a fourfold increase in the concentration of quartz dust in the intake air causes a fourfold increase in the intensity of wear of the first piston ring of the D-54A engine, regardless of the size of the dust grains (Figure 24).

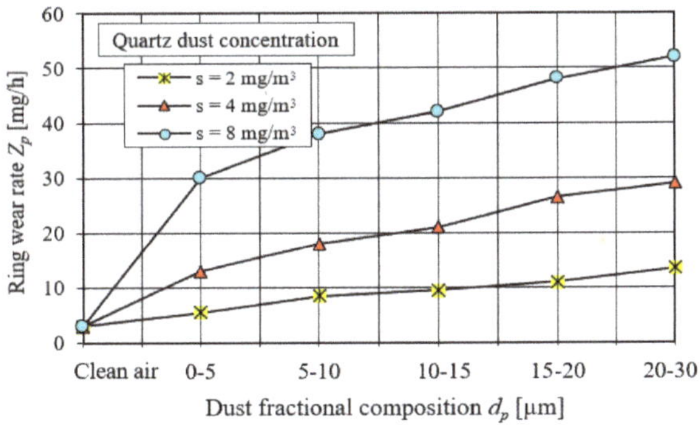

Figure 24. Wear intensity of the first piston ring of the D-54A engine for different values of quartz dust concentration in the intake air and different fractional compositions. Figure made by the authors using data from [108].

Figure 25 shows the effect of particle size and dust concentration on piston ring and cylinder liner wear for two cases. In the first case, the dust particle size is reduced by 50% from the reference value and is 3.4 µm, and in the second case, the dust concentration per engine cylinder is reduced by 50% from the reference value and is 3.525 mg/cylinder.

A 50% reduction in particle size or dust concentration reduces the wear rate of both the cylinder liner and piston ring compared to the reference wear by 40% and 50%, respectively (Figure 25). Thus, reducing the amount of dust per engine cylinder has a greater effect on the wear rate compared to reducing the particle size.

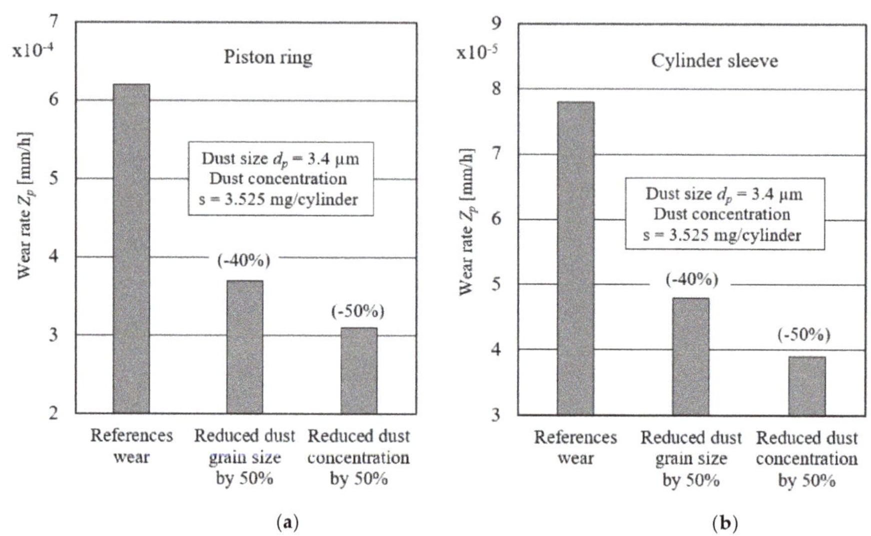

Figure 25. Wear rate: (**a**) piston ring, (**b**) cylinder liner for 50% reduction in dust particle size (case 1) and 50% reduction in dust concentration per cylinder (case 2). Figure made by the authors using data from [103].

However, both factors have a significant effect on the wear rate of the piston ring and cylinder liner and are interrelated. Therefore, it is important to use air filters with high filtration efficiency and accuracy, which can reduce the influence of both factors especially during the operation of vehicles in dusty conditions, and thus ensure longer engine life.

Figure 26 shows the abrasive wear rate of the cylinder liner and piston ring when three air filters with different filtration efficiencies are used compared to when no air filter is installed in the engine.

It can be seen that installing inlet air filters for the engine significantly reduces the wear rate of its components. For example, a filter with 97.8% filtration efficiency reduces the abrasive wear of the cylinder liner from 107×10^{-5} to 2.02×10^{-5} µm/h, i.e., by almost 98%. Any air filter with higher efficiency definitely reduces the wear of both engine components, with the wear rate of piston rings being several times higher than that of the cylinder liner.

Assuming the abrasive wear of both the cylinder liner and piston ring without air filter as 100% (Figure 26), the use of an air filter with 97.8% efficiency reduces the wear of both elements to 1.89%. After using an air filter with higher efficiency (99.45%), the wear of the cylinder liner and piston ring is only 0.41% compared to the wear when there was no air filter in the engine.

An increase in the engine oil filtration efficiency from $\varphi_1 = 66.7\%$ to $\varphi_2 = 95\%$ results in a more than 50% decrease in the wear rate of the piston rings of the loaded Diesel engine—Figure 27 [109].

Figure 28 shows the dependence of piston ring wear magnitude on slip velocity for different ring loads at 24-h tests. As the slip velocity increases, the ring wear increases the higher the load. In contrast, Figure 28b shows the dependence of ring wear on load for 24-h test runs at 150 °C and for two different slip speeds. Both at slip speeds of 1 and 4 m/s the wear increases almost proportionally to the ring load, with the higher speed having higher wear values.

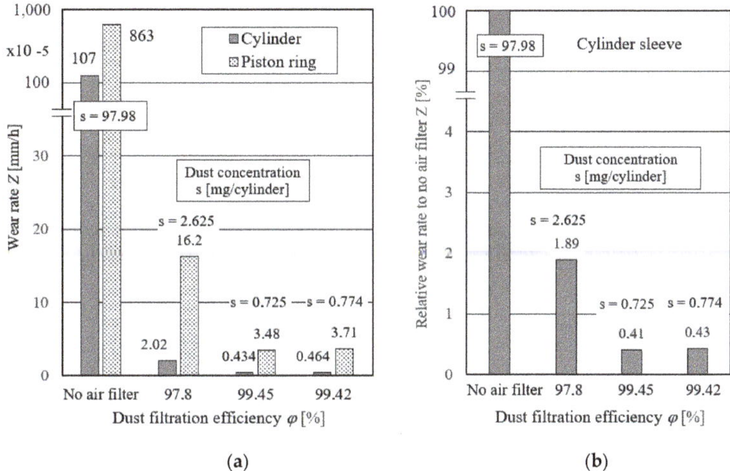

Figure 26. Effect of air filter efficiency on: (**a**) wear rate of cylinder liner and piston rings, (**b**) relative wear rate of both compared to no air filter. Figure made by the authors using data from [103].

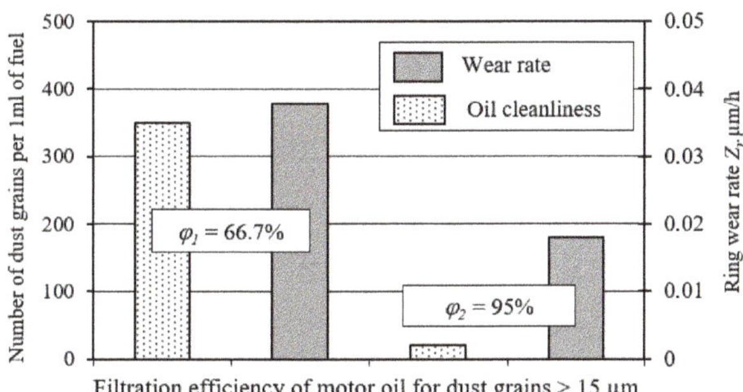

Figure 27. Effect of filtration efficiency of dust grains larger than 15 μm contained in engine oil on the wear rate of piston rings of a Diesel engine. Figure made by the authors using data from [109].

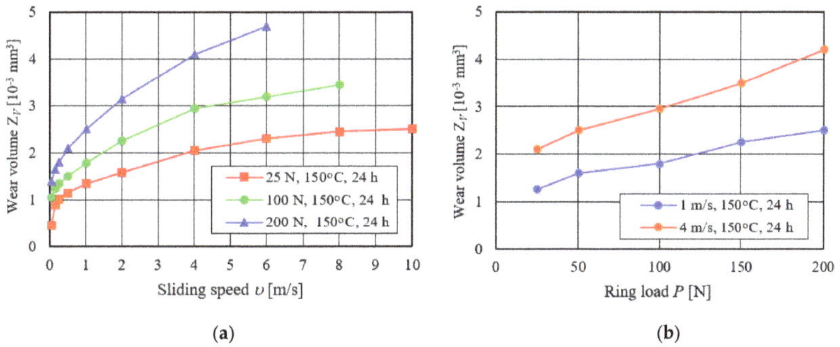

Figure 28. Ring wear as a function of: (**a**) slip rate at 24-h test time, average temperature 150 °C, and ring load of 25, 100, and 200 N, (**b**) as a function of load. Figure made by the authors using data from [51].

The paper [110] presents an operational study of the wear of the piston-cylinder group of a Jamz 242N truck engine. The vehicles were operated on unpaved roads while transporting iron ore. The engines were equipped with two different air filters. The wear of the piston-cylinder group of the engine with the paper cartridge filter was 50% lower than when the engine was equipped with the inertia-wash filter, which had a lower filtration efficiency. The cylinder liner wear intensity at TDC (at the level of the first piston ring) had an average value of 1.6 and 3 μm/1000 km, respectively (Figure 29). When these cars were used on asphalt roads, the wear intensity of the piston–cylinder group elements decreased six times [110].

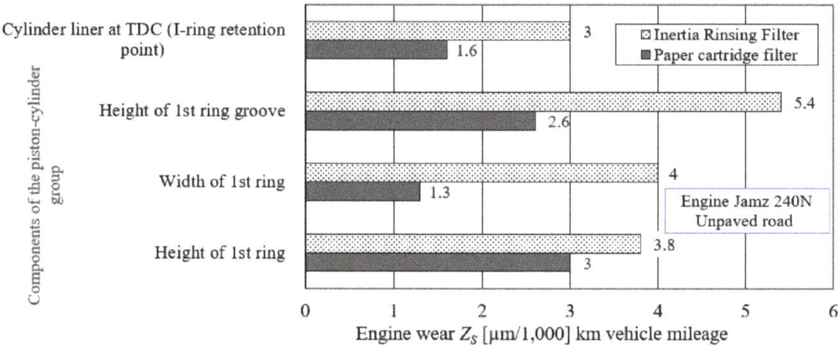

Figure 29. Wear of engine piston–cylinder group components during use on an unpaved road. Figure made by the authors using data from [110].

The results of piston ring wear tests are presented in [45]. The test object was a 4-cylinder turbocharged compression ignition engine with a displacement of 1.3 dm^3, with charge air cooling and exhaust gas recirculation. The engine was equipped with a common rail direct injection system and had a maximum power of 66 kW at 4000 rpm and a maximum torque of 200 Nm in the range 1750–2250 rpm. The engine had a typical ring arrangement which included a first rectangular sealing ring with a barrel-shaped, chrome-plated face, a second conical sealing ring and a double lip scraper ring with chrome-plated faces and with a helical spring. The aluminum piston had a cast iron insert under the first sealing ring and cooling channels.

The tests were conducted during a long-term durability test of the engine operating under heavy load conditions according to a special cycle that was repeated 336 times. The total operating time of the engine during the durability tests was about 1200 h [45]. The wear of individual piston rings was determined as an increase in ring lock clearance (Figure 30a) and as a decrease in ring height (Figure 30b).

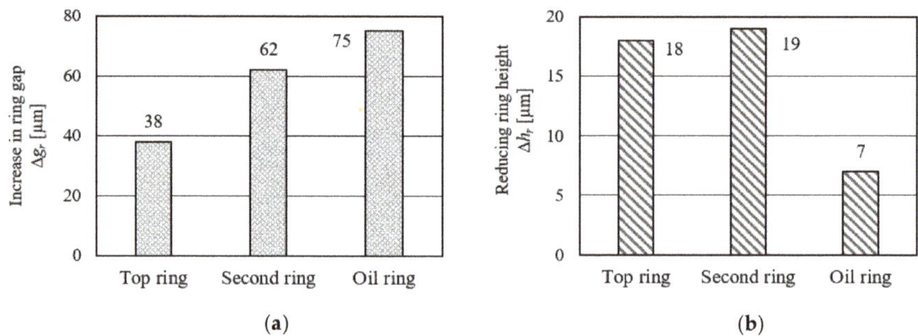

Figure 30. Piston ring wear during durability test: (**a**) increase in lock clearance for individual rings, (**b**) decrease in ring height. Figure made by the authors using data from [45].

According to the authors of the paper [111], the intensity of cylinder liner wear at TDC (at the level of the first piston ring) was on average 1.38 μm/1000 km of the mileage of a vehicle used in sandy desert conditions and equipped with a two-stage filter (monocyclone paper cartridge) and 3.8 μm/1000 km of the mileage for an engine equipped with an inertia-washout filter, i.e., having lower filtration efficiency. The wear of the cylinder liner in the lower part was twice less [111].

The accelerated wear of the cylinder liner, resulting from the operation of a truck engine with a faulty air filtration system, is shown in Figure 31. The abrasive wear of the cylinder liner in the form of parallel continuous bands of scratches along the liner formation, located in the upper zone of the cylinder liner on approximately 1/5 of its circumference, is clearly visible [112]. The scratch bands are so intensive and deep that no traces of the final honing treatment of the cylinder liner surface can be seen. The view of the cylinder face of a truck engine operated with an efficient air filtration system is shown in Figure 31c. There are clear honing marks and combustion product deposits on the cylinder liner above the top dead center of the first piston ring.

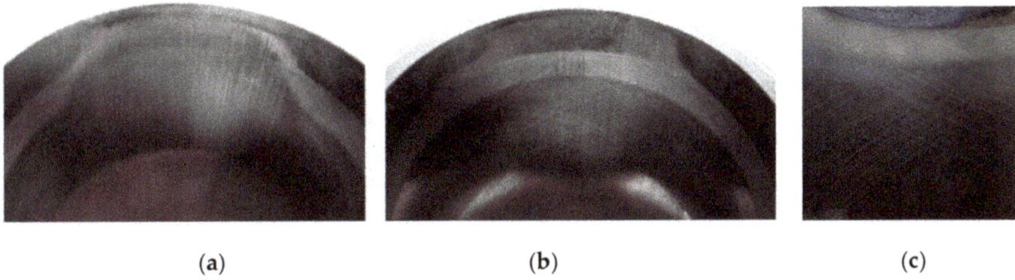

(a)   (b)   (c)

Figure 31. Wear view of the cylinder liner of a truck engine operated with an inoperative and an operative air filtration system and efficient air filtration system: (**a**) clear strands of scratches without any honing traces, (**b**) visible single scratches against the background of surface treatment traces, (**c**) efficient air filtration system [112].

The loss of material in the form of scratches (grooves) results from the mechanical impact of large-size foreign particles on the surface of the cylinder liner and the hardness is much higher than the hardness of the metallic liner matrix. The cylinder liner has scratches up to 70 μm wide (Figure 32). The width of the cracks in the cylinder face is a representation of the size of the foreign particles between the mating parts.

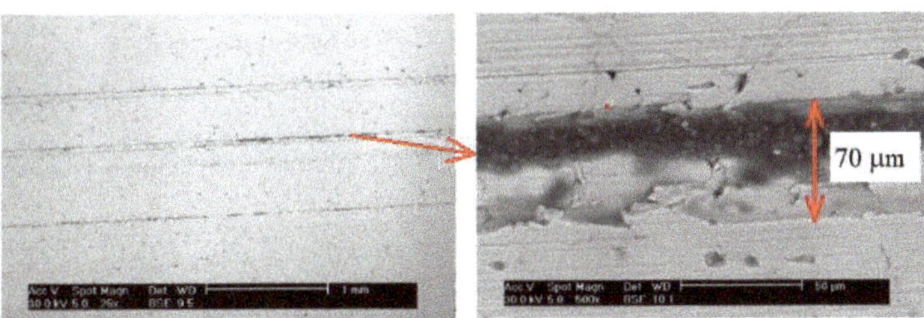

Figure 32. Scratches in the cylinder face caused by foreign particles of high hardness [112].

A paper cartridge newly installed in the filter only after exceeding 20–30% (Figure 33) of the predicted mileage ensures the required efficiency (99.5%) and accuracy of air filtration [72,87]. This is due to the phenomena occurring in the initial period of the filtration process of the fibrous material. Therefore, the frequency of air filter servicing (filter cartridge replacement with a new one) may affect engine wear and durability.

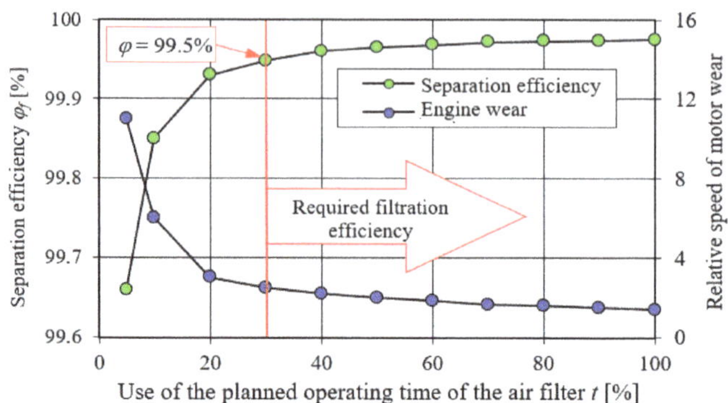

Figure 33. Separation efficiency and engine wear rate as a function of air filter run time. Figure made by the authors using data from [89,106].

The length of this period depends on the type and parameters of the filter material, the concentration and granulometric composition of the dust, and the filtration rate. Figure 34 shows the results of experimental tests of several filter materials that differ in the length (duration) of the initial filtration period [113]. With the increase of the mass of dust retained in the filtration layer (increase of the km coefficient), the filtration efficiency of the tested cartridges assumes increasingly higher values. The required value of filtration efficiency (φ_w = 99.9%) is achieved by filter cartridges working in the same conditions (the same value of dust concentration and air flow) after different time. Insert A achieves an efficiency of φ_w = 99.9% after obtaining a dust absorption coefficient k_{mA} = 110.7 g/m². For inserts B, C, D, and E made of other filtration materials the initial period (time of obtaining the required filtration efficiency) is much shorter. For inserts E (polyester + PTFE membrane) and D (cellulose + polyester + nanofiber), this period ends at the earliest when the dust absorption coefficient k_{mE} = 5.77 and k_{mD} = 7.22 g/m², respectively, is reached. For cartridge B (polyester), the initial period ends at a dust absorption coefficient of k_{mB} = 31.9 g/m², and cartridge C (cellulose + polyester) for a coefficient of k_{mC} = 48.5 g/m² (Figure 34).

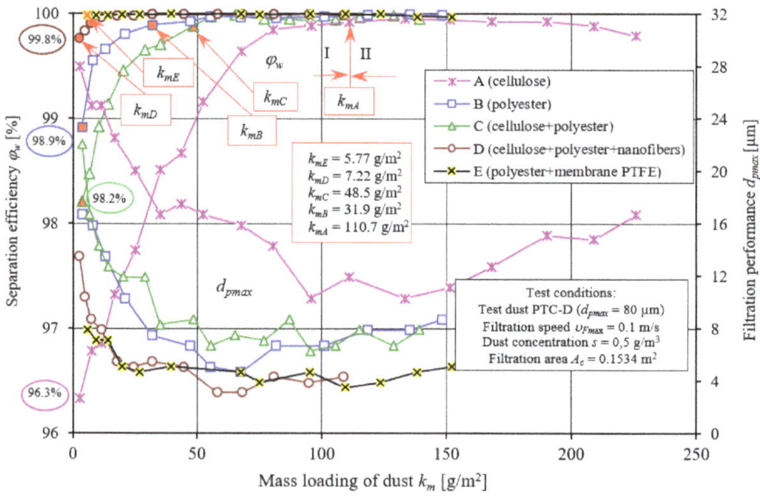

Figure 34. Separation efficiency φ_w and filtration performance d_{pmax} depending on the dust mass loading k_m of the tested filter cartridges [113].

At this time, there are large dust grains in the air behind the filters. The largest dust grain sizes (d_{pmax} = 28 µm) were recorded downstream of cartridge A (cellulose), and the smallest (d_{pmax} = 7.9 µm), downstream of cartridge E, which is closely related to the initial filtration efficiency of these cartridges. This is due to the higher grammage, and smaller pore size of the filter materials tested.

The initial filtration efficiencies of the tested cartridges take on different values. The lowest value (φ_{w0A} = 96.3%) was recorded for the filter cartridge made of filtering material A (cellulose). Inserts B, C, D, and E have higher values of initial filtration efficiency, respectively: φ_{w0B} = 98.9%, φ_{w0C} = 98.2%, φ_{w0D} = 99.8%, φ_{w0E} = 99.97% (Figure 34).

Low initial filtration efficiency and the presence of large dust grains in the cleaned air during the initial period of filter operation (after replacing the contaminated filter insert with a new one) can affect the accelerated wear of mainly P-PR-CL association. Thus, frequent, unjustified filter cartridge replacement may be the cause of accelerated wear of engine components.

On the other hand, the authors of the paper [114] believe that the wear of the friction pair "cylinder–piston rings" and thus the durability of the internal combustion engine depends not only on the efficiency of the inlet air filtration and the wear resistance of the parts, but also on the design of the intake manifold. In the intake manifold of a multi-cylinder car engine (Figure 35a), due to the rapid twisting of the flowing air in the ducts and the associated effect of the inertial force on the dust particles, uneven distribution of the dust particles to the individual engine cylinders can occur, resulting in uneven wear of the piston–cylinder group.

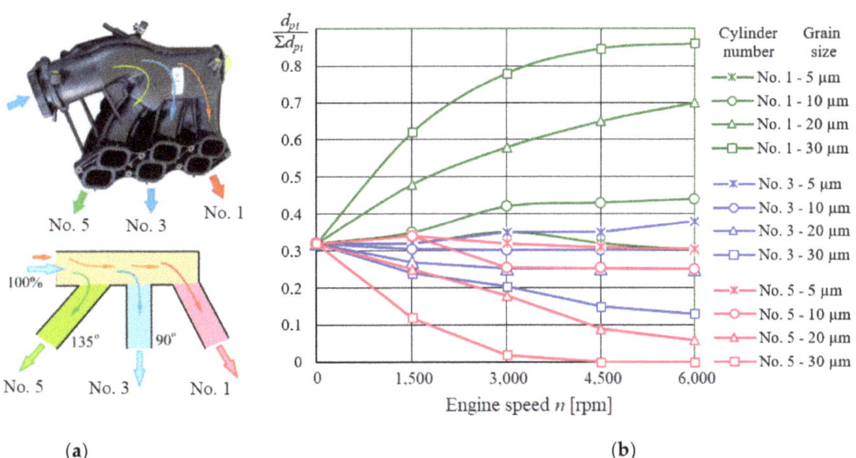

Figure 35. Intake manifold with different outlet angles (1, 3, and 5—engine cylinder numbers): (**a**) construction and functional diagram, (**b**) the ratio of the number of dust particles d_{pi} the size of 5, 10, 20, and 30 µm in the cylinders to the total number of particles Σd_{pi} at the inlet at different rotation frequency n of the crankshaft. The figure was made by the authors based on data from [114].

Simulation studies were carried out in the air velocity range of 5–20 m/s in branched manifold channels with diversion angles of 45°, 90°, and 135° for the most characteristic particle sizes of 5–30 µm [114,115]. The computational results showed that the dust particles deviate from the air flow line due to inertia and can move through the side outlet the larger the particle size, channel deflection angle and air velocity—Figure 35b. The uneven distribution of particles in some modes of operation is so great that as much as 75–85% of the dust can enter the outermost cylinder, furthest from the inlet to the intake manifold. This explains the reason for the uneven local abrasive wear of the cylinder–piston group and valve mechanism components in individual cylinders.

According to the author of the paper [116], the excessive presence of sulfur in the fuel may be the cause of additional corrosion wear of piston rings and cylinder liners, i.e., those engine elements which are in contact with exhaust gases. This type of corrosion wear caused by sulfuric acid intensifies when the engine is operated at low temperatures and when exhaust gas recirculation is used. It can be seen from Figure 36 that the rate of cylinder liner wear increases with increasing particle size in the range of 15–30 μm and engine speed in the range of 1000–2500 rpm. The dust samples used in this study were obtained from the Jordanian desert. The wear rate of the cylinder liner increased significantly when sulfur was cycled into the fuel at 0.03%, 0.6%, and 0.1% (Figure 36b). A maximum wear rate of 2.41 μm/h was obtained when the engine was operated at n = 2500 rpm, for maximum particle size (30 μm) and maximum sulfur content (0.1%) in the fuel.

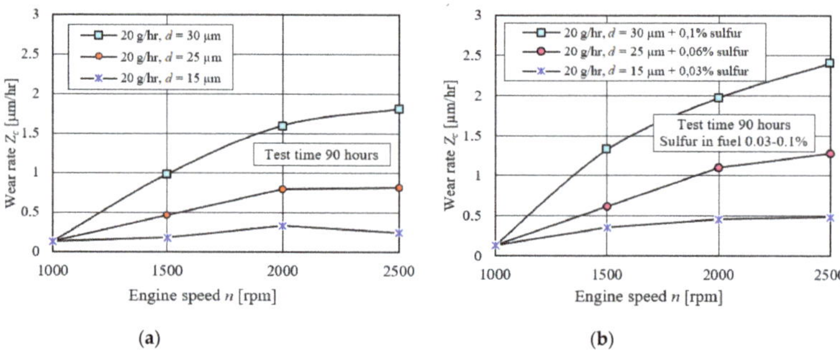

Figure 36. Cylinder liner wear rate as a function of engine speed: (**a**) for different sizes of dust grains in the 15–30 μm range, (**b**) for different sizes of dust grains and with the addition of sulfur in the fuel. Figure made by the authors using data from [116].

Figure 37 shows the effect of dust particles dosed with air and sulfur added to fuel on engine cylinder liner wear along its formation. It can be clearly seen that the abrasive wear takes maximum values at the top dead center (TDC), slightly smaller values at the bottom dead center (BDC) and very little values at the other locations of the bushing (Figure 37a). When the engine burns fuel with added sulfur, cylinder liner wear increases because corrosive wear occurs in addition to abrasive wear caused by dust particles. The total wear near TDC is higher than elsewhere in the cylinder liner [116]. This is probably due to the difficult operating conditions of the rings in this zone: very low piston and ring velocity, which disappears to zero at TDC. The upper ring is exposed to high temperature gases (proximity to the combustion chamber), which causes the oil near the upper ring to evaporate and reduce its viscosity. As a consequence, the lubricating wedge disappears, so that most of the time there is metal-to-metal contact which is the cause of abrasive wear.

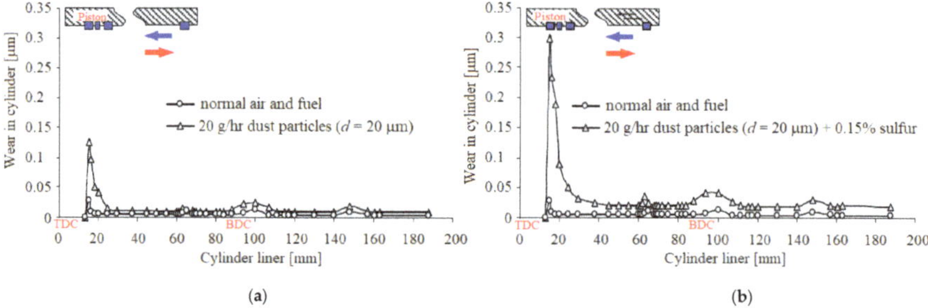

Figure 37. Cylinder liner wear along its length at 2000 rpm and after 120 h of engine operation: (**a**) effect of dust particles, (**b**) effect of dust and sulfur particles. Figure made by the authors using data from [116].

Figure 37a shows the relationship between the wear of the first and second piston rings and the mass of dust dispensed into the cylinder. The wear of the first ring, due to high temperature and pressure, lower oil viscosity, is higher than that of the second ring. The wear increases linearly as the intensity of the dust mass flowing into the cylinder increases [116].

The relationship between the wear of the first and second piston ring and the degree of sulfur content in the fuel is shown in Figure 38b. The course of ring wear is similar (linear) to that shown in Figure 38a, but changes with less intensity. This shows that dust particles have a greater effect on piston ring wear than sulfur in fuel. That is, the abrasive wear rate of engine components is higher than the corrosive wear rate.

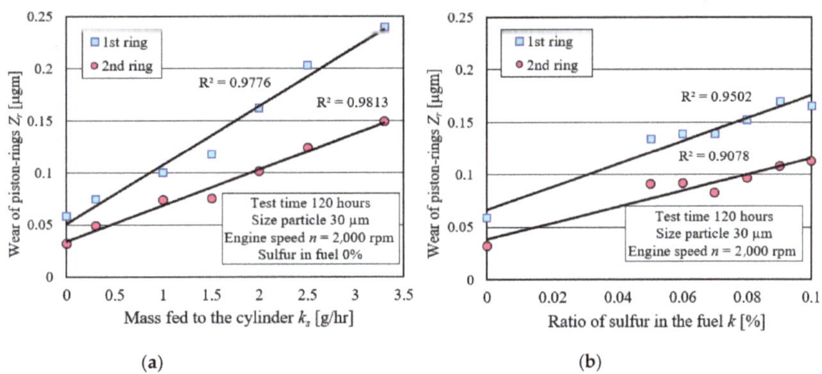

Figure 38. Piston ring wear rate at engine speed n = 2000 rpm depending on: (**a**) the mass of dust dispensed into the cylinder, (**b**) the percentage of sulfur in the fuel. The figure was made by the authors based on data from [116].

Experimental studies of the wear of piston-cylinder assembly components of five 4-cylinder 85.3 kW engines were presented in [117]. Each of the five tested engines was equipped with a completely new set consisting of: honed sleeve, piston, rings, pins, and valves. The first stage of the study involved running the engines during the run-in period for 10 h at 1400–2800 rpm and varying load and with increased air dust. The engine load and speed were increased systematically from minimum to maximum value. The second stage of testing consisted of seven 3-h cycles of 2 h 50 min full load at 2800 rpm and 10 min idle to obtain a total of 21 h of engine operation. In this stage, test dust containing mainly 73.7% SiO_2 and 14.8 Al_2O_3 was precisely metered into the engine intake manifold at a rate of 1.3 g/h, corresponding to an ambient dust concentration of about 8 mg/m^3. To avoid the accumulation of dust and wear products inside the engine lubrication system, an additional external fine oil filtration system was added.

The wear of the pistons was expressed by the change in their average radius, while the wear of the top, middle, and bottom ring was expressed by the change in the size of the end gaps: PI, PII, and PIII. The lapping operation causes only 1.3–3.9 µm of cylinder liner wear due to the change in the roughness parameters of the cylinder liners. The ring wear during engine run-in averages 53.2 µm for top rings, 153.2 µm for middle rings, and 369 µm for bottom rings (Figure 39a). During the second stage of testing, the average wear is respectively—for cylinder liners: 4.0–24.3 µm, for pistons: 2.6–4.4 µm, and for rings: ΔPI = 340 µm, ΔPII = 560 µm, ΔPIII = 960 µm (Figure 39b).

From the above analysis, it can be seen that wear of the elements of the friction pair "cylinder liner—piston rings" and other friction associations in the engine is inevitable due to the presence of hard dust grains in the engine intake air. This is due to the fact that standard cellulose-based filter materials used for engine intake air filters have filtration efficiencies in excess of = 99.5% and dust grain retention accuracy of more than 2–5 µm [113]. It follows that all dust grains above 2–5 µm enter the engine cylinders. From available literature data, all dust grains above $d_p \geq 1\ \mu m$ cause accelerated wear of internal combustion engine components [76,94,95,106]. The only way to protect engines from excessive wear of frictionally cooperating surfaces is to use materials with an engine inlet air filtration accuracy above 1 μm. Such possibilities are created by polymer nanofibers, i.e., fibers with a diameter below 1 μm obtained by electrospinning [118–123].

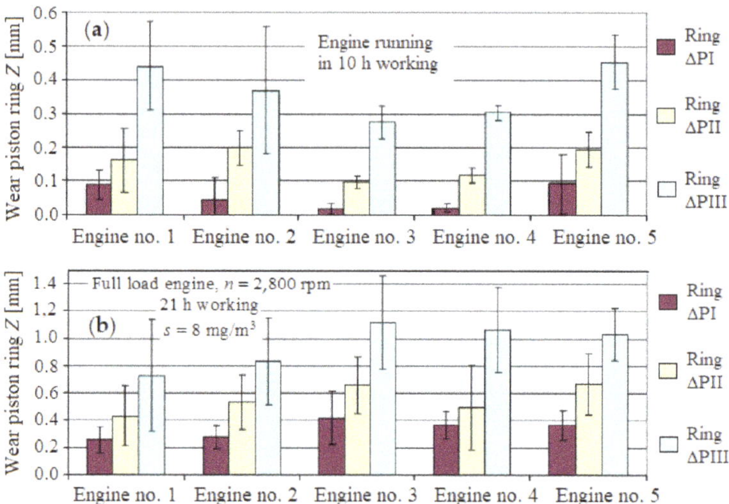

Figure 39. Piston ring wear: (**a**) after 10 h of engine run-in test, (**b**) after 21 h of engine operation at full load, at 2800 rpm, and dust concentration in intake air $s = 8\ \mathrm{mg/m^3}$. Figure made by the authors using data from [117].

Nanofibers have completely different and new properties compared to standard fibers. First of all, they have a large surface area in relation to their weight and a significantly higher strength, and they also have a higher chemical activity and higher moisture sorption. In automotive technology, nanofibers with very small diameters of about 50–800 nm are used. Companies producing filter media using nanofibers have developed their own technologies for this purpose, such as Donaldson's Ultra-Web® and Fibra-Web® technology, Finetex Mats™ from Finetex Technology Inc. and AM-SOIL Ea Air Filters. Due to the limited mechanical and strength properties of the nanofiber thin film (1–5 μm), it is applied over a substrate of conventional filter materials that have greater thickness and strength.

A thin nanofiber layer applied on the inlet side of a standard filter bed, which can be cellulose, nylon, or polyester (Figure 40) traps contaminant particles before they penetrate deep into the filter material [124–126].

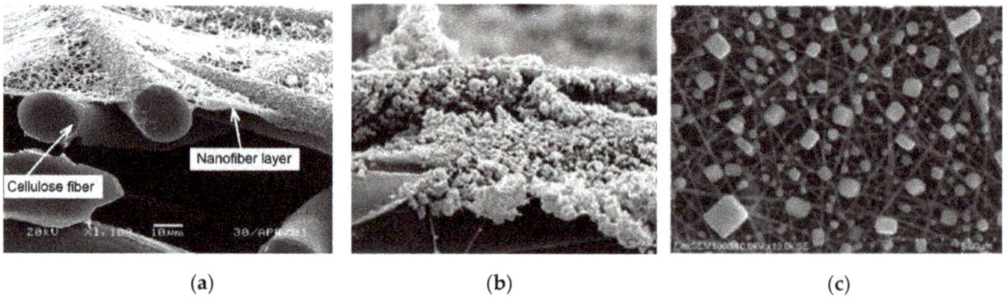

Figure 40. Filter bed made of cellulose and nanofiber layer: (**a**) nanofiber layer applied to the substrate—cross-sectional view of the bed [124], (**b**) dust layer retained on the nanofiber layer [125], (**c**) dust grains retained on the nanofiber layer [126].

The use of nanofibers as an additional layer applied to standard air filter materials used in motor vehicles significantly improves filtration efficiency and accuracy, but unfortunately increases flow resistance. Figure 41 shows the filtration efficiency of a nanofiber filter medium with a cellulose substrate on which a layer of nanofibers with a thickness of 300 μm, a weight $g_m = 0.1$ g/m^2 and a fiber diameter in the range of 40–800 nm was placed [127]. For filtration velocity $v_F = 0.03$ m/s and dust grains in the range $d_p = 0.2$–4.5 μm, the filtration efficiency of this bed reaches values $\varphi = 64$–99%, respectively. For a much higher filtration velocity $v_F = 0.2$ m/s, the filtration efficiency reaches a slightly lower level. These efficiency values are much higher than those of cellulose-based and commercial nanofiber materials (Figure 40).

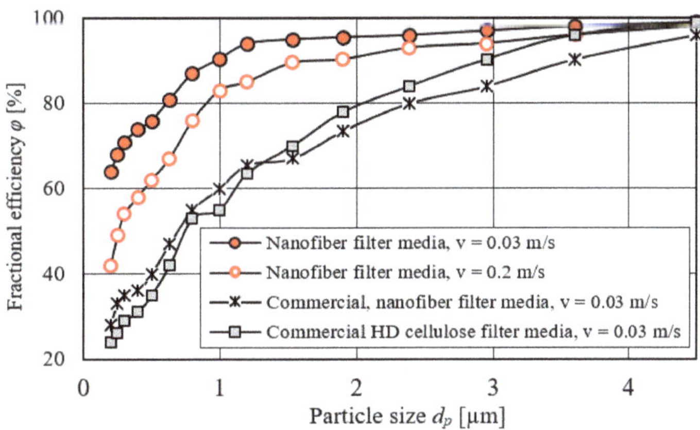

Figure 41. Pleated filter elements made of cellulose fibers, nanofiber layer, and cellulose fibers filtration efficiency. Figure made by the authors using data from [127].

The second area of activity aimed at reducing friction and wear of elements of the friction pair "cylinder liner—piston rings" and other friction bonds is focused on the application of innovative technologies. Surface modification and coating, suitable lubricating oil or surface texturing are important techniques for reducing friction between two friction surfaces. One of the factors preventing excessive wear of engine elements is the application of antiwear coatings on the most thermally stressed and wear-prone elements of the piston–ring–cylinder unit. The results of research on the influence of introduction of these coatings for various parameters of internal combustion engines, both car and aircraft, are presented in [128–135].

Examples include thermal application of advanced cylinder liner coatings such as nickel and diamond-like carbon nanocomposite [131], coating of piston rings (PR) with molybdenum (Mo) powder [132], coating of diamond-like carbon (DLC) on piston rings and Fe coating on cylinder liner [133], coating of alumina nanopowder on piston rings and CrN coating on cylinder liner [134], and thermal spraying of coatings on ring faces [135].

Surface texturing of piston rings and cylinder liners is an innovative technology to reduce friction and improve wear resistance of these most heat and strength stressed components. After conventional honing, the surface texture of cylinder liners can be changed by a number of techniques such as flat honing, helical honing, laser honing, and laser texturing [136]. The honing angle or the angle of the transverse notch created by the honing process is believed to affect the tribological behavior of the sleeve–ring tribosystem. Surface texturing, as an effective method to improve the tribological properties of P-PR-CL friction pairs, has received much attention in the relevant scientific literature [137–141]. Studies have included the influence of texture parameters: the shape of indentations, their size, and surface ratio on the reduction of friction and wear of sliding elements in an engine [137], the influence of honing angles (20–100°) under boundary lubrication conditions on the tribological behavior of the "cylinder liner–piston and piston ring" pair [138], effect of topography of DLC coating applied on piston ring and cylinder liner on tribological characteristics of friction pair [139], effect of texture of thread grooves with different widths on cylinder liner surfaces and texture of circular indentation on piston ring on diesel engine performance [140], and effect of normal load on changes of liner surface texture [141].

According to the authors [142], well-chosen surface texturing parameters can improve the tribological properties of lubricated surfaces under relative sliding motion conditions by creating a hydrodynamic lift phenomenon at the interface between the ring and cylinder liner surfaces. The test results of the authors of paper [143] showed that under different lubrication conditions, laser texturing of the cylinder liner can effectively reduce the friction force in the reciprocating motion of the ring-sleeve pair. The average friction coefficients of the laser textured cylinder liner-ring pair are reduced by about (18.6–37.6%). The authors of the paper [144] optimized textures with diameters of 100 μm and 120 μm on a parabolic piston ring and showed a reduction in friction of 27% and 21% compared to a ring without texture.

The authors of the paper [145] stated that reduction of friction and wear can affect engine performance and can be achieved by using proper honing technique.

In [146], the effect of microstructure laid on the cylinder liner of an internal combustion engine on the lubrication condition of the control ring was investigated. Measurements using a floating sleeve engine showed that the microstructure improved lubrication conditions by reducing hydrodynamic friction.

Among all these methods, surface texturing has become one of the main methods to improve the tribological properties of mechanical pairs undergoing relative sliding motion.

In conclusion of this section of the analysis, it should be stated that a large number of factors influence the wear of engine friction pair components. These are dust parameters (grain size, hardness, and shape), properties of the material from which engine elements are made, engine operating conditions (rotational speed, load size, operating temperature), efficiency and accuracy of inlet air filtration, quality of servicing air and oil filters, type and quality (sulfur content) of fuel used, and conditions (value of dust concentration in the air) in which a car is used. It is impossible to carry out research comprising all these factors. Therefore, for a number of years, experimental tests have been carried out, which evaluate the wear of only single or several factors simultaneously. Each researcher uses a different methodology and test conditions. Moreover, the researchers use different wear indicators. Therefore, the comparison of results is not always unambiguous.

3. Effect of Wear of P-PR-CL Junction Elements on Engine Operation

Wear and tear of the cylinder liner and piston rings caused by contaminants that penetrate the engine cylinders with the intake air as well as contaminants contained in the oil causes a loss of compressed medium and thus a pressure drop at the end of the compression stroke. The result is a decrease in engine power and an increase in specific fuel consumption. Excessive clearance in the P-PR-CL connection is the cause of increased flow of exhaust gases into the oil sump, which increases the temperature of the lubricating oil, decreases its lubricating properties and causes the oil to be blown out by the exhaust gases. The effect of this phenomenon is the disappearance of the "oil film", resulting in the system moving from fluid friction conditions to boundary friction. The increased clearance in the P-PR-CL combination intensifies the phenomenon of pumping action of the piston rings, thereby increasing oil consumption and exhaust gas toxicity.

Figures 42–47 show the results of the P-PR-CL assembly wear and its effect on changes in power, torque, specific fuel consumption, and exhaust gas blow-by into the crankcase of a 1.3 dm^3 turbocharged 4-cylinder ZS engine with charge air cooling and exhaust gas recirculation [45]. This is a passenger car engine with a maximum power of 66 kW achieved at 4000 rpm and a maximum torque of 200 Nm in the range 1750–2250 rpm. The honed cylinder liners had narrow, deep grooves (lubricating oil reservoirs) in the upper part made using laser machining.

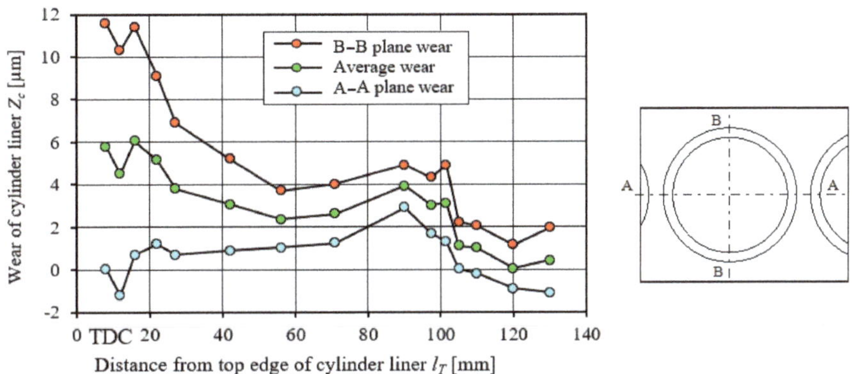

Figure 42. Average cylinder liner wear of a 4-cylinder turbocharged Diesel engine with N_{emax} = 66 kW in the plane perpendicular (B-B) and parallel (A-A) to the engine axis [45].

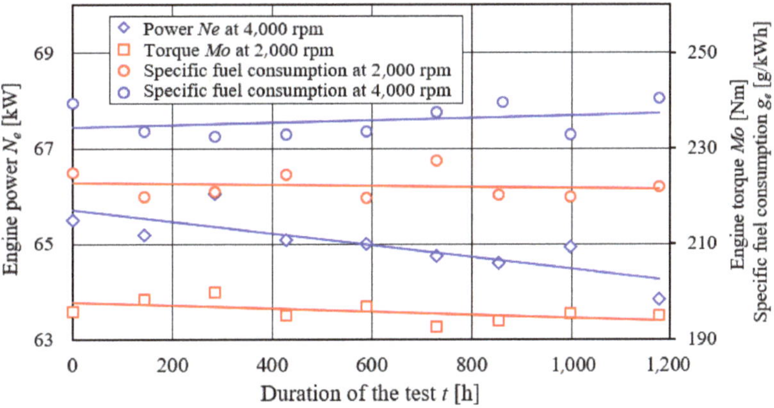

Figure 43. Variation of power and specific fuel consumption at 4000 rpm and torque and specific fuel consumption at 2000 rpm. Figure made by the authors using data from [45].

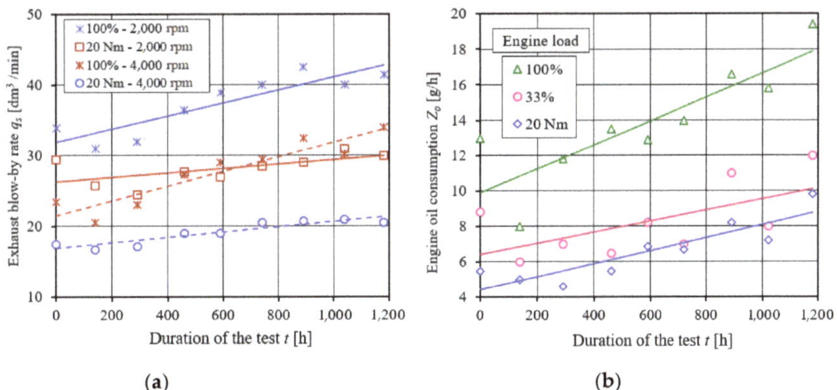

Figure 44. Effect of test duration on (**a**) exhaust blow-by volume at full and light engine loads, (**b**) engine oil consumption at different engine reliefs. Figure made by the authors using data from [45].

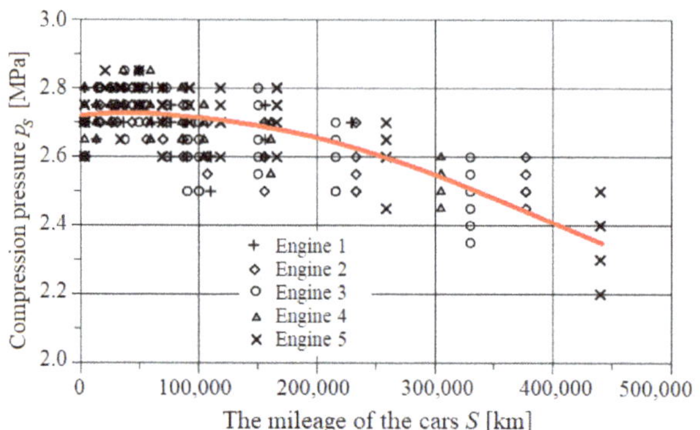

Figure 45. Changes in compression pressure in engine cylinders as a function of car mileage. Figure made by the authors using data from [147].

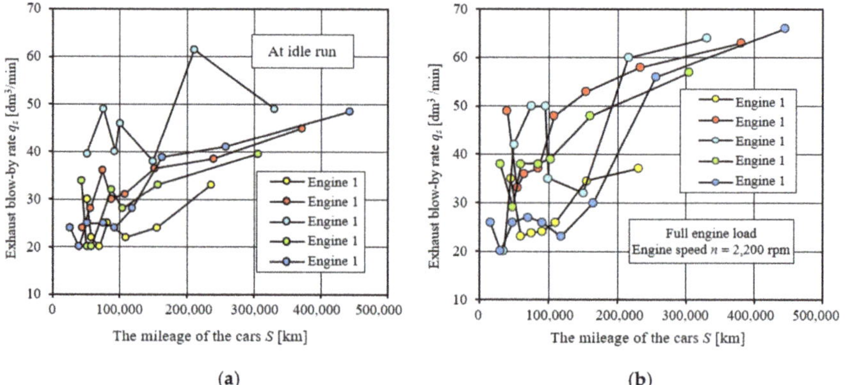

(**a**) (**b**)

Figure 46. Changes in the intensity of exhaust gas blow-by depending on the course of cars during engine operation: (**a**) at idle, (**b**) at full load with 2200 rpm. Figure made by the authors using data from [147].

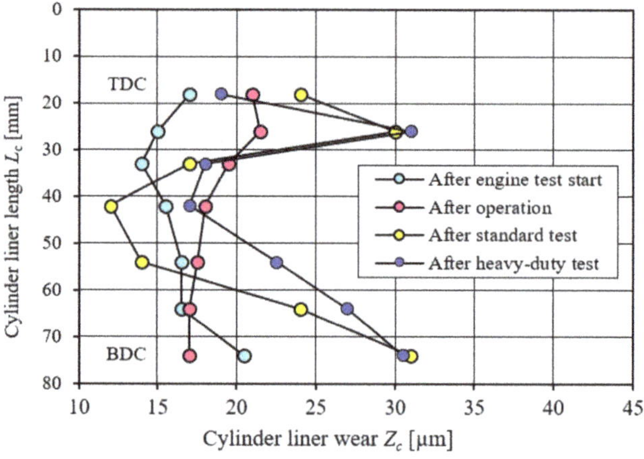

Figure 47. Cylinder liner wear profiles under different engine operating conditions and during engine start-up. Figure made by the authors using data from [148].

Long-term durability tests were conducted on the dynamometer bench, during which each engine was operated under heavy load conditions according to a special cycle that was repeated 336 times. The total operating time of each engine during the tests was 1200 h. Every 42 test cycles or so, speed characteristics and other measurements were made to determine motor performance.

Figure 42 shows the cylinder liner wear averaged over all cylinders of both engines. The cylinder liner diameters in the direction parallel to the engine axis changed little, while in the perpendicular direction the diameter increments were larger, especially in the upper part of the liner. It should be assumed that the above changes in diameters are a consequence of both wear of the sleeve surfaces and their deformation. The average values for both directions indicate greater wear of the sleeves near the turning positions of the ring pack, especially the upper turning position.

The wear of the cylinder liner and piston rings caused by impurities penetrating into the cylinder liner together with the inlet air, as well as impurities contained in the oil, resulted in a decrease in the tightness of the piston crown space. As a result, a loss of the compressed medium occurred, and thus a pressure drop at the end of the compression stroke, and in consequence a decrease in the power of the tested engine by about 2.5% and an increase in specific fuel consumption by 3.4% (Figure 43).

Wear of the P-PR-CL association is at the same time an increase in the intensity of exhaust gas blow-by into the crankcase (Figure 44a), which causes an increase in the temperature of the lubricating oil, a decrease in its lubricating properties, and oil blow-by of the exhaust gases. The effect of this phenomenon is the disappearance of the "lubricating wedge," resulting in the system moving from fluid friction conditions to boundary friction.

After 1200 h of engine operation at full load and 4000 rpm, the blow-by rate increased by 61%. This greatly accelerates the degradation of engine oil. The increased clearance in the P-PR-CL combination intensifies the phenomenon of pumping action of piston rings, thereby increasing oil consumption and exhaust toxicity. At the same time, engine oil consumption increased by 108%, 96%, and 113% at 100%, 33%, and 20 Nm loads, respectively (Figure 44b).

The results of testing the integrity of the combustion chamber of 5 examples of six-cylinder Diesel engines are presented in the paper [147]. These were engines with displacement V_{ss} = 6.8 dm^3 and power N_e = 110 kW used for driving trucks, which were exploited in similar conditions with an average mileage of 10,000 km per month. Changes in combustion chamber tightness as a result of increased wear of engine cylinder liners were determined on the basis of compression pressure measurements and the intensity of exhaust gas blow-by into the crankcase. The measurements of the intensity of the blow-by gases to the crankcase were carried out when the engine was idling and on the chassis dynamometer at full engine load in the range of the crankshaft rotational speed n = 1,570–2800 rpm. Until the car reached the mileage of 100,000 km, the measurements were performed every 15,000 km, and after exceeding 100,000 km, every 50,000 km.

As a result of increased wear of the engine cylinder liners, the P-PR-CL junction tightness decreases, and thus the pressure at the end of the p_s compression stroke decreases and the exhaust gas flow rate (blow-by) to the crankcase increases. With the mileage of vehicles in the range of 0–500,000 km, the values of engine compression pressure decrease, but a definite decrease occurs only after the mileage of about 80,000 km (Figure 45). The measurement results were characterized by significant scatter.

The average compression pressure value after 500,000 km decreased by only 16% compared to the value at zero mileage. Exhaust blow-by intensities increased significantly: by 116% at idle, while the increase at full load was: 117% at 1570 rpm, 149% at 1880 rpm, 194% at 2200 rpm and 51% at 2800 rpm (Figure 46)

This was most probably due to the fact that the engines were being run-in during their first period of operation. Above 50,000 km, the blow-by volume increased almost linearly with the vehicle mileage. The values of the intensity of the exhaust gas blowing into the crankcase of a fully loaded engine (Figure 46b) have higher values, especially after the mileage of $S = 100{,}000$ km.

Figure 47 shows the results of engine cylinder liner wear after bench and operational tests [148]. The values of wear of cylinder liners during three reliable bench tests were compared with the results of long-term operation of the same type of compression-ignition internal combustion engine in a vehicle. Wear tests of cylinder liners of a 4CT90 internal combustion Diesel engine were carried out for the following operating conditions [148]:

- Tests during long term operation (5 engines);
- Bench testing according to the "standard test" described in BN-79/1374-04;
- Tests on a laboratory test bench according to "heavy load test";
- Tests on a laboratory test bench according to "start-up test".

The object of this research was a 4-cylinder internal combustion engine with ZS 4CT90, which is the driving unit of a Lublin delivery truck manufactured by WSK "Andoria". This is an engine with indirect injection, displacement $V_{ss} = 2.417$ dm^3, maximum power 63.5 kW at 4100 rpm, and maximum torque of 195 Nm at 2500 rpm.

In order to compare the obtained results of wear and tear of the test bench tests with the wear and tear observed during operation of the vehicle (mileage of about 100,000 km), it was assumed that 1 h of operation of the engine on the dynamometer during the "heavy load test" and the "standard test" corresponds to the mileage of the vehicle of 60 km, while start-ups take place every 7 km on average.

The average wear of the engine cylinder liners on the test bench after the "starting test" represents 85.5% of the total average wear during engine operation in a vehicle that has performed the same number of starts as during the starting test. For the "standard test" and "heavy load test", the average wear is higher and represents 103.4% and 107.5% of the average wear during engine operation in the vehicle, respectively.

A comparison of the cylinder liner wear profiles shows that for both the "standard test" and the "heavy load test" the highest wear values occurred near the upper and lower reciprocating positions of the piston rings. Of all three bench tests compared, the wear profile that was closest to the operating profile was obtained during the "start-up test".

The increased wear of the cylinder liners in the TDC area is due to several reasons:

- The action of high exhaust gas temperature on the piston bottom and the first piston ring, which causes a decrease in oil viscosity and a decrease in oil film thickness, and consequently a transition from liquid to boundary lubrication;
- Low value of piston velocity (in TDC piston velocity is 0) which is the cause of decrease of oil film thickness;
- Penetration of dust grains between mating surfaces of piston liner and piston rings, which get into the upper part of the cylinder liner together with the intake air;
- Increased wear in the plane perpendicular to the engine axis.

The research on the influence of wear of piston-cylinder unit elements of five 4-cylinder engines ($N_e = 85.3$ kW) on the changes of power, torque, specific fuel consumption and exhaust gas toxicity was presented in the paper [44]. The study was conducted in two stages: the first stage of the study involved running the engines in the run-in period for 10 h at 1400–2800 rpm and variable load, the second stage lasted 21 h of engine operation and consisted of seven 3-h cycles of 2 h 50 min full load at 2800 rpm and 10 min idle. During the tests, test dust was dosed into the engine manifold at a rate of 1.3 g/h, corresponding to an ambient dust concentration of approximately 8 mg/m^3.

The tests showed noticeable differences between the engines in the obtained parameters. Statistically, the average power output of the five engines after running-in and after the full test slightly decreased by 2.5% from 83 to 80.9 kW. The same trend was observed for the average torque, which changed from 285.5 to 277.9 Nm, also a reduction of 2.5%. Fuel consumption increased by 3.6% from 364.3 to 377.4 g/(kWh), while total engine efficiency decreased by 2.2% from 22.6% to 22.1%. A larger change was seen for exhaust emission factors: CO increased by 35% and CH by 40.8%.

The most reliable in the field of wear of engine elements are experimental tests conducted on a real object. In the available literature, the results of wear of P-PR-CL engine elements are presented, however, each researcher used an engine with different parameters and applied a different testing methodology. Engine bench tests and in-service tests were conducted. A common feature of these tests is the determination of the wear degree of the P-PR-CL friction pair and the effects of wear in the form of blow-by in the crankcase, a decrease in engine power and increased exhaust emissions. In engineering practice, the parameter that determines the degree of wear of the P-PR-CL association and evaluation of engine technical efficiency is the "pressure value at the end of the compression stroke".

4. The Effect of Oil Contaminants on Engine Wear and Operation

Exterior contaminants, mainly mineral dust particles, are introduced into the engine oil via the engine's supply system together with air and fuel as well as during maintenance work. As a result of the piston's movement in the BDC direction, the rings scrape oil together with mineral dust particles from the cylinder head into the oil sump. The engine oil also contains internal contaminants such as particles of dust and metals that were not removed during production, wear products of engine components, products of incomplete combustion and products of chemical conversion. The concentration of contaminants in oil is a function of the operating time of the oil in an engine and depends on: oil type and properties, amount of oil added, type of oil filtration system, as well as operating conditions. Contaminants in the oil of the lubricating system are distributed through the oil system to those tribological areas of the engine that are lubricated, e.g., to the journal–crankshaft, journal–camshaft, valve guide–valve stem combinations. Contaminants in the oil cause scratches and damage to the mating surfaces (Figure 48). They may settle in the material of bearing shells, and the effects of their presence in the form of abrasive wear will be felt even after oil change, which is the basic form of removing contaminants from the lubrication system.

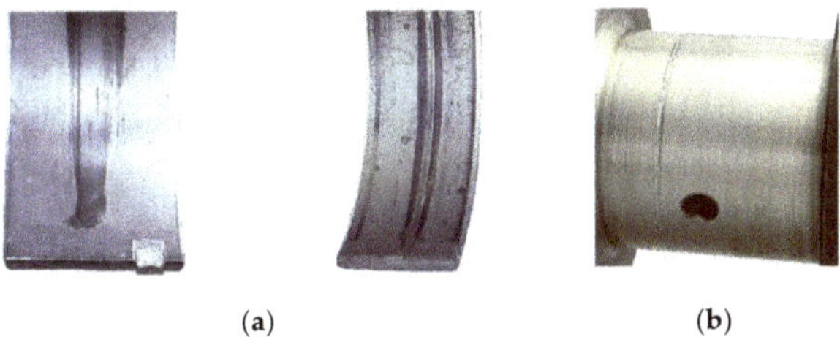

(a)   (b)

Figure 48. Wear view of pan (**a**) and crankshaft journal (**b**) of a car engine operated with a faulty air filtration system [149].

The effect of oil change frequency on the intensity of wear of the P-PR-CL junction and, consequently, on the decrease in pressure at the end of the compression stroke and the decrease in power and increase in specific fuel consumption of the Cummins N 14 engine was presented in [81]. The engine in which oil change was performed cyclically, every 25,000 miles, registered more than 18% decrease in power compared to the engine new and the engine in which oil change was performed twice as often (Figure 49).

Along with oil and filter changes, contaminants that caused accelerated wear of P-PR-CL association components were removed. Changing the oil at a lower frequency caused the accumulation of contaminants and an increase in their concentration, and thus increased wear of this association. This resulted in increased charge loss with leaks of the P-PR-CL association during its compression in the cylinder. The loss of charge mass from the cylinders is a decrease in engine inflation and power.

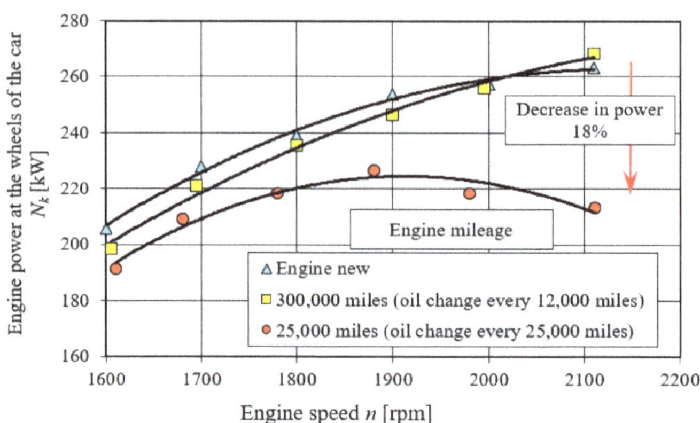

Figure 49. Power change of Cummins N14 engine new and after various vehicle mileages with oil changes every 12,000 and 25,000 miles. Figure made by the authors using data from [81].

The main source of contamination in engine oil are the dust particles that are carried with the air into the engine's cylinders and then transported to the oil sump by the piston rings. There are also other contaminants in the engine oil: metallic wear products, carbon deposits, and soot from the exhaust gases, fuel, and coolant. The grains that came into contact between the surfaces of the piston rings and the cylinder liner and were involved in the wear process may have crumbled into smaller grains and been smoothed out. Therefore, their influence on the wear of those engine components that are lubricated with oil may be different. The only way to reduce the effect of contaminants on engine components is to use oil filters and to change oil and filters regularly.

5. Effects of Contaminants on Air Flow Meter Performance

One of the most significant sensors that are on an internal combustion engine is the mass airflow (MAF) sensor used to measure the air entering the engine and determine the exact amount of fuel to be injected. The MAF sensor is a device that directly measures the mass of air entering the engine and sends a voltage signal to the engine control unit. Electronic engine fuel injection systems use the MAF sensor to control the air/fuel ratio. In doing so, they maintain the desired stoichiometric composition of the fuel mixture to achieve the best balance between fuel economy and emission reduction. Therefore, a high level of operational accuracy is required from the air flow meter sensor. MAF sensors are calibrated under steady flow conditions without turbulence or vortices with a long straight pipe to provide a turbulence-free, fully developed velocity profile. Unfortunately, these flow conditions are not maintained when MAF sensors are used in engines. Air flow causes turbulence around the sensor and interference with the sensor signal, resulting in unstable mass flow measurements [150].

To measure the mass of air flowing into the engine, mainly two types of mass air flow meters are used: with wire thermo-anemometer HLM (hot wire) and with layer thermo-anemometer HFM (hot layer). As air flows past the hot wire, the wire cools, reducing its resistance, which in turn allows more current to flow through the circuit. As the current flows, the temperature of the wire increases until the resistance reaches equilibrium again. The amount of current needed to maintain the temperature is directly proportional to the mass of air flowing through the wire.

The basic element of the flowmeter is a cylindrical housing (Figure 50) through which the air flow to the engine passes. The measuring element of the Bosch wire thermo-anemometer flowmeter is a 70-μm-thick platinum wire stretched between the walls of the housing perpendicular to the direction of the air stream flow.

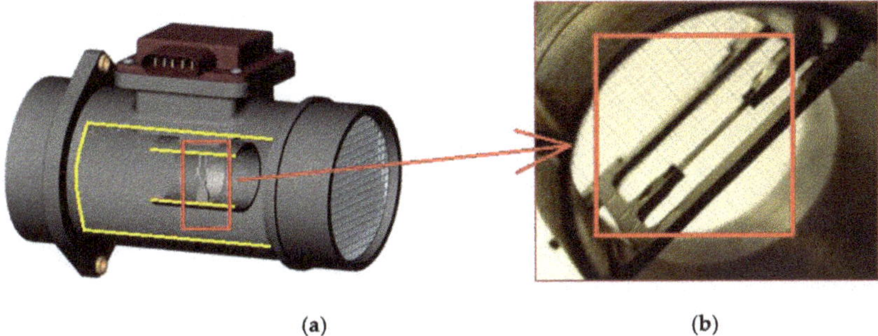

(a)   (b)

Figure 50. Air flow meter with wire thermo-anemometer: (**a**) general view, (**b**) measurement system The figure was made by the authors based on data from [151].

Layer thermo-anemometer (hot layer) air flow meters are built with three main components (Figure 51):

- The housing (measuring tube), which is built into the intake system after the air filter;
- A sensor (measuring element and electronics) located in the measuring tube;
- An inlet grid located at the inlet of the housing that protects against contaminants and forces turbulent flow, similar to a wire thermo-anemometer flowmeter.

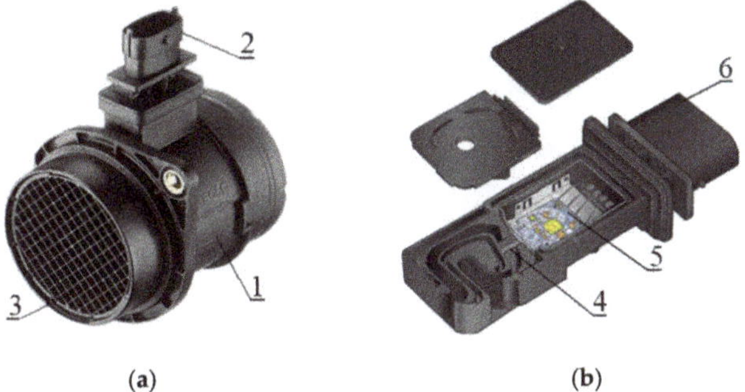

(a)   (b)

Figure 51. Bosch HFM6 type air flow meter: (**a**) general view (1—housing, 2—sensor, 3—inlet grille), (**b**) sensor (4—measuring element, 5—electronics, 6—digital interface). The figure was made by the authors based on data from [152].

The dust grains which are not stopped by the air filter move with the inlet air stream into the engine cylinders with an average speed $v_z = 15$–25 m/s, depending on the cross-sectional area of the inlet duct. Some of the grains impact and some are deposited on the measuring element, forming a kind of thermal shield. The "wire flow meters" in particular are exposed to this type of contamination. The dust grains that are not stopped by the air filter settle on the surface of the wire, which traps the dust grains like a filter fiber in a porous filter baffle.

At the same time, the grains of SiO_2 and Al_2O_3 minerals in the flowing air stream, which are characterized by sharp edges and very high hardness, strike the gauge wire, crippling and scratching its surface (Figure 52) and thus weakening the entire wire. A dust grain with a diameter $d_p = 5$ μm moving with a velocity $v_z = 15$ m/s hits the measuring element with a kinetic energy $E_k = 1.56 \times 10^{-10}$ J. A dust grain with a diameter $d_p = 80$ μm (in the case of a faulty filter) moving with the same velocity has a kinetic energy more than 4000 times higher.

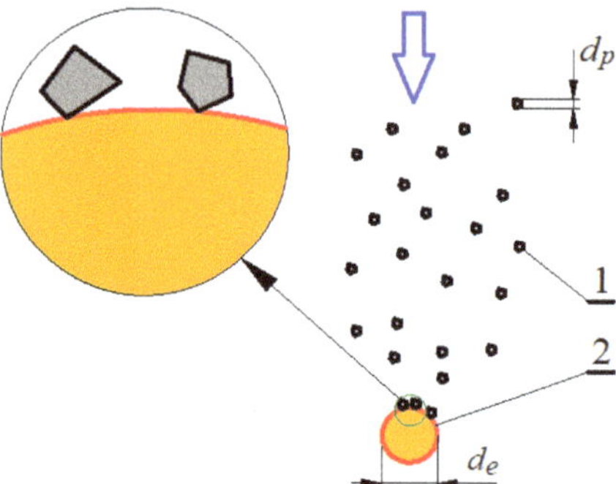

Figure 52. Flow diagram of dust grains hitting the heating wire of HLM air flow meter: 1—dust grains with diameter $d_p < 5$ μm, 2—heating element, $d_e = 70$ μm—diameter of the heating wire.

A malfunctioning filter or a leak in the air supply system causes accelerated wear of the flow meter heating wire, which leads to its failure even after several thousand kilometers. Over time, some of the dust grains are retained on the wire and form a layer whose thickness increases with the number of incoming grains—Figure 53.

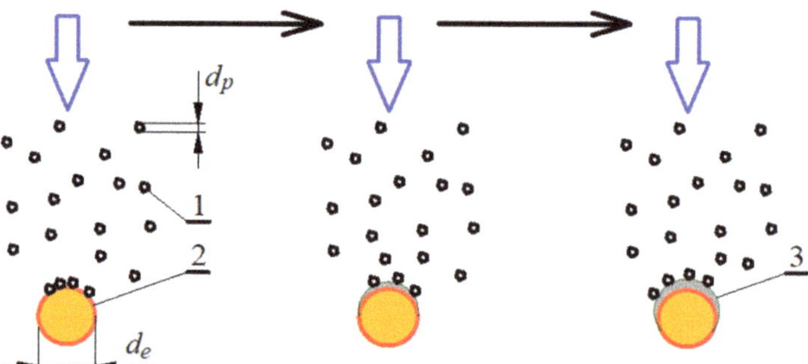

Figure 53. Diagram of dust grains deposition on the heating element (wire) of the LFM air flow meter: 1—dust grains with diameter $d_p < 5$ μm, 2—heating element, 3—layer formed by dust grains, $d_e = 70$ μm—diameter of heating element (wire).

The resulting layer, whose basic component is silica SiO_2, is a good insulator that has a thermal conductivity coefficient of $\lambda_s = 0.651$ W/m-K, which is more than 100 times lower than the thermal conductivity coefficient of the material (platinum $\lambda_p = 71$ W/mK) from which the heating wire is made.

Moisture contained in atmospheric air can also settle on the measuring element. The flowing air cools the dust-laden heating wire much less intensively. Thus, the value of the current needed to heat the wire is lower. The information about the value of the air stream sent to the on-board computer is different than if the heating wire was completely clean.

The engine intake air stream also contains contaminants in the form of:

- Oil particles which are fed into the air supply system via the crankcase ventilation system, e.g., as a result of a defective oil separator;
- Carbon deposited in the flow meter due to a defective exhaust gas recirculation system.

These deposits on the measuring element of the flow meter are usually caused by the reverse flow of the air stream in the direction from the engine to the air filter. These backflows are most often caused by pulsations caused by cyclic intake of air to individual engine cylinders during intake valve opening periods.

Layer flow meters are more resistant to contaminations, since only a part of the air stream flows over the measuring element. At the inlet and outlet of the casing there is a mesh (Figure 54) whose task is to protect the inside of the flowmeter from contaminants and also to force a turbulent air flow through the flowmeter throat.

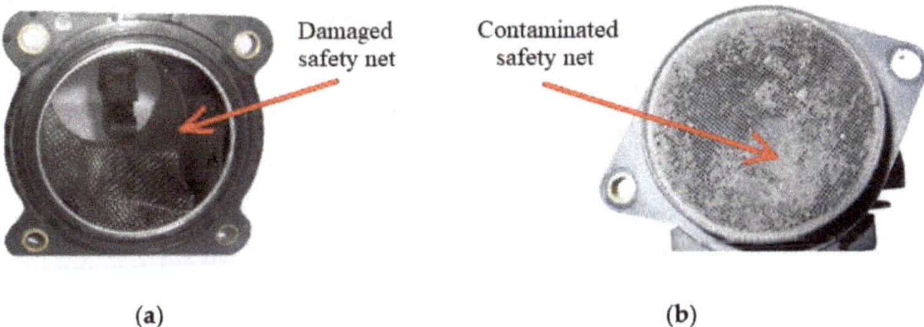

Figure 54. Protective mesh of flow meter with layered anemometer: (**a**) damaged, (**b**) contaminated with dust. The figure was made by the authors based on data from [153].

A view of the layered anemometer flow meter measuring element with mechanical damage caused by mineral dust grain impacts is shown in Figure 55.

Figure 55. Mechanical damage to the measuring element of the layered anemometer flow meter due to impacts of mineral dust grains. Figure made by the authors using data from [153].

Salt found in the intake air and coming from the ground can also be deposited on the surface of the sensor layer, especially in winter (Figure 56).

Figure 56. Salt contaminated flow meter sensor measurement element with layered anemometer. Figure made by the authors using data from [154].

Oil mist may settle on the gauge sensor (Figure 57). The presence of engine oil in the air stream (load) is a consequence of the crankcase ventilation system. The exhaust gases always contain a small amount of oil despite passing through the filtration system. If the oil separator of the ventilation system is damaged, the amount of incoming exhaust gas and thus oil is particularly high. The accumulation of such contaminants is generally caused by a backflow of the air stream in the direction from the head intake ducts to the air filter. This backflow is usually caused by pulsation of the air stream with insufficiently closed intake valves (or worn intake valves), through which the medium from the engine cylinders enters during the compression stroke.

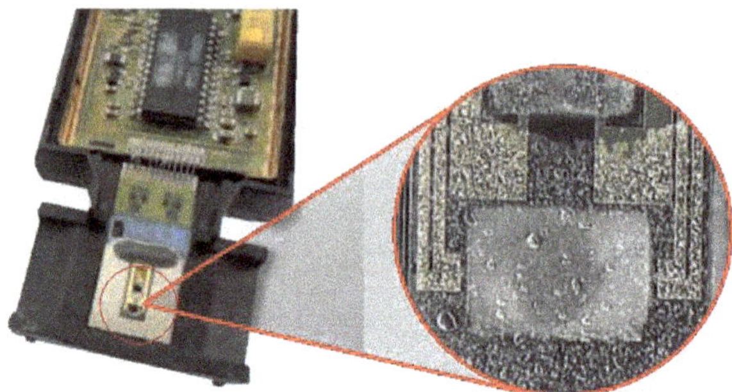

Figure 57. Oil-contaminated flow meter sensor measuring element with layered anemometer. Figure made by the authors using data from [153].

Deposition of this type of contamination is generally caused by backflow of the airflow in the direction from the head intake ducts to the air filter.

Figure 58 shows the results of laboratory testing of the $U_w = f(Q_m)$ characteristic of a Bosch HFM5 stratified air flow meter whose measuring element was contaminated with dust, followed by oil and dust. When the measuring element was contaminated with road dust there was an approximately 12% (Figure 58), drop in the output voltage U_w over the entire range of air flow Q_m. In the case when the measuring element was contaminated with oil and dust, the drop in output voltage U_w is much greater and for the maximum air flow $Q_m = 526$ m^3/h is 50%.

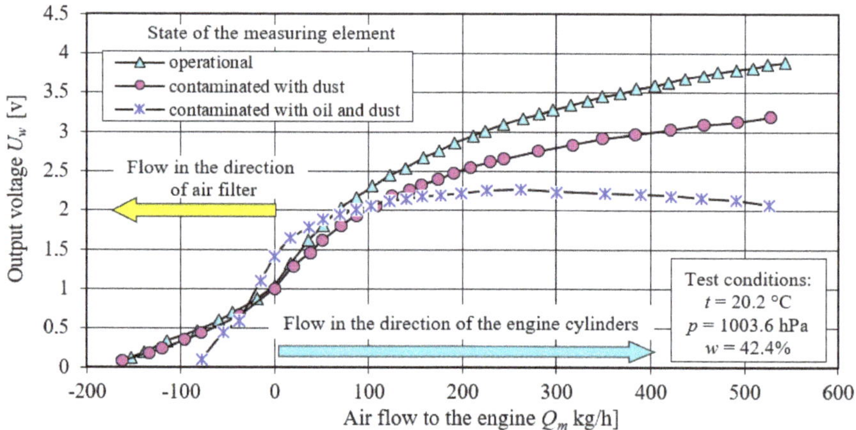

Figure 58. Characteristics $U_w = f(Q_m)$ of HFM5 layer flowmeter made by Bosch, used in Nissan cars, in working order and with the measuring element contaminated with dust.

Contamination of the flowmeter measuring element with road dust causes its isolation because silica SiO_2, which is the main component of dust, is a good insulator both electrically and thermally. The layer of dust covering the measuring resistors is the cause of reduced intensity of their cooling, which results in lower value of the current needed to maintain a constant temperature difference between them. As a result, there is a decrease in the output voltage U_w in relation to the output signal of the flowmeter in good condition. The higher value of the voltage drop U_w in the case of oil and dust contamination is probably due to a thicker layer of dust deposited on the measuring element, where oil acts as a binder.

In real engine operating conditions for lower values of output voltage U_w the on-board computer reads lower values of the air flow and in order to maintain the appropriate composition of the mixture doses a smaller mass of fuel. As a result, the engine power decreases.

Dust grains which are sucked in together with the air have a destructive effect not only on the friction couples of the engine but also on the measuring element of the air flow meter. In the air supply system of the modern engine there is a turbocharger. The basic element of this device are two rotors connected by a shaft. Rotor located in the exhaust system of the engine acquires a rotational speed (about 240,000 rpm) from the exhaust stream, which is transferred to the rotor located in the inlet system of the engine. High velocities of the air stream flowing onto the compressor blades (50–80 m/s) and of the exhaust stream (over 300 m/s), high peripheral velocities of rotor units (200–500 m/s) cause that the dust grains, upon contact with the surfaces of these elements, have significant kinetic energy which results in high impact force of the grains. This results in tearing out metal microparticles from the surface of the parts, violation of their surface structure and geometric shapes. The main effect of the impact of the dust grains on the rotor is the accelerated erosive wear of the rotor in the form of a change in geometry and deterioration of the smoothness of their surfaces and, consequently, a decrease in the efficiency and durability of the engine. The authors did not analyze this problem, as it is a topic for a separate article.

6. Conclusions

This paper provides a systematic review of the research progress in the field of the influence of engine intake air pollutants on the accelerated wear of the elements of the combination: piston, piston rings, cylinder (P-PR-CL) and plain bearing (journal–panel). Commonly harmful for internal combustion engines in operation is road dust, which enters the engine cylinders mainly with the intake air, causing accelerated wear, and thus reducing their reliability and durability. The main components of road dust are silica (SiO_2) and corundum Al_2O_3, whose proportion in dust reaches 60–95. The hardness of these components evaluated on the basis of the ten grade Mohs scale is 7 and 9, respectively, and significantly exceeds the hardness of construction materials used in engine construction.

Only 10–20% of the dust mass which enters the engine with the air through the intake system is deposited on the oil-coated cylinder liner walls. These are mineral dust grains with a size of less than 2–5 µm, as this is the accuracy provided by the cellulose-based filter materials commonly used. The dust together with the oil forms a kind of abrasive paste, which penetrates between the mating surfaces of the engine parts, e.g., piston-piston rings-cylinder wall, causing their abrasive wear.

As a result of the movement of the piston towards its bottom dead center (BTC) the rings scrape oil and contaminants from the cylinder floor into the oil sump. After reaching the lubrication system, the contaminants are distributed via the oil system to those tribological areas of the engine that are lubricated by oil and thus subject to accelerated wear. These include, for example, friction pairs: the "journal–cup" plain bearings of the crankshaft and camshaft, and the valve guide–valve stem assembly.

This paper presents an extensive analysis of the effect of three basic dust parameters (grain size and hardness and dust concentration in air) on the accelerated wear of the friction pair: piston, piston rings, cylinder (P-PR-CL) and plain bearing (journal–cup). It was pointed out that the wear values of the same component were obtained by different researchers using different testing techniques and under different conditions and evaluated by different indices. However, regardless of the different methodologies, the researchers' conclusions are similar.

(1) The most dangerous for two mating parts are dust particles whose diameter d_p is equal to the oil film thickness h_{min} between two surfaces at a given moment. In an internal combustion engine, there are many mating parts lubricated by engine oil, where the oil film thickness depends on the conditions and parameters of engine operation and oil properties, and therefore takes on different values in the range h_{min} = 0–50 μm;

(2) Abrasive wear of engine components is mainly caused by particles of 1–40 μm, with dust grains of 1–20 μm being the most dangerous. Grains smaller and larger than this range cause the same amount of wear;

(3) Depending on the type of association, its components experience maximum wear by a different range of particle sizes. It can be caused by different dynamic clearances between moving parts as well as different operating conditions and different measuring techniques;

(4) The upper piston ring and cylinder liner are most susceptible to abrasive wear. Maximum wear occurs at the top of the cylinder liner, in a plane perpendicular to the crankshaft axis, where the ring reaches TDC. This is due to the particularly unfavorable conditions for liner-ring interaction at this point: high temperature (oil viscosity drop) and low piston speed, the effect of normal force on the cylinder walls is greatest, high pressure of exhaust gases on the rear wall of the rings. As a result, the value of the oil film decreases significantly or disappears completely. This is the reason for the occurrence of boundary friction and increased wear.

However, more reliable tests of wear of engine elements are those carried out during experimental bench tests or during operation of engines in real car exploitation. Such research is labor-consuming and expensive; hence, the number of research results in the literature is insignificant. Excessive wear of the P-PR-CL junction elements causes several unfavorable phenomena in engine operation. First, there is increased blowing of the compressed charge and hot exhaust gases into the crankcase. As a result, the pressure decreases at the end of the compression stroke, and as a consequence, the torque and power of the engine decrease and the specific fuel consumption increases. Increased blow-by of hot exhaust gases into the oil sump causes an increase in the temperature of the P-PR-CL bonding elements and the oil, which in turn lowers its viscosity. As a result, the thickness of the oil film decreases, and consumption increases. This phenomenon becomes particularly important when the piston is located near TDC, where the conditions of oil film formation, due to low piston speed, high temperature and high load, are extremely unfavorable. In addition, hot exhaust gases containing soot and other contaminants flow into the crankcase and adversely affect (degrade) the oil there.

An inherent phenomenon of complicated piston ring movements during their operation is pumping oil over the piston bottom into the combustion chamber. The burning oil increases the emission of toxic exhaust components. It should be noted that blow-through of compressed cargo and exhaust gases to the crankcase occurs in every engine, but with the mileage of the vehicle and the progressive wear of the P-PR-CL combination elements, this phenomenon increases and may have an impact on the increased emission of toxic components of exhaust gases.

The unfavorable phenomenon of friction and wear of friction components of an engine, resulting mainly from the presence of dust, is an inherent process of car engine operation. This phenomenon can be minimized by:

- use of filtration materials with the use of nanofibers, which increase the effectiveness and accuracy of engine inlet air filtration—grains above 1 μm,
- use of wear-resistant ring and cylinder surfaces by spraying
- Texturing of ring and cylinder surfaces,
- use of suitable lubricating oils and fuels.

Dust in the air drawn into the engine is trapped and deposited on the measuring element of the air mass sensor. Other contaminants (moisture, salt, oil) also settle on this element, but dust grains are particularly dangerous due to their hardness and sharp edges. Firstly, the deposited layer of dust is mainly silica, which is a good insulator and deteriorates the heat exchange between the measuring element material and the flowing air, which is the essence of the flow meter operation. The voltage generated by the flowmeter then has a lower value than an efficient flowmeter, which the computer reads as a smaller mass of air and dispenses a smaller mass of fuel. As a result, engine power decreases. In addition, dust grains coming in at high speed scratch the surface of the airflow meter's heating wire and reduce its life, which combined with vibrations from the engine can cause it to break.

Author Contributions: Conceptualization, T.D.; methodology, T.D.; software, T.D. and S.D.D.; validation, T.D. and S.D.D.; formal analysis, T.D. and S.D.D.; investigation, T.D. and S.D.D.; resources, T.D.; data curation, T.D.; writing—original draft preparation, T.D. and S.D.D. All authors have read and agreed to the published version of the manuscript.

Funding: This research received no external funding.

Institutional Review Board Statement: Not applicable.

Informed Consent Statement: Not applicable.

Data Availability Statement: Data is contained within the article.

Conflicts of Interest: The authors declare no conflict of interest.

References

1. Nikas, G.K. A state-of-the-art review on the effects of particulate contamination and related topics in machine-element contacts. *Proc. Inst. Mech. Eng. Part J J. Eng. Tribol.* **2010**, *224*, 453–479. [CrossRef]
2. Shang, X.; Lee, M.; Lim, S.; Gustafsson, Ö.; Lee, G.; Chang, L. Dust Criteria Derived from Long-Term Filter and Online Observations at Gosan in South Korea. *Atmosphere* **2021**, *12*, 1419. [CrossRef]
3. Chun, Y.; Kim, J.; Choi, J.C.; Boo, K.O.; Oh, S.N.; Lee, M. Characteristic number size distribution of aerosol during Asian dust period in Korea. *Atmos. Environ.* **2001**, *35*, 2715–2721. [CrossRef]
4. Luo, H.; Zhou, W.; Jiskani, I.M.; Wang, Z. Analyzing Characteristics of Particulate Matter Pollution in Open-Pit Coal Mines: Implications for Green Mining. *Energies* **2021**, *14*, 2680. [CrossRef]
5. Csavina, J.; Field, J.; Taylor, M.P.; Gao, S.; Landázuri, A.; Betterton, E.A.; Sáez, A.E. A review on the importance of metals and metalloids in atmospheric dust and aerosol from mining operations. *Sci. Total Environ.* **2012**, *433*, 58–73. [CrossRef]
6. Li, L.; Zhang, R.; Sun, J.; He, Q.; Kong, L.; Liu, X. Monitoring and prediction of dust concentration in an open-pit mine using a deep-learning algorithm. *J. Environ. Health Sci. Eng.* **2021**, *19*, 401–414. [CrossRef] [PubMed]
7. Baddocka, M.C.; Zobeck, T.M.; Scot Van Pelt, R.; Fredrickson, E.L. Dust emissions from undisturbed and disturbed, crusted playa surfaces: Cattle trampling effects. *Aeolian Res.* **2011**, *3*, 31–41. [CrossRef]
8. Norman, M.; Sundvor, I.; Denby, B.R.; Johansson, C.; Gustafsson, M.; Blomqvist, G.; Janhäll, S. Modelling road dust emission abatement measures using the NORTRIP model: Vehicle speed and studded tyre reduction. *Atmos. Environ.* **2016**, *134*, 96–108. [CrossRef]
9. Rivera, B.H.; Rodriguez, M.G. Characterization of Airborne Particles Collected from Car Engine Air Filters Using SEM and EDX Techniques. *Int. J. Environ. Res. Public Health* **2016**, *13*, 985. [CrossRef]
10. Chan, D.; Stachowiak, G.W. Review of automotive brake friction materials. *Proc. Inst. Mech. Eng. Part D J. Automob. Eng.* **2004**, *218*, 953–966. [CrossRef]
11. Haider, A.; Haider, S.; Kang, I.K. A comprehensive review summarizing the effect of electrospinning parameters and potential applications of nanofibers in biomedical and biotechnology. *Arab. J. Chem.* **2018**, *11*, 1165–1188. [CrossRef]
12. Thorpe, A.; Harrison, R.H. Sources and properties of non-exhaust particulate matter from road traffic: A review. *Sci. Total Environ.* **2008**, *400*, 270–282. [CrossRef]
13. Barr, B.C.; Andradóttir, H.Ó.; Thorsteinsson, T.; Erlingsson, S. Mitigation of Suspendable Road Dust in a Subpolar, Oceanic Climate. *Sustainability* **2021**, *13*, 9607. [CrossRef]
14. Kreider, M.L.; Panko, J.M.; McAtee, B.L.; Sweet, L.I.; Finley, B.L. Physical and chemical characterization of tire-related particles: Comparison of particles generated using different methodologies. *Sci. Total Environ.* **2010**, *408*, 652–659. [CrossRef]
15. Kupiainen, K.J.; Tervahattu, H.; Räisänen, M.; Mäkelä, T.; Aurela, M.; Hillamo, R. Size and composition of airborne particles from pavement wear, tires, and traction sanding. *Environ. Sci. Technol.* **2005**, *39*, 699–706. [CrossRef] [PubMed]
16. Panko, J.; Kreider, M.; Unice, K. Review of tire wear emissions: A review of tire emission measurement studies: Identification of gaps and future needs. In *Non-Exhaust Emissions*; Elsevier: Amsterdam, The Netherlands, 2018; Chapter 7; pp. 147–160. [CrossRef]
17. Oroumiyeh, F.; Zhu, Y. Brake and tire particles measured from on-road vehicles: Effects of vehicle mass and braking intensity. *Atmos. Environ. X* **2021**, *12*, 100121. [CrossRef]

18. Kovochich, M.; Parker, J.A.; Oh, S.C.; Lee, J.P.; Wagner, S.; Reemtsma, T.; Unice, K.M. Characterization of Individual Tire and Road Wear Particles in Environmental Road Dust, Tunnel Dust, and Sediment. *Environ. Sci. Technol. Lett.* **2021**, *8*, 1057–1064. [CrossRef]
19. Merkisz, J.; Pielecha, J.; Radzimirski, S. *Emisja Zanieczyszczeń Motoryzacyjnych w Świecie Nowych Przepisów Unii Europejskiej*; WKŁ: Warszawa, Poland, 2012. (In Polish)
20. Mrozik, M.; Merkisz-Guranowska, A. Environmental Assessment of the Vehicle Operation Process. *Energies* **2020**, *14*, 76. [CrossRef]
21. Lähde, T.; Giechaskiel, B.; Pavlovic, J.; Suarez-Bertoa, R.; Valverde, V.; Clairotte, M.; Martini, G. Solid particle number emissions of 56 light-duty Euro 5 and Euro 6 vehicles. *J. Aerosol Sci.* **2022**, *159*, 105873. [CrossRef]
22. Giechaskiel, B.; Woodburn, J.; Szczotka, A.; Bielaczyc, P. *Particulate Matter (PM) Emissions of Euro 5 and Euro 6 Vehicles Using Systems with Evaporation Tube or Catalytic Stripper and 23 nm or 10 nm Counters*; SAE International: Warrendale, PA, USA, 2020. [CrossRef]
23. Muthu, M.; Gopal, J.; Kim, D.-H.; Sivanesan, I. Reviewing the Impact of Vehicular Pollution on Road-Side Plants-Future Perspectives. *Sustainability* **2021**, *13*, 5114. [CrossRef]
24. Zhao, J.; Peng, P.; Song, J.; Ma, S.; Sheng, G.; Fu, J. Research on flux of dry atmospheric falling dust and its characterization in a subtropical city, Guangzhou, South China. Air Quality. *Atmos. Health* **2010**, *3*, 139–147. [CrossRef]
25. Smialek, J.L.; Archer, F.A.; Garlick, R.G. Turbine Airfoil Degradation in the Persian Gulf War. *JOM* **1994**, *46*, 39–40. [CrossRef]
26. Dzierżanowski, P.; Kordziński, W.; Otyś, J.; Szczeciński, S.; Wiatrek, R. *Napędy Lotnicze. Turbinowe Silniki Śmigłowe i Śmigłowcowe*; WKŁ: Warszawa, Poland, 1985. (In Polish)
27. Cardozo, J.I.H.; Sánchez, D.F.P. An experimental and numerical study of air pollution near unpaved Roads. *Air Qual. Atmos. Health* **2019**, *12*, 471–489. [CrossRef]
28. Rienda, I.C.; Alves, C.A. Road dust resuspension: A review. *Atmos. Res.* **2021**, *261*, 105740. [CrossRef]
29. Fujiwara, F.; Rebagliati, R.J.; Dawidowski, L.; Gómez, D.; Polla, G.; Pereyra, V.; Smichowski, P. Spatial and chemical patterns of size fractionated road dust collected in a megacitiy. *Atmos. Environ.* **2011**, *45*, 1497–1505. [CrossRef]
30. Schaeffer, J.W.; Olson, L.M. Air Filtration Media for Transportation Applications. *Filtr. Sep.* **1998**, *35*, 124–129.
31. Elis, M.; Bojdo, N.; Covey-Crump, S.; Jones, M.; Filippone, A.; Pawley, A. Generalized Predictions of Particle-Vane Retention Probability in Gas Turbine Engines. *J. Turbomach.* **2021**, *143*, 111008. [CrossRef]
32. Jaroszczyk, T.; Pardue, B.A.; Heckel, S.P.; Kallsen, K.J. Engine air cleaner filtration performance—Theoretical and experimental background of testing. In Proceedings of the AFS Fourteenth Annual Technical Conference and Exposition, Tampa, FL, USA, 1 May 2001.
33. Pinnick, R.G.; Fernandez, G.; Hinds, B.D.; Bruce, C.W.; Schaefer, K.W.; Pendelton, J.D. Dust Generated by Vehicular Traffic on Unpaved Roadways: Sizes and Infrared Extinction Characteristics. *Aerosol Sci. Technol.* **1985**, *4*, 99–121. [CrossRef]
34. Barbolini, M.; Di Pauli, F.; Traina, M. Simulation der luftfiltration zur auslegung von filterelementen. *MTZ-Mot. Z.* **2014**, *5*, 52–57. [CrossRef]
35. De Fré, R.; Bruynseraede, P.; Kretzschmar, J.G. Air pollution measurements in traffic tunnels. *Environ. Health Perspect.* **1994**, *102* (Suppl. S4), 31–37. [CrossRef] [PubMed]
36. Dziubak, T. Zapylenie powietrza wokół pojazdu terenowego. *Wojsk. Przegląd Tech.* **1990**, *3*, 154–157. (In Polish)
37. Burda, S.; Chodnikiewicz, Z. Konstrukcja i badania pyłowe filtrów powietrza silnika czołgowego. *Bull. Mil. Univ. Technol.* **1962**, *3*, 12–34. (In Polish)
38. Szczepankowski, A.; Szymczak, J.; Przysowa, R. The Effect of a Dusty Environment Upon Performance and Operating Parameters of Aircraft Gas Turbine Engines. Available online: https://www.sto.nato.int/publications/STO%20Meeting%20Proceedings/STO-MP-AVT-272/MP-AVT-272-18P.pdf (accessed on 2 January 2022).
39. Ptak, T.J.; Fallon, S.L. Particulate Concentration In Automobile Passenger Compartments. *Part. Sci. Technol.* **1994**, *12*, 313–322. [CrossRef]
40. Long, J.; Tang, M.; Sun, Z.; Liang, Y.; Hu, J. Dust Loading Performance of a Novel Submicro-Fiber Composite Filter Medium for Engine. *Materials* **2018**, *11*, 17. [CrossRef]
41. Bojdo, N.; Filipone, A. Effect of desert particulate composition on helicopter engine degradation rate. In Proceedings of the 40th European Rotorcraft Forum, Southampton, UK, 2–5 September 2014. [CrossRef]
42. Wróblewski, P. Effect of asymmetric elliptical shapes of the sealing ring sliding surface on the main parameters of the oil film. *Combust. Engines* **2017**, *168*, 84–93. [CrossRef]
43. Wróblewski, P.; Iskra, A. Problems of Reducing Friction Losses of a Piston-Ring-Cylinder Configuration in a Combustion Piston Engine with an Increased Isochoric Pressure Gain. In Proceedings of the SAE Powertrains, Fuels & Lubricants Meeting, Online, 22–24 September 2020; SAE Technical Paper 2020-01-2227; SAE International: Warrendale, PA, USA, 2020. [CrossRef]
44. Woś, P.; Michalski, J. Effect of Initial Cylinder Liner Honing Surface Roughness on Aircraft Piston Engine Performances. *Tribol. Lett.* **2011**, *41*, 555–567. [CrossRef]
45. Koszałka, G.; Suchecki, A. Changes in performance and wear of small diesel engine during durability test. *Combust. Engines* **2015**, *162*, 34–40. [CrossRef]
46. Koszałka, G.; Suchecki, A. Changes in blow-by and compression pressure of a diesel engine during a bench durability test. *Combust. Engines* **2013**, *154*, 34–39. [CrossRef]

47. Wróblewski, P. Analysis of Torque Waveforms in Two-Cylinder Engines for Ultralight Aircraft Propulsion Operating on 0W-8 and 0W-16 Oils at High Thermal Loads Using the Diamond-Like Carbon Composite Coating. *SAE Int. J. Engines* **2022**, *15*, 129–146. [CrossRef]
48. Rao, X.; Sheng, C.; Guo, Z. The Influence of Different Surface Textures on Wears in Cylinder Liner Piston Rings. *Surf. Topogr. Metrol. Prop.* **2019**, *7*, 045011. [CrossRef]
49. Meng, Y.; Xu, J.; Jin, Z.; Prakash, B.; Hu, Y. A review of recent advances in tribology. *Friction* **2020**, *8*, 221–300. [CrossRef]
50. Ferreira, R.; Martins, J.; Carvalho, O.; Sobral, L.; Carvalho, S.; Silva, F. Tribological solutions for engine piston ring surfaces: An overview on the materials and manufacturing. *Mater. Manuf. Process.* **2020**, *35*, 498–520. [CrossRef]
51. Biberger, J.; Füßer, H.-J. Development of a test method for a realistic, single parameter-dependent analysis of piston ring versus cylinder liner contacts with a rotational tribometer. *Tribol. Int.* **2017**, *113*, 111–124. [CrossRef]
52. Priest, M.; Dowson, D.; Taylor, C.M. Predictive wear modelling of lubricated piston rings in a diesel engine. *Wear* **1999**, *231*, 89–101. [CrossRef]
53. Zabala, B.; Igartua, A.; Fernández, X.; Priestner, C.; Ofner, H.; Knaus, O.; Nevshupa, R. Friction and wear of a piston ring/cylinder liner at the top dead centre: Experimental study and modelling. *Tribol. Int.* **2017**, *106*, 23–33. [CrossRef]
54. Delprete, C.; Razavykia, A. Piston ring-liner lubrication and tribological performance evaluation: A review. *Proc. Inst. Mech. Eng. Part J J. Eng. Tribol.* **2017**, *232*, 193–209. [CrossRef]
55. Etsion, I.; Sher, E. Improving fuel efficiency with laser surface textured piston rings. *Tribol. Int.* **2009**, *42*, 542–547. [CrossRef]
56. Meng, X.; Gu, C.; Zhang, D. Modeling the wear process of the ring/liner conjunction considering the evaluation of asperity height distribution. *Tribol. Int.* **2017**, *112*, 20–32. [CrossRef]
57. Michelberger, B.; Jaitner, D.; Hagel, A.; Striemann, P.; Kröger, B.; Leson, A.; Lasagni, A.F. Combined measurement and simulation of piston ring cylinder liner contacts with a reciprocating long-stroke tribometer. *Tribol. Int.* **2021**, *163*, 107146. [CrossRef]
58. Trivedi, H.K.; Bhatt, D.V. Effect of Lubricating Oil on Tribological behaviour in Pin on Disc Test Rig. *Tribol. Ind.* **2017**, *39*, 90–99. [CrossRef]
59. Reitz, R.D.; Ogawa, H.; Payri, R.; Fansler, T.; Kokjohn, S.; Moriyoshi, Y.; Agarwal, A.; Arcoumanis, D.; Assanis, D.; Bae, C.; et al. IJER editorial: The future of the internal combustion engine. *Int. J. Engine Res.* **2020**, *21*, 3–10. [CrossRef]
60. Grabon, W.; Pawlus, P.; Wos, S.; Koszela, W.; Wieczorowski, M. Effects of cylinder liner surface topography on friction and wear of liner-ring system at low temperature. *Tribol. Int.* **2018**, *121*, 148–160. [CrossRef]
61. Indudhar, M.R.; Banapurmath, N.R.; Rajulu, K.G.; Patil, A.Y.; Javed, S.; Khan, T.M.Y. Optimization of Piston Grooves, Bridges on Cylinder Head, and Inlet Valve Masking of Home-Fueled Diesel Engine by Response Surface Methodology. *Sustainability* **2021**, *13*, 11411. [CrossRef]
62. Di Blasio, G.; Ianniello, R.; Beatrice, C. Hydrotreated vegetable oil as enabler for high-efficient and ultra-low emission vehicles in the view of 2030 targets. *Fuel* **2021**, *310*, 122206. [CrossRef]
63. Agarwal, A.K.; Dhar, A.; Gupta, J.G.; Kim, W.I.; Choi, K.; Lee, C.S.; Park, S. Effect of fuel injection pressure and injection timing of Karanja biodiesel blends on fuel spray, engine performance, emissions and combustion characteristics. *Energy Convers. Manag.* **2015**, *91*, 302–314. [CrossRef]
64. Kim, M.-S.; Akpudo, U.E.; Hur, J.-W. A Study on Water-Induced Damage Severity on Diesel Engine Injection System Using Emulsified Diesel Fuels. *Electronics* **2021**, *10*, 2285. [CrossRef]
65. Su, Y.; Zhang, Y.; Xie, F.; Duan, J.; Li, X.; Liu, Y. Influence of ethanol blending ratios on in-cylinder soot processes and particulate matter emissions in an optical direct injection spark ignition engine. *Fuel* **2022**, *308*, 121944. [CrossRef]
66. Mujtaba, M.A.; Masjuki, H.H.; Kalam, M.A.; Noor, F.; Farooq, M.; Ong, H.C.; Gul, M.; Soudagar, M.E.M.; Bashir, S.; Rizwanul Fattah, I.M.; et al. Effect of Additivized Biodiesel Blends on Diesel Engine Performance, Emission, Tribological Characteristics, and Lubricant Tribology. *Energies* **2020**, *13*, 3375. [CrossRef]
67. Karczewski, M.; Szczęch, L. Influence of the F-34 unified battlefield fuel with bio components on usable parameters of the IC engine. *Eksploat. Niezawodn.—Maint. Reliab.* **2016**, *18*, 358–366. [CrossRef]
68. Rimkus, A.; Vipartas, T.; Matijošius, J.; Stravinskas, S.; Kriaučiunas, D. Study of Indicators of CI Engine Running on Conventional Diesel and Chicken Fat Mixtures Changing EGR. *Appl. Sci.* **2021**, *11*, 1411. [CrossRef]
69. Agarwal, D.; Singh, S.K.; Agarwal, A.K. Effect of Exhaust Gas Recirculation (EGR) on performance, emissions, deposits and durability of a constant speed compression ignition engine. *Appl. Energy* **2011**, *88*, 2900–2907. [CrossRef]
70. Duan, X.; Liu, Y.; Liu, J.; Lai, M.-C.; Jansons, M.; Guo, G.; Zhang, S.; Tang, Q. Experimental and numerical investigation of the effects of low-pressure, high-pressure and internal EGR configurations on the performance, combustion and emission characteristics in a hydrogen-enriched heavy-duty lean-burn natural gas SI engine. *Energy Convers. Manag.* **2019**, *195*, 1319–1333. [CrossRef]
71. George, S.; Balla, S.; Gautam, M. Effect of diesel soot contaminated oil on engine wear. *Wear* **2007**, *262*, 1113–1122. [CrossRef]
72. Oungpakornkaew, P.; Karin, P.; Tongsri, R.; Hanamura, K. Characterization of biodiesel and soot contamination on four-ball wear mechanisms using electron microscopy and confocal laser scanning microscopy. *Wear* **2020**, *458–459*, 203407. [CrossRef]
73. Al Sheikh Omar, A.; Motamen Salehi, F.; Farooq, U.; Morina, A.; Neville, A. Chemical and physical assessment of engine oils degradation and additive depletion by soot. *Tribol. Int.* **2021**, *160*, 107054. [CrossRef]
74. Matarrese, R. Catalytic Materials for Gasoline Particulate Filters Soot Oxidation. *Catalysts* **2021**, *11*, 890. [CrossRef]

75. Awad, O.I.; Ma, X.; Kamil, M.; Ali, O.M.; Zhang, Z.; Shuai, S. Particulate emissions from gasoline direct injection engines: A review of how current emission regulations are being met by automobile manufacturers. *Sci. Total Environ.* **2020**, *718*, 137302. [CrossRef]

76. Durst, M.; Klein, G.; Moser, N. *Filtration in Fahrzeugen*; Mann+Hummel GMBH: Ludwigsburg, Germany, 2005.

77. Maddineni, A.K.; Das, D.; Damodaran, R.M. Numerical investigation of pressure and flow characteristics of pleated air filter system for automotive engine intake application. *Sep. Purif. Technol.* **2019**, *212*, 126–134. [CrossRef]

78. Halim, M.A.A.; Mohd, N.A.R.N.; Nasir, M.N.M.; Dahalan, M.N. Experimental and Numerical Analysis of a Motorcycle Air Intake System Aerodynamics and Performance. *Int. J. Automot. Eng.* **2020**, *17*, 7607–7617.

79. Kaźmierczak, A.; Tkaczyk, M. Team durability test of a 1.3 MW locomotive diesel engine with prototype piston rings. *Combust. Engines* **2021**, *187*, 77–82. [CrossRef]

80. Schäffer, J.; Wachtmeister, G. Analyse der Kolbengruppenreibung und -schmierung von Ottomotoren—Die Bedeutung des Ölabstreifrings. *Forsch. Ing.* **2021**, *4*, 1065–1075. [CrossRef]

81. Avis, M. Particles: Friend or Foe? Understanding the Value of Particles in Oil Analysis–Machinery Lubrication 6. Available online: https://www.machinerylubrication.com/Read/28974/particles-friend-foe (accessed on 2 January 2022).

82. Fitch, J. Clean Oil Reduces Engine Fuel Consumption—Practicing Oil Analysis Magazine. Available online: https://www.machinerylubrication.com/Read/401/oil-engine-fuel-consumption. (accessed on 2 January 2022).

83. McClelland, J.E.; Billett, S.M. Filter Life versus Engine Wear. *SAE Trans.* **1966**, *74*, 922–936.

84. Gupta, N.; Singh, S.K.; Pandey, S.M. Tribological characterisation of thermal sprayed CrC alloyed coating—A review. *Adv. Mater. Process. Technol.* **2020**, *7*, 660–683. [CrossRef]

85. Ali, M.K.A.; Xianjun, H.; Turkson, F.R.; Ezzat, M. An analytical study of tribological parameters between piston ring and cylinder liner in internal combustion engines. *Proc. Inst. Mech. Eng. Part K J. Multi-Body Dyn.* **2016**, *230*, 329–349. [CrossRef]

86. Fatjo, G.G.A.; Smith, E.H.; Sherrington, I. Piston-ring film thickness: Theory and experiment compared. *Proc. Inst. Mech. Eng. Part J J. Eng. Tribol.* **2017**, *232*, 550–567. [CrossRef]

87. Needelman, W.M.; Madhavan, P.M. Review of Lubricant Contamination and Diesel Engine Wear—SAE Technical Paper Series 881827. In Proceedings of the Truck and Bus Meeting and Exposition, Indianapolis, IN, USA, 7–10 November 1988.

88. Grafe, T.; Gogins, M.; Barris, M.; Schaefer, J.; Canepa, R. Nanofibers in filtration applications in transportation, filtration. In Proceedings of the International Conference and Exposition, Chicago, IL, USA, 3–5 December 2001.

89. Kosmider, K.; Scott, J. Polymeric Nanofibres Exhibit an Enhanced Air Filtration Performance. *Filtr. Sep.* **2002**, *39*, 20–22. [CrossRef]

90. Wei, L.; Qin, X. Nanofiber bundles and nanofiber yarn device and their mechanical properties: A review. *Text. Res. J.* **2016**, *86*, 1885–1898. [CrossRef]

91. Leung, W.W.F.; Hau, C.W.Y.; Choy, H.F. Microfiber-nanofiber composite filter for high-efficiency and low pressure drop under nano-aerosol loading. *Sep. Purif. Technol.* **2018**, *206*, 26–38. [CrossRef]

92. Schmid, B.; Kreiner, A.; Poljak, I.; Klein, G.M. Luftfilter Mit Nanofaserbeschichtungen. *MTZ—Mot. Z.* **2012**, *73*, 592–597. [CrossRef]

93. Jaroszczyk, T. Air filtration in heavy-duty motor vehicle applications. In Proceedings of the Dust Symposium III, Vicksburg, MS, USA, 15–17 September 1987.

94. Nagy, J. Filtrowanie a żywotność silnika. *Siln. Spalinowe* **1973**, *3*, 43–47. (In Polish)

95. Jaroszczyk, T.; Fallon, S.L.; Liu, Z.G.; Heckel, S.P. *Development of a Method to Measure Engine Air Cleaner Fractional Efficiency*; SAE Technical Paper 1999-01-0002; SAE International: Warrendale, PA, USA, 1999.

96. Watson, C.E.; Hanly, F.J.; Burchell, R.W. Abrasive Wear of Piston Rings. *SAE Trans.* **1955**, *63*, 717–728.

97. Borawska, I. Papierowe przegrody filtracyjne–Własności, metody badań i zastosowanie. In Proceedings of the I Ogólnopolska Konferencja Naukowo—Techniczna: Filtry i Filtracja Płynów, Rzeszów, Poland, October 1974. (In Polish)

98. Thomas, G.E.; Culbert, R.M. Ingested dust, filters, and diesel engine ring wear. In Proceedings of the National West Coast Meeting, San Francisco, CA, USA, 12–15 August 1968; Society Of Automotive Engineers, Inc.: Warrendale, PA, USA, 1968. Available online: https://saemobilus.sae.org/content/680536/ (accessed on 29 December 2021).

99. Addison, J.A.; Needelman, W.M. *Diesel Engine Lubricant Contamination and Wear*; Scientific and Laboratory Services Department; Pall Corporation: Glen Cove, NY, USA, 1986. Available online: https://p2infohouse.org/ref/31/30453.pdf (accessed on 5 November 2021).

100. Jones, G.W.; Eleftherakis, J.G. *Correlating Engine Wear with Filter Multipass Testing*; SAE Technical Paper Series; SAE International: Warrendale, PA, USA, 1995. [CrossRef]

101. Treuhaft, M.B. *The Use of Radioactive Tracer Technology to Measure Engine Ring Wear in Response to Dust Ingestion*; SAE Technical Paper Series; SAE International: Warrendale, PA, USA, 1993. [CrossRef]

102. Milder, F.L.; Armini, A.J.; Jones, G.W. *The Use of Surface Layer Activation Wear Monitoring for Filter Design and Evaluation*; SAE Technical Paper Series; SAE International: Warrendale, PA, USA, 1981. [CrossRef]

103. Alfadhli, A.; Alazemi, A.; Khorshid, E. Numerical minimisation of abrasive-dust wear in internal combustion engines. *Int. J. Surf. Sci. Eng.* **2020**, *14*, 68–88. [CrossRef]

104. Eckardt, C. Einfluß von Fremdkörpern im Schmiermittel auf das Betriebsverhalten von Motorengleitlagern. *MTZ—Mot. Z.* **1983**, *44*, 393–401.

105. Khorshid, E.A.; Nawwar, A.M. A review of the effect of sand dust and filtration on automobile engine wear. *Wear* **1991**, *141*, 349–371. [CrossRef]
106. Barris, M.A. *Total Filtration™: The Influence of Filter Selection on Engine Wear, Emissions, and Performance*; SAE Technical Paper Series 952557; SAE International: Warrendale, PA, USA, 1995.
107. Ptak, T.J.; Richberg, P.; Vasseur, T. *Discriminating Tests for Automotive Engine Air Filters*; SAE Technical Paper Series; SAE International: Warrendale, PA, USA, 2001. [CrossRef]
108. Łysoń, F.; Siwula, Z. Analiza konstrukcji filtrów powietrza do tłokowych silników spalinowych. In Proceedings of the Materiały I Ogólnokrajowej Konferencji Naukowo—Technicznej: Filtry i Filtracja Płynów, Rzeszów, Poland, October 1974. (In Polish).
109. Truhan, J. Filter Performance as the Engine Sees It. *Filtr. Sep.* **1997**, *34*, 1019–1022.
110. Slabov, E.; Antropov, B.; Nilov, V. Influence of Air Quality on Engine Life. *Automobile Transport.* **1978**, *4*, 41–42.
111. Ruzaev, I.G.; Strykowski, F.R. Investigation of the Combined Engine Air Purification System. *Automotive Industry.* **1979**, *8*, 8–10.
112. Dziubak, T.; Wysocki, T. *Analiza Wpływu Konstrukcji Wkładu Filtra Powietrza na Wystąpienie Uszkodzeń w Postaci Zarysowań Wzdłużnych Gładzi Tulei Cylindrowych w Silniku Ciągnika Siodłowego MAN F19. Opinia Techniczna*; WAT: Warszawa, Poland, 2009. (In Polish)
113. Dziubak, T.; Dziubak, S.D. Experimental Study of Filtration Materials Used in the Car Air Intake. *Materials* **2020**, *13*, 3498. [CrossRef] [PubMed]
114. Saraiev, O.; Khrulev, A. Devising a model of the airflow with dust particles in the intake system of a vehicle's internal combustion engine. *East.-Eur. J. Enterp. Technol.* **2021**, *2*, 61–69. [CrossRef]
115. Khrulev, A.E.; Dmitriev, S.A. Influence of the inlet system design on dust centrifugation and the parts wear of the modern internal combustion engines. *Intern. Combust. Engines* **2020**, *2*, 73–84. [CrossRef]
116. Al-Rousan, A.A. Effect of dust and sulfur content on the rate of wear of diesel engines working in the Jordanian desert. *Alex. Eng. J.* **2006**, *45*, 527–536.
117. Michalski, J.; Woś, P. The effect of cylinder liner surface topography on abrasive wear of piston–cylinder assembly in combustion engine. *Wear* **2011**, *271*, 582–589. [CrossRef]
118. Xu, Y.; Zhang, X.; Hao, X.; Teng, D.; Zhao, T.; Zeng, Y. Micro/nanofibrous nonwovens with high filtration performance and radiative heat dissipation property for personal protective face mask. *Chem. Eng. J.* **2021**, *423*, 130175. [CrossRef]
119. Deng, Y.; Lu, T.; Cui, J.; Samal, S.K.; Xiong, R.; Huang, C. Bio-based electrospun nanofiber as building blocks for a novel eco-friendly air filtration membrane: A review. *Sep. Purif. Technol.* **2021**, *277*, 119623. [CrossRef]
120. Li, Y.; Yin, X.; Yu, J.; Ding, B. Electrospun nanofibers for high-performance air filtration. *Compos. Commun.* **2019**, *15*, 6–19. [CrossRef]
121. Sofi, H.S.; Rashid, R.; Amna, T.; Hamid, R.; Sheikh, F.A. Recent advances in formulating electrospun nanofiber membranes: Delivering active phytoconstituents. *J. Drug Deliv. Sci. Technol.* **2020**, *60*, 102038. [CrossRef]
122. Malviya, R.M. *Nano-Fiber Filters for Automotive Applications*; SAE Technical Paper Series; SAE International: Warrendale, PA, USA, 2018. [CrossRef]
123. Zhu, W.H.; Poudyal, A.; Martin, P.M.; Tatarchuk, B.J. Characterization of Dirt Holding Capacity of Microfiber-Based Filter Media Using Thermal Impedance Spectroscopy. *ACS Appl. Mater. Interfaces* **2020**, *12*, 15737–15747. [CrossRef]
124. Graham, K.; Ouyang, M.; Raether, T.; Grafe, T.; Mc Donald, B.; Knauf, P. Polymeric Nanofibers in Air Filtration Applications. In Proceedings of the 5th Annual Technical Conference & Expo of the American Filtration & Separations Society, Galveston, TX, USA, 9–12 April 2002.
125. Technologia Medium Ultra-Web. Available online: https://www.donaldson.com/pl-pl/industrial-dust-fume-mist/technical-articles/ultra-web-media-technology/ (accessed on 24 January 2022).
126. Zhang, H.; Xie, Y.; Song, Y.; Qin, X. Preparation of high-temperature resistant poly (m-phenylene isophthalamide)/polyacrylonitrile composite nanofibers membrane for air filtration. *Colloids Surf. A Physicochem. Eng. Asp.* **2021**, *624*, 126831. [CrossRef]
127. Jaroszczyk, T.; Fallon, S.L.; Schwartz, S.W. Development Of High Dust Capacity, High Efficiency Engine Air Filter With Nanofibers. *J. KONES Powertrain Transp.* **2008**, *15*, 215–224.
128. Wróblewski, P.; Koszalka, G. An Experimental Study on Frictional Losses of Coated Piston Rings with Symmetric and Asymmetric Geometry. *SAE Int. J. Engines* **2021**, *14*, 853–866. [CrossRef]
129. Wróblewski, P.; Rogólski, R. Experimental Analysis of the Influence of the Application of TiN, TiAlN, CrN and DLC1 Coatings on the Friction Losses in an Aviation Internal Combustion Engine Intended for the Propulsion of Ultralight Aircraft. *Materials* **2021**, *14*, 6839. [CrossRef] [PubMed]
130. Wróblewski, P. Technology for Obtaining Asymmetries of Stereometric Shapes of the Sealing Rings Sliding Surfaces for Selected Anti-Wear Coatings. In Proceedings of the SAE Powertrains, Fuels & Lubricants Meeting, Online, 22–24 September 2020; SAE Technical Paper 2020-01-2229. SAE International: Warrendale, PA, USA, 2020. [CrossRef]
131. Dolatabadi, N.; Forder, M.; Morris, N.; Rahmani, R.; Rahnejat, H.; Howell-Smith, S. Influence of advanced cylinder coatings on vehicular fuel economy and emissions in piston compression ring conjunction. *Appl. Energy* **2020**, *259*, 114129. [CrossRef]
132. Ficici, F.; Kurgun, S. Analysis of Weight loss in Reciprocating Wear Test of Cylinder Liner and Piston Ring Coated with Molybdenum. *Arab. J. Sci. Eng.* **2021**, *46*, 7801–7813. [CrossRef]
133. Liu, Z.; Liang, F.; Zhai, L.; Meng, X. A comprehensive experimental study on tribological performance of piston ring–cylinder liner pair. *Proc. Inst. Mech. Eng. Part J J. Eng. Tribol.* **2021**, *236*, 184–204. [CrossRef]

134. Singh, S.K.; Chattopadhyaya, S.; Pramanik, A.; Kumar, S.; Pandey, S.M.; Walia, R.; Sharma, S.; Khan, A.M.; Dwivedi, S.P.; Singh, S.; et al. Effect of alumina oxide nano-powder on the wear behaviour of CrN coating against cylinder liner using response surface methodology: Processing and characterizations. *J. Mater. Res. Technol.* **2022**, *16*, 1102–1113. [CrossRef]
135. Sonia; Walia, R.; Suri, N.; Chaudhary, S.; Tyagi, A. Potential applications of thermal spray coating for I.C. engine tribology: A Review. *J. Phys. Conf. Ser.* **2021**, *1950*, 012041. [CrossRef]
136. Pawlus, P.; Reizer, R.; Wieczorowski, M. Analysis of surface texture of plateau-honed cylinder liner—A review. *Precis. Eng.* **2021**, *72*, 807–822. [CrossRef]
137. Patil, A.S.; Shirsat, U.M. Effect of laser textured dimples on tribological behavior of piston ring and cylinder liner contact at varying load. *Mater. Today Proc.* **2021**, *44*, 1005–1020. [CrossRef]
138. Ajith Kurian, B.; Rajendrakumar, P.K.; Deepak, K.L. Influence of honing angle on tribological behaviour of cylinder liner–piston ring pair: Experimental investigation. *Tribol. Int.* **2022**, *167*, 107355.
139. Ferreira, R.; Almeida, R.; Carvalho, O.; Sobral, L.; Carvalho, S.; Silva, F. Influence of a DLC coating topography in the piston ring/cylinder liner tribological performance. *J. Manuf. Process.* **2021**, *66*, 483–493. [CrossRef]
140. Rao, X.; Sheng, C.; Guo, Z.; Zhang, X.; Yin, H.; Xu, C.; Yuan, C. Effects of textured cylinder liner piston ring on performances of diesel engine under hot engine tests. *Renew. Sustain. Energy Rev.* **2021**, *146*, 111193. [CrossRef]
141. Grabon, W.; Pawlus, P.; Wos, S.; Koszela, W.; Wieczorowski, M. Evolutions of cylinder liner surface texture and tribological performance of piston ring-liner assembly. *Tribol. Int.* **2018**, *127*, 545–556. [CrossRef]
142. Venkateswara Babu, P.; Syed, I.; Ben Beera, S. Experimental investigation on effects of positive texturing on friction and wear reduction of piston ring/cylinder liner system. *Mater. Today Proc.* **2020**, *24*, 1112–1121. [CrossRef]
143. Yin, B.; Xu, B.; Jia, H.; Hua, X.; Fu, Y. Experimental research on the frictional performance of real laser-textured cylinder liner under different lubrication conditions. *Int. J. Engine Res.* **2021**, 146808742199529. [CrossRef]
144. Rajput, H.; Atulkar, A.; Porwal, R. Optimization of the surface texture on piston ring in four-stroke IC engine. *Mater. Today Proc.* **2021**, *44*, 428–433. [CrossRef]
145. Pawlus, P. Effects of honed cylinder surface topography on the wear of piston-piston ring-cylinder assemblies under artificially increased dustiness conditions. *Tribol. Int.* **1993**, *26*, 49–55. [CrossRef]
146. Kikuhara, K.; Koeser, P.S.; Tian, T. Effects of a Cylinder Liner Microstructure on Lubrication Condition of a Twin-Land Oil Control Ring and a Piston Skirt of an Internal Combustion Engine. *Tribol. Lett.* **2022**, *70*, 6. [CrossRef]
147. Koszałka, G. Model of operational changes in the combustion chamber tightness of a diesel engine. *Eksploat. Niezawodn.—Maint. Reliab.* **2014**, *16*, 133–139.
148. Kordos, P. The influence of the combustion engine work conditions on the cylinder wear in the stand test researches. *Eksploat. Niezawodn.—Maint. Reliab.* **2006**, *32*, 11–15.
149. Oil and Your Engine. Available online: https://sos.hastingsdeering.com.au/oil/HDLabServices/pubs/SEBD0640-04_OilAndYourEngine.pdf (accessed on 5 November 2021).
150. Savcı, I.H.; Şener, R.; Duman, I. A study of signal noise reduction of the mass air flow sensor using the flow conditioner on the air induction system of heavy-duty truck. *Flow Meas. Instrum.* **2022**, *83*, 102121. [CrossRef]
151. Dziubak, T. Właściwości eksploatacyjne przepływomierzy powietrza wlotowego silników spalinowych pojazdów mechanicznych. *Logistyka* **2015**, *3*, 1217–1226. (In Polish)
152. *Przepływomierz Powietrza z Gorącą Warstwą HFM 6*; Zeszyt do Samodzielnego Kształcenia nr 358; Volkswagen AG: Wolfsburg, Germany, 2007. (Paper version available in Poland)
153. Przepływomierze Masowe Powietrza. Rozwiązywanie Problemów, Usterki i Testy. Pierburg. Available online: https://static.webdealauto.com/images//medias/pdf/0005/pg-si-0079-pl-web.pdf (accessed on 5 November 2021).
154. Myszkowski, S. Przepływomierze powietrza—Rodzaje i uszkodzenia. *Serwis Motoryz.* **2008**, *1*, 10–15. (In Polish)

 *energies*

 MDPI

Article

Experimental Study of the Effect of Air Filter Pressure Drop on Internal Combustion Engine Performance

Tadeusz Dziubak * and Mirosław Karczewski

Faculty of Mechanical Engineering, Military University of Technology, 2 Gen, Sylwestra Kaliskiego Street, 00-908 Warsaw, Poland; miroslaw.karczewski@wat.edu.pl
* Correspondence: tadeusz.dziubak@wat.edu.pl; Tel.: +48-261837121

Abstract: The paper presents the problem of the effect of air filter pressure drop on the operating parameters of a modern internal combustion engine with compression ignition. A literature analysis of the results of investigations of the effect of air filter pressure drop on the filling, power and fuel consumption of carburetor and diesel engines with classical injection system was carried out. It was shown that each increase in the air filter pressure drop Δp_f by 1 kPa results in an average decrease in engine power by SI 1–1.5% and an increase in specific fuel consumption by about 0.7. For compression ignition engines, the values are 0.4–0.6% decrease in power and 0.3–0.5% increase in specific fuel consumption. The values of the permissible resistance of the air filter flow Δp_{fdop} determined from the condition of 3% decrease in engine power are given, which are at the level of 2.5–4.0 kPa—passenger car engines, 4–7 kPa—truck engines and 9–12 kPa—special purpose vehicles. Possibilities of decreasing the pressure drop of the inlet system, which result in the increase of the engine filling and power, were analyzed. The program and conditions of dynamometer engine tests were worked out in respect to the influence of the air filter pressure drop on the operation parameters of the six-cylinder engine of the swept volume V_{ss} = 15.8 dm³ and power rating of 226 kW. Three technical states of the air filter were modeled by increasing the pressure drop of the filter element. For each technical state of the air filter, measurements and calculations of engine operating parameters, including power, hourly and specific fuel consumption, boost pressure and temperature, were carried out in the speed range n = 1000–2100 rpm. It was shown that the increase in air filter pressure drop causes a decrease in power (9.31%), hourly fuel consumption (7.87%), exhaust temperature (5.1%) and boost pressure (3.11%). At the same time, there is an increase in specific fuel consumption (2.52%) and the smoke of exhaust gases, which does not exceed the permissible values resulting from the technical conditions for admission of vehicles to traffic.

Keywords: internal combustion engines; air filter pressure drop; engine power; hourly and specific fuel consumption; smoke opacity

Citation: Dziubak, T.; Karczewski, M. Experimental Study of the Effect of Air Filter Pressure Drop on Internal Combustion Engine Performance. *Energies* **2022**, *15*, 3285. https://doi.org/10.3390/en15093285

Academic Editor: Constantine D. Rakopoulos

Received: 28 March 2022
Accepted: 28 April 2022
Published: 30 April 2022

1. Introduction

The basic task of the intake system of an internal combustion engine of a motor vehicle is to supply air from the environment to the engine cylinders in appropriate quantities and with appropriate pressure, temperature and density, so as to ensure the correct course of the fuel combustion process in the cylinders and optimize the engine performance [1–4]. An engine requires at least 14.5 kg of air to burn 1 kg of fuel. When operating at rated conditions, passenger car engines draw 150–400 m³/h of air per hour. For truck engines this value is 900–2000 m³/h, and for a special vehicle engine e.g., Leopard 2 tank—more than 6000 m³/h. The air is added to the combustion process in order to obtain the required engine performance.

Various pollutants—gaseous and solid—get into internal combustion engines together with the air. The basic component of mechanical impurities is mineral dust, which is lifted from the ground to the height of several meters by the movement of vehicles or the wind,

---

and then sucked in by the intake air intake [5,6]. The basic components of mineral dust are SiO_2 and Al_2O_3 grains, which are characterized by sharp edges, very high hardness and account for 65–95% of the total dust mass depending on the type of substrate atmospheric conditions [7,8].

In internal combustion engines, mineral dust grains are carried with the air into the piston crown where they settle onto the surface of the cylinder liner and piston. Together with the oil, the dust forms an abrasive paste which, when it comes into contact with the surfaces of the engine components P-PR-C (piston–piston ring–cylinder), causes abrasive wear. As a result of excessive wear of the "piston–cylinder" pair, there is a decrease in the leak-tightness of the piston crown and the resulting drop in charge pressure at the end of the compression stroke, and consequently a decrease in the engine's compression ratio and power as well as an increased blow-by of hot combustion gases into the oil sump, which cause an increase in the temperature of the P-PR-C bonding elements and the oil. This leads to a decrease in oil viscosity and to excessive wear of cylinder rings and cylinder liner [9]. To counteract this phenomenon, special composite (diamond-like and titanium) coatings are applied to the piston ring surfaces [10–12].

From the point of view of abrasive wear, the most dangerous are dust particles, whose diameter dp is equal to the thickness of the oil film hmin between two surfaces at a given moment. In typical combustion engine associations, the oil film thickness hmin does not exceed the value of 50 µm [9,12].

In order to minimize the wear of engine components, an air filter is installed in the intake system, which ensures that the air supplied to the engine cylinders is of suitable quality (purity) in terms of mechanical components. Air filters differ in principle and method of operation, construction, type of filtration baffle and efficiency of operation. The passenger car engines, which are operated at low concentrations of air dust, are equipped with single-stage (baffle) filters, where the filtering element is a commonly used cartridge (rectangular panel), made of pleated filter paper or non-woven filter cloth [13–15]. Engines of trucks, special vehicles (tanks, armored personnel carriers, infantry fighting vehicles) and working machines operating in conditions of high dust concentrations in the air are usually equipped with two-stage filters. The first stage of air filtration is a multicyclone and the second one is a cylindrical filtering insert which, due to space limitations and efforts to increase the surface area of the filtering material, is made of pleated filter paper [16,17].

The other tasks of the air intake system are damping of the noise caused by the air flow [18] and forcing the wave phenomena, causing so-called dynamic (resonant) supercharging in the desired ranges of engine work, and thus increasing the cylinder filling and engine power [19]. The other elements of the air supply system of the internal combustion engine which are involved in the supply of air to the engine cylinders are: the flow meter, the turbocharger, the charge air cooler, the air damper, the intake manifold, the intake ducts in the head and the intake valves. The flow meter, located downstream from the air filter, continuously determines the mass of air flowing and transmits this information to the ECU (Engine Control Unit).

The purpose of the turbocharger (standard equipment in diesel engines) is to increase the mass of working medium supplied to the engine's cylinders by increasing the pressure (and temperature) of the air sucked into the system. The air cooler lowers the temperature of the charge air (its density increases) and thus increases the mass flow of the air supplied to the engine cylinders. In order to measure the permissible resistance of the air filters of trucks and special-purpose vehicles, sensors are used, which are mounted on the air filter outlet duct. The crankcase ventilation system and the EGR (exhaust gas recirculation) system are related to the air supply system.

A characteristic feature of baffle filters is that, during operation, as a result of deposition and accumulation of dust particles in the filter bed, the filter pressure drop pf defined as the static pressure drop behind the filter increases systematically (Figure 1). The intensity of the increase of the pressure drop depends on the conditions in which the vehicle is operated, and mainly on the dust concentration in the air and the engine operating time. The higher the value of the dust concentration in the air sucked into the engine, the faster the filter reaches the permissible value—Δp_{fdop}. The dust retention process is strongly dependent on the size of the incoming particles [20,21]. As the number of fine particles arriving with the air on the filter bed increases, which are characterized by a larger surface area per unit mass than particles of large size, a layer of difficult permeability forms on the filter layer. Therefore, the pressure drop of the filtration bed increases rapidly and the filter service life, limited by the Δp_{fdop} value, becomes shorter [22,23].

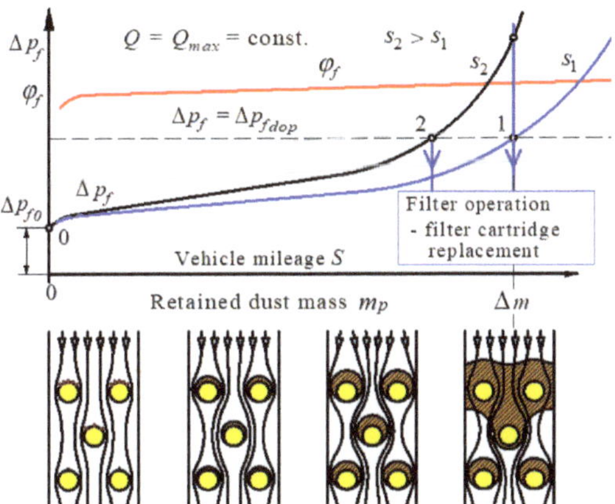

Figure 1. Changes in filtration efficiency and pressure drop of baffled air filter during operation: 1—low air dustiness, 2—high air dustiness.

The filtration process is also strongly dependent on the type of pollutants. Soot coming mainly from engine exhaust emissions very quickly forms a tight, greasy layer on the surface of the fibrous filter bed, which results in a more intensive increase in pressure drop and a faster attainment of the Δp_{fdop} value [24].

When the dust absorption capacity of the filter bed is exhausted (Point 1 in Figure 1), there is a sharp increase in the pressure drop of the air filter. The filter pressure drop Δp_f corresponding to the pf value is called the permittivity resistance Δp_{fdop}.

As dust settles inside or on the surface of the filter material, the air flow rate tends to decrease due to increasing resistance to air flow through the filtration system. Operating an engine with an air filter with increased pressure drop causes a decrease in fill and torque, resulting in a decrease in maximum engine power. The result is a reduction in the dynamic properties of the vehicle. The high resistance of the air filter and the significant negative pressure created behind the filter can cause the forces of detachment to exceed the forces of adhesion of the grains to the filter substrate, resulting in the dust being detached in an avalanche (so-called secondary emission) and being sucked into the engine cylinders along with the air, causing accelerated wear of the P-PR-C elements. In extreme cases, a high pressure drop may cause mechanical damage (rupture, cracking) of the filter insert (Figure 2), which will result in intensive wear of the engine elements.

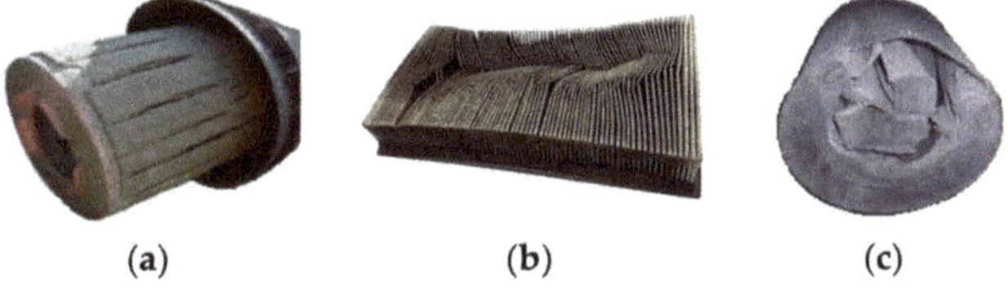

(**a**)   (**b**)   (**c**)

Figure 2. Effects of operating an air filter beyond the permissible resistance: (**a**) excessively contaminated filter element, (**b**,**c**) filter element failure.

For this reason, when the filter reaches a certain resistance value, it is necessary to service the air filter by replacing the filter cartridge. In passenger cars, due to low values of dust concentration in the air and thus small increments of the filter pressure drop, this operation is performed depending on the mileage of the vehicle or its operation time. These values vary considerably, depending on the design specifics of the engines and the car manufacturer, ranging from 15,000–90,000 km or from one to 4 years [25,26].

According to the authors of [27,28], in the case of passenger cars, the air filter should be replaced when the increase in the pressure drop reaches 1.0–2.5 kPa. The research presented in [29,30] has shown that the replacement of the filter insert, depending on the mileage of the vehicle or its operation time, may be premature—before the full dust absorption capacity and the permissible pressure drop are reached, or too late. In this case, the increased filter pressure drop will cause a decrease in engine filling and deterioration of engine characteristics—a decrease in power output.

For trucks and special vehicles, which are operated in conditions of high and variable concentrations of dust in the air, thus, the increase of the pressure drop is significant (by 5–8 kPa) and occurs with different intensities. Failure to replace the filter cartridge on a regular basis, prolonging the mileage of the vehicle, results in an additional increase in the pressure drop and, as a consequence, a significant drop in the engine power and reduction of the vehicle's dynamics. At the same time, the filter still provides high efficiency and filtration accuracy.

For this reason, after the air filter of the engines of vehicles operating in conditions of high dust concentration in the air, sensors are installed, signaling the achievement of the set value of the permissible pressure drop Δp_{fdop}, which is a signal to operate the air filter. The value of the permissible resistance Δp_{fdop} is determined from the condition of a 3% decrease in engine power and is 2.5–4.0 kPa—passenger car engines—4–7 kPa—truck engines [31]—and 9–12 kPa—special purpose vehicles [32].

Technically, air filter life is commonly defined as the level of restriction that causes the pressure on a passenger car filter to drop by about 2.5 kPa above the pressure drop of a new (clean) filter. For trucks and special vehicles, Δp_{fdop} values are assumed to be about 6.25–7.5 kPa above the pressure drop of a clean air filter [33]. Therefore, efforts are made to minimize the pressure drop of clean air filters to reduce engine energy loss and extend vehicle mileage. Air filter performance is a technical trade-off between pressure drop (vehicle mileage), filtration efficiency and accuracy, and engine wear and durability and vehicle reliability [34,35].

In light of the comments made, it can be thought that:

(1) There are no unambiguously defined values of the engine power drop resulting from the need to overcome the air pressure drop. Thus, it is impossible to determine the resulting permissible values of the air filter pressure drop.

(2) The literature does not give unequivocally the nature of changes in the engine performance—decrease in useful power, increase in smoke, increase in specific fuel consumption and smoke of the engine exhaust, depending on the air pressure drop.

In the available literature, it is difficult to find a sufficiently complete picture of the influence of the air pressure drop on the engine operation indicators, and in particular on its performance characteristics. Such an assessment may be obtained during experimental tests of the engine on an engine dynamometer or of the whole vehicle on a chassis dynamometer. However, these tests are costly and labour-intensive, which explains the limited number of results available. Nevertheless, it is the most reliable research method. In the available literature, there are few experimental results available on the effect of the pressure drop on the performance parameters of naturally aspirated carburetor and diesel engines with classical injection system—with an in-line, sectional injection pump.

Electronically controlled direct petrol injection systems are commonly used to prepare the mixture in modern high speed spark ignition internal combustion engines. They enable precise control of the fuel dose and the beginning of fuel injection, which, together with the information from the air flow meter, makes it possible to determine the optimum fuel mixture composition for the operating conditions, and, as a result, reduce the emission of toxic exhaust components. Passenger cars are mostly equipped with spark-ignition engines with direct injection of petrol and electronic control system, while truck engines are equipped with high-pressure diesel injection systems such as common rail, electronically controlled pump-injector systems. The engine ECU controls all the quantities that affect the value of the generated torque produced by the engine, while at the same time meeting the requirements in the area of exhaust emissions and fuel consumption throughout the life of the vehicle.

In a modern passenger vehicle engine air supply system, there is a flow meter, most often an HFM layer thermoanemometer, which measures the mass of air drawn in by the engine and transmits the appropriate signal to the engine control system. In the exhaust system, on the other hand, there is a λ probe, which continuously monitors the amount of oxygen in the exhaust gases, and its signal is used to optimize the composition of the mixture with regard to the content of toxic compounds in the exhaust gases.

Increasing resistance of the air filter during operation decreases the value of the mass flow of air, which may cause disturbances in the process of mixture preparation and combustion. In the available literature, there are no results of investigations of the influence of the air pressure drop on the work of modern SI engine (spark ignition engine) as well as compression ignition engine. In this paper it was decided to partly fill this gap by carrying out experimental tests on an engine dynamometer on the influence of the air pressure drop on the work parameters of a modern compression ignition engine used to drive a truck tractor. The basic operating parameters of the engine were determined as a function of the engine speed, including: power, hourly and specific fuel consumption, boost pressure. The experimental study was preceded by a literature analysis of the influence of the pressure drop of the inlet system, mainly the air filter pressure drop, on the engine filling and operating parameters. The constructional possibilities of reducing the air filter pressure drop and increasing the filling and power of the internal combustion engine were analyzed.

2. Literature Analysis on the Influence of the Inlet System Pressure Drop on the Engine Operation

2.1. Factors Affecting the Value of the Engine Filling Factor

The effective power of a reciprocating internal combustion engine can be expressed by the relation [36]:

$$N_e = \frac{W_d \cdot \rho_{pow} \cdot V_s \cdot i \cdot n}{L_t \cdot k} \cdot \frac{\eta_i}{\lambda} \cdot \eta_m \cdot \eta_v, \tag{1}$$

where: W_d—fuel calorific value, ρ_{pow}—air density, vs.—cylinder displacement, i—number of cylinders, n—engine speed, L_t—theoretical air demand, k—stroke number factor, λ—excess air factor, η_i—indexed efficiency, η_m—mechanical efficiency, η_v—filling factor.

From the above expression, it follows that the effective power of the reciprocating internal combustion engine depends on many factors, and mainly on the filling factor η_v, which is determined from the relation:

$$\eta_v = \frac{\dot{m}_{rz}}{\dot{m}_t}, \tag{2}$$

where: $\dot{m}_{rz}$—mean real flow rate of air supplied to engine cylinders in determined operating conditions and in time interval long enough to eliminate the influence of pressure pulsations in intake duct, $\dot{m}_t$—theoretical mass flow rate of air supplied to engine cylinders in determined operating conditions.

The value of the engine filling factor is influenced by the following factors: thermodynamic, structural and operational.

The thermodynamic factors of the charge in the cylinder are related to each other by the relation [37]:

$$\eta_v = \frac{T_H}{p_H(\varepsilon - 1)} \cdot \left(\varepsilon \frac{p_N}{T_N} - \frac{p_r}{T_r} \right), \tag{3}$$

where: p_H, T_H—ambient pressure and temperature, p_N, T_N pressure and temperature at the end of the filling stroke, p_r, T_r—pressure and temperature of the rest of the exhaust gas, ε—compression ratio.

The various thermodynamic parameters have different and not always unambiguous effects on the value of the filling ratio. However, the influence of various factors on the value of the filling ratio according to relation (3) does not lead to unambiguous conclusions.

An increase in the ambient temperature T_H causes an increase in the temperature T_N at the end of the filling stroke, which in the presence of a constant ambient pressure results in a decrease in air density and thus a lower mass filling of the engine cylinders.

The ambient pressure p_H has a significant effect on the filling ratio. This is clearly visible as the altitude at which the engine operates increases, which is associated with a decrease in air density, as well as with a decrease in ambient temperature [37]. For example, at an altitude of 1000 m above sea level, the pressure is $p_H = 893$ hPa and is lower than the pressure at sea level (1013.25 hPa) by 11%, and the ambient temperature T_H by 2.4%, resulting in a decrease in air density ρ_p by nearly 11%. The filling of the engine cylinders also deteriorates by this amount.

According to the authors of [37,38], the filling ratio is affected more by thermodynamic parameters at the inlet than by those at the outlet, especially by the end-of-fill pressure p_N. For fixed valve timing phases, the end-fill pressure p_N depends mainly on the pressure drop of the intake system Δp_U according to the relation:

$$p_N = p_H - \Delta p_U. \tag{4}$$

Among the constructional factors, the pressure drop of the intake system has the greatest influence on the filling factor of the internal combustion engine, which depends on:

(1) Air filter pressure drop Δp_f—its value depends on the type of filter (baffle, cyclone) and the degree of dust contamination of the filter element,
(2) Pressure drop of external inlet ducts Δp_D—its value depends on the cross-sectional area and the quality (smoothness) of internal duct walls, the number and radii of bends in the air path,
(3) Pressure drop of the assembly "intake seat—intake valve" Δp_{GK}—located in the head channel, and its value will depend primarily on the lift of the intake valve h_z, which is assigned by the design angle of rotation of the crankshaft α. The value of the pressure drop of the assembly "intake seat—intake valve" can be described by the function $\Delta p_G = \mathrm{f}(\alpha)$.
(4) Pressure drop of carburetor and throttle Δp_G—its value depends on the opening angle of the mixture throttle.

Then the above relation takes the form:

$$p_N = p_H - (\Delta p_f + \Delta p_D + \Delta p_{GK} + \Delta p_G). \tag{5}$$

For a given engine construction (valve timing, valve lift, piston speed, cylinder capacity, compressor), at a constant speed and load, the maximum filling of the engine cylinders will depend mainly on the air filter pressure drop pf, which increases its value during the vehicle operation (Figure 1). The air filter pressure drop Δp_f, which increases with the operating conditions as a result of accumulation and deposition of pollutants on the filter cartridge, will cause a drop in pressure p_N at the end of the filling stroke, and thus in the filling level. As a consequence, the mass of air delivered to the engine cylinders decreases. In naturally aspirated compression ignition engines equipped with a classic in-line injection pump, there is a decrease in the excess air coefficient λ (at the same fuel dose $G_e = $ const). For the compression ignition engines meeting the requirements not higher than EURO II operating in the rated conditions, the coefficient should be equal to $\lambda \cong 1.4$ [39], whereas for the engines meeting the standards above EURO II, it is necessary to increase the value of the coefficient to the value above 1.5–1.6 due to the necessity of reducing the particulate matter emission [40]. At this value of the coefficient λ in the cylinders, there are good conditions for the preparation of the mixture, the initiation of the ignition, the course of combustion and the release of heat, and thus the power achieved by the engine is the highest. Reduced air mass (oxygen deficiency in the combustion chamber) causes the fuel not to burn completely, which is one of the reasons for the formation of complete and incomplete fuel combustion products. These are primarily: PM (particulate matter) and CO (carbon monoxide), HC (hydrocarbons), NO_x, (nitrogen oxides) which are toxic components of the exhaust gas. There is a decrease in the efficiency of the engine and thus its torque M_o and power N_e.

The most important operational factors affecting the engine filling are the engine speed and its load, as well as the air pressure drop pf increasing during operation. In most of the currently operated reciprocating engines, an increase and then a decrease in the filling ratio η_v is observed as the engine speed n increases. A typical waveform $\eta_v = f(n)$ for a naturally aspirated spark-ignition and compression-ignition engine is shown in Figure 3.

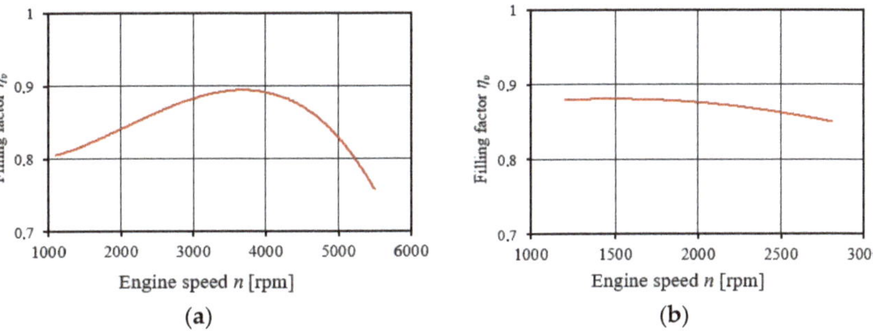

Figure 3. Dependence of the filling ratio on the rotational speed n for naturally aspirated engines: (**a**) spark-ignition, (**b**) compression-ignition.

Increasing of the filling level with increasing engine rotational speed of the SI engine results from high kinetic energy of the flowing charge and dynamic supercharging. Decreasing filling with further increase of engine speed occurs due to increasing pressure drop of the intake system with increasing speed of the flowing medium according to the relation:

$$\Delta p_U = \zeta \frac{\rho_L \cdot v_L^2}{2},\tag{6}$$

where: ζ—dimensionless pressure drop coefficient of the inlet system, ρ_L—charge density, v_L—average charge velocity in the inlet system.

The influence of the engine load on the filling ratio varies and depends on the way the fuel supply equipment is regulated. In the case of an SI engine, as the engine load decreases and the throttle is closed, the resistance of the inlet system increases. This results in a decrease in the filling level and thus a decrease in engine power.

2.2. Influence of Air Pressure Drop on Performance of an Internal Combustion Engine

In the available literature, the dependencies, which unambiguously define the nature of the changes in the decrease of engine power depending on the increasing resistance of the air filter, are not very often found. The results of experimental research on carburetor or compression-ignition engines with classical injection systems—with a piston (sectional) in-line injection pump—are the most common [41–48].

The characteristics of the filling ratio $\eta_v = f(n)$, power $N_e = f(n)$ and torque $M_o = f(n)$ of the eight-cylinder naturally aspirated (V_{ss} = 6.842 dm^3) 359M compression-ignition engine with the classical injection system are presented in Figures 4–6, for three different values of the air filter pressure drop Δp_f [41]. At engine speed n = 2800 rpm, the filter pressure drop had the following values:

- air filter pressure drop with clean filter insert—$\Delta p_f = \Delta p_{fo} = 2.3$ kPa.
- permissible air filter pressure drop—$\Delta p_f = \Delta p_{fdop} = 6$ kPa,
- air filter pressure drop—$\Delta p_f = 2\Delta p_{fdop} = 12$ kPa.

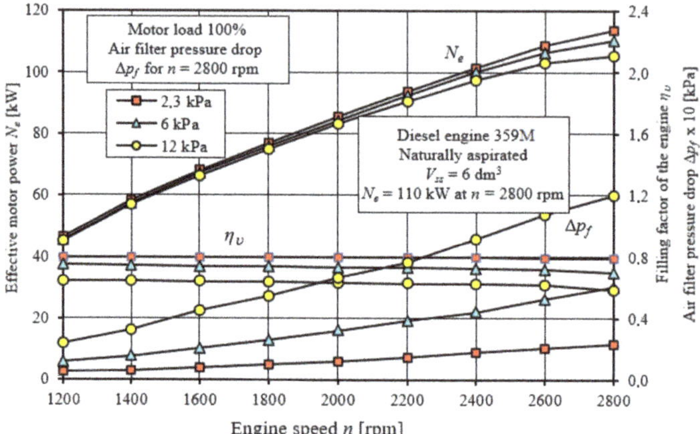

Figure 4. Dependence of fill factor and power of a naturally aspirated 359M diesel engine with classic injection system for different values of air filter pressure drop. The figure was made by the authors based on data from the paper [41].

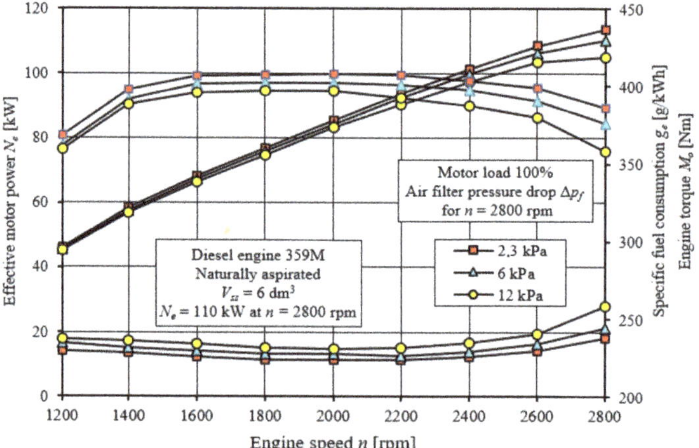

Figure 5. External characteristics of a naturally aspirated 359M compression ignition engine with a classic injection system for different air filter pressure drop (at constant speed $n = 2800$ rpm). The figure was made by the authors based on data from the paper [41].

As the engine speed increases, regardless of the value of the air pressure drop, for lower and medium engine speeds, the filling factor assumes almost constant values. At higher engine speeds, a slight decrease of the filling ratio η_v can be observed, resulting from the increasing resistance to the flow caused by the increasing speed of the air flow. This is, therefore, a typical characteristic curve of the filling factor $\eta_v = f(n)$ for the naturally aspirated diesel engine. As the value of the pressure drop of the air filter increases from $\Delta p_{fo} = 2.3$ kPa to $\Delta p_{fdop} = 6$ kPa, and then to $\Delta p_f = 2\Delta p_{fdop} = 12$ kPa, the filling characteristic $\eta_v = f(n)$ shifts almost in parallel towards lower values η_v (Figure 4). At engine speed $n = 2800$ rpm, the filling factor takes the values: $\eta_v = 0.785, 0.695, 0.583$. An increase in air pressure drop by 1 kPa results in a decrease in the fill factor by 2.65% on average.

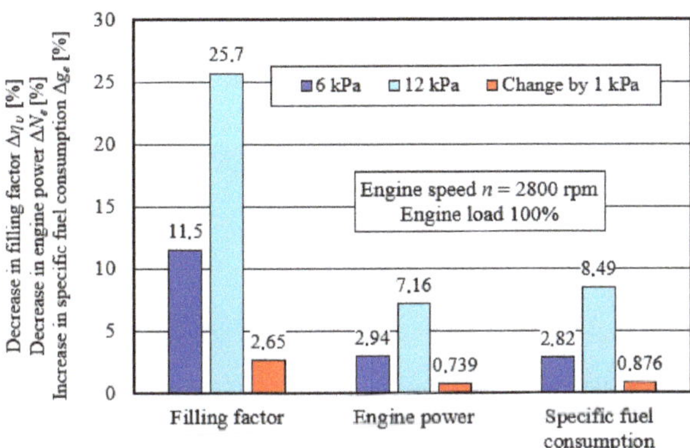

Figure 6. Influence of the air pressure drop of the naturally aspirated 359M engine with classic injection system on its operating parameters.

The decrease in the filling ratio causes a deterioration of the engine performance: a decrease in torque M_0 and power N_e and an increase in specific fuel consumption g_e. The family of curves $M_o = f(n)$, $N_e = f(n)$ and $g_e = f(n)$ for pressure drop values: $\Delta p_{f0} = 2.3$ kPa, $\Delta p_{fdop} = 6$ kPa, $2\Delta p_{fdop} = 12$ kPa are shown in Figure 5.

An increase in air filter pressure drop in the range of 2.3–12 kPa, during engine operation at engine speed n = 2800 rpm and 100% load, causes a decrease in the filling factor by 25.7%, in power by 7.16%, and an increase in specific fuel consumption by 8.49%. An increase in the air pressure drop by 1 kPa causes an average decrease in the filling factor by 2.65%, a decrease in power by 0.739% and an increase in specific fuel consumption by 0.876% (Figure 6).

Reducing the coefficient η_v, and thus of the mass of air supplied to the engine cylinders at the same fuel dose (n = const, G_e = const.) resulted in the decrease of the excess air coefficient λ from 1.36 at $\Delta p_{f0} = 2.3$ kPa to $\lambda = 0.96$ (at $\Delta p_f = 12$ kPa). For the diesel engine working in the rated conditions (in such conditions the tested engine worked), this coefficient should have the value $\lambda = 1.4$ [39]. At this value of the coefficient in the engine cylinders there are good conditions of the mixture preparation, ignition initiation, combustion process and heat release, and thus the engine power is the highest. Increase of the air pressure drop disturbs the mentioned phenomena. Lack of air causes the fuel not to burn completely, resulting in a decrease in engine efficiency, and consequently a decrease in torque M_o and power N_e.

Figures 7 and 8 show the results of the study of the effect of the air filter pressure drop Δp_f on the external characteristics of the effective power N_e and specific fuel consumption g_e of the diesel engine of a special vehicle [42].

This was a 12-cylinder (V-system) naturally aspirated engine with a displacement of 38.88 dm^3 and a rated power of 430 kW (580 hp) at n = 2000 rpm, with a classical injection system and a multi-range speed controller (direct fuel injection, 4 valves per cylinder). It can be seen from the presented graphs that the effect of pressure drop is only visible when $\Delta p_f = 6$ kPa is reached. Further increase of the air pressure drop already causes a significant decrease in engine power and an increase in specific fuel consumption, as well as a parallel shift of the external characteristics of power and specific fuel consumption towards lower values of N_e power, as well as fuel consumption g_e, with a simultaneous shift towards lower rotational speeds. At pressure drop $\Delta p_f = 26.7$ kPa, the engine operates in the speed range 1000–1800 rpm [42].

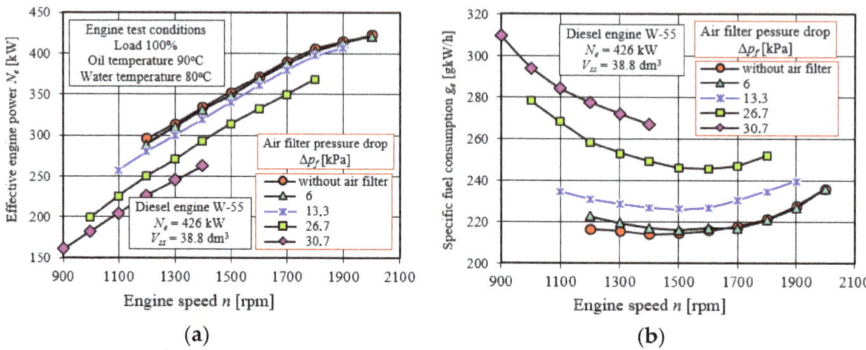

Figure 7. Influence of air filter pressure drop Δp_f on the characteristics of: (**a**) specific fuel consumption, (**b**) effective power of a twelve-cylinder, naturally aspirated W-55 compression ignition engine. The figure was made by the authors based on data from the paper [42].

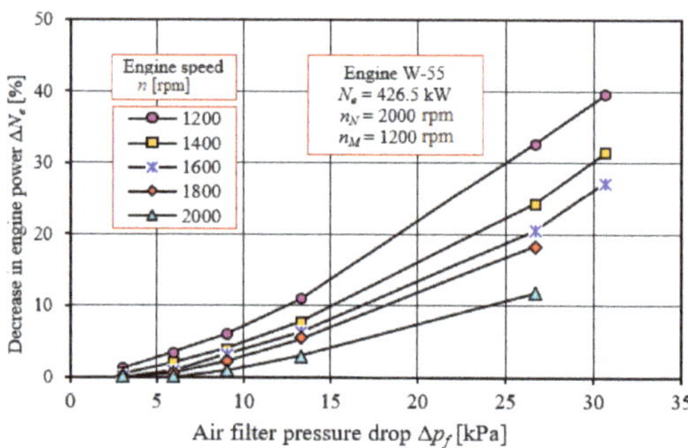

Figure 8. The effect of air filter pressure drop Δp_f on the power drop of a naturally aspirated diesel engine W-55 different speeds.

As the air filter pressure drop Δp_f increases (at a constant speed on the external characteristic curve), the percentage decreases in ΔN_e engine power increase parabolically. The decrease in power is greater the lower the motor speed is. For pressure drop $\Delta p_f = 26.7$ kPa, the power drops ΔN_e take the values: 11.75% at 2000 rpm and 20.6% at $n = 1400$ rpm and 32.7% for 1200 rpm (Figure 8).

The paper [43] presents an experimental study of the effect of air pressure drop Δp_f on the filling factor η_v and smoke opacity of a turbocharged, six-cylinder ($V_{ss} = 6$ dm^3), diesel engine T359E with a classical injection system.

The tested engine, being a driving unit of a truck, was equipped with an air filter with a cylindrical paper cartridge. The effect of four technical states of the air filter, differing in pressure drop, was studied. State number one—$\Delta p_f = 3.1$ kPa—filter with a clean paper insert. Number two, three, four—$\Delta p_f = 11, 18.7, 24.7$ kPa, respectively (Figure 9). The pressure drop values were due to the different contamination state of the three paper filter cartridges.

The engine filling factor η_v was determined using relation (2). To measure the real mass flow rate of the air $\dot{m}_{rz}$ supplied to the engine cylinders, the HFAM-1000 flowmeter (with hot wire) with the measurement range of 10–1200 kg/h and measurement accuracy of 5% was used. For measurement of the engine smoke opacity, the MDO-2 absorption opacimeter was used, with the measuring range of the light absorption coefficient of 0–10 m^{-1} and accuracy of 0.01 m^{-1}.

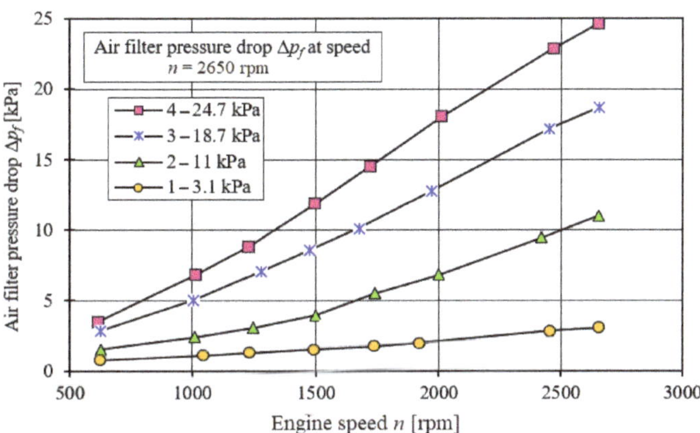

Figure 9. Flow characteristics $\Delta p_f = f(n)$ for different air filter conditions in the engine speed range n = 650–2650 rpm and without external engine load [43].

The tests were carried out using the method of free acceleration of the engine loaded with its own resistance moment and its total moment of inertia, hereinafter referred to as the "dynamic characteristics" method [44]. This method enables the determination of the instantaneous value of the engine torque without loading it on the dynamometer bench. The determination of this engine parameter consists of the precise determination of the course of the crankshaft angular accelerations during the engine acceleration process, resulting from the step change of the fuel supply to its cylinders.

The effect of the air pressure drop Δp_f on the variations of the air flow rate supplied to the engine cylinders is shown in Figure 10. The maximum value of the air stream (at the end of the acceleration process) is about 685 kg/h for the air filter pressure drop Δp_f = 3.1 kPa. For the remaining states (two, three, four) of the air filter, the air flux values were observed to be respectively smaller for each state. For state number four, the air flux m_{pow} = 500 kg/h. This is 25% less than for the air filter with a clean filter cartridge.

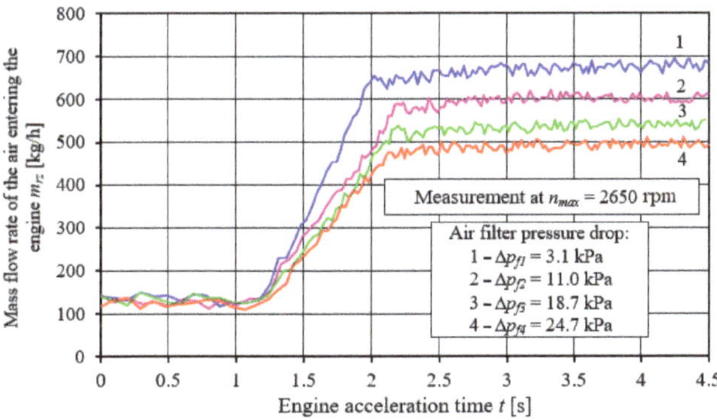

Figure 10. Mass flow rate of air $\dot{m}_{rz}$ entering the engine at different air filter condition as a function of time t of engine acceleration. The figure was made by the authors based on data from the paper [43].

The decrease in the amount of air supplied to the engine cylinders primarily results in a decrease in the filling ratio η_v, and thus in power output. The changes of the filling degree η_v for the recorded values of the air flow $\dot{m}_{rz}$ are shown in Figure 11. At the final stage of the engine acceleration process for the air filter with a clean filter insert (state number one) the filling degree is $\eta_v \approx 1.02$. With the increase of the pressure drop the filling degree takes on smaller and smaller values, respectively: $\eta_v \approx 0.90; 0.81; 0.75$—Figure 11. Thus, an increase in the resistance Δp_f by 1 kPa causes a decrease in the filling degree by 1.49, 1.29 and 1.23%, on average.

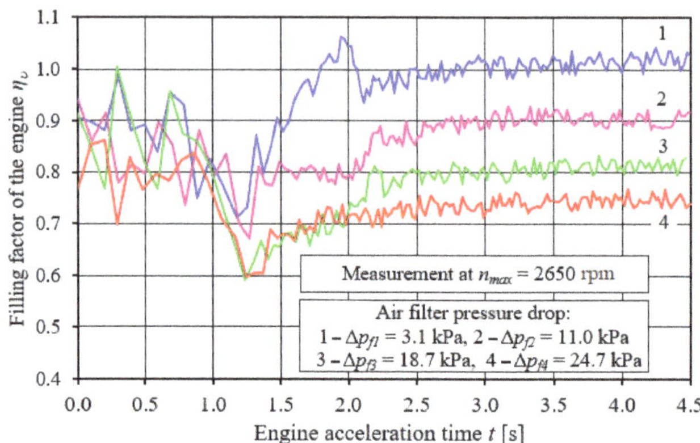

Figure 11. Changes in fill factor η_v as a function of time t of acceleration of the T359E engine at different values of air filter pressure drop—filter element contamination degree. The figure was made by the authors based on data from the paper [43].

Decreasing the filling degree, causes at the same fuel dose a decrease in the value of the excess air coefficient λ. It causes disturbance in engine work. It causes deterioration of conditions of mixture preparation, ignition initiation, combustion process and heat release, and thus decrease of engine efficiency, its torque and power.

An eight-fold increase in the resistance Δp_f over the initial resistance Δp_{f0} results in a two-fold increase in the smoke opacity of the T359E engine (increase in the light absorption coefficient) to $k = 0.81$ m^{-1}. For the T359E engine, this does not cause exceeding its maximum value—$k_{max} = 3.0$ m^{-1}.

The paper [45] presents fuel consumption and exhaust emissions during operation of a carburetor engine with and without an air filter (Figure 12). When the engine is operated with a constant load in the speed range from 1500 to 2500 rpm, the hourly fuel consumption with an air filter increases from 0.687 to 1.028 dm^3/h, i.e., by 49.6%. On the other hand, when operating the engine under the same conditions but without air filter, the hourly fuel consumption increases by only 35.2%.

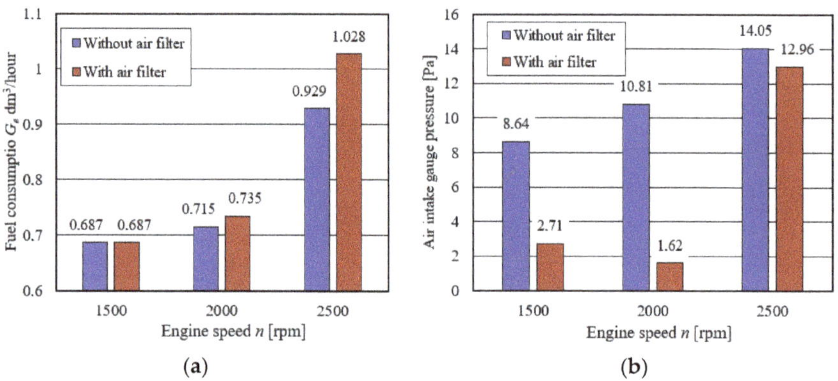

Figure 12. Effect of air filter pressure drop on: (**a**) fuel consumption at constant load, (**b**) air intake gauge pressure at constant load. Figure made by the authors using data from [45].

The air inlet pressure for the variant with air filter assumes slightly smaller values than that without air filter, especially when the engine is running at n = 1500 rpm and

n = 2000 rpm—Figure 12. For the first case it is 31.2%, and in the second case it is 15% of the pressure without air filter. At n = 2500 rpm the inlet pressure with air filter has only a slightly lower value than the inlet pressure without filter. The higher inlet pressure promotes more efficient filling of the engine cylinders. Therefore, it is necessary that the air filter element be changed regularly. A clean filter cartridge guarantees a higher inlet pressure.

At the same time, the authors' study [45] shows that engine operation without an air filter is characterized by a higher volume percentage of CO_2 and NO_x in the exhaust gas. The operation of the engine with a constant load, in the rotational speed range 1500–2000 rpm (without filter), causes that the volumetric share of CO_2 in the exhaust gases being only 7.7 and 12% higher than for the operation with an air filter. For the speed of 2500 rpm it is 25% more. This is because, due to the lack of an air filter, more air mass flows into the engine cylinders. The excess air causes the fuel to burn completely and completely, producing more CO_2 as a result.

The effect of the air filter on the NO_x concentration in the engine exhaust is similar. Running the engine with a constant load at 1500, 2000 and 2500 rpm (without filter), results in a volume fraction of NO_x in the exhaust, respectively, of 51.7, 8.1 and 20.2%, higher than for during operation with air filter. The formation of NO_x in the engine is influenced by the maximum temperature and pressure of the combustion process. Without air filter the combustion process is more efficient, which leads to higher temperature and pressure of the gases produced during the process. The higher temperature and pressure in the cylinder, which are caused by more efficient combustion, promotes the formation of more NO_x.

The paper [46] presents the results of research on the effect of three variants of the baffle filter on the characteristics of the UTD-20 internal combustion engine of a special vehicle. This is a compression-ignition engine with displacement V_{ss} = 15.8 dm^3 and power rating of 226 kW with a classical fuel injection system. Tests performed on a dynamometer stand included the comparison of effective parameters obtained for maximum load of the engine equipped with a standard air filter (S variant), with two other variants of the filter and engine operation:

S—standard filter, Δp_{fdop} = 13.2 kPa—at engine speed n = 2600 rpm,

A—upgraded filter (lower pressure drop by about 8 kPa), Δp_{fdop} = 4.9 kPa—at engine speed n = 2600 rpm,

B—modernized filter—operating with the engine with increased (about 7%) fuel dose.

The test results in the observations of these characteristics: external effective power N_e, torque M_o, hourly G_e and specific fuel consumption g_e, which are shown in Figure 13.

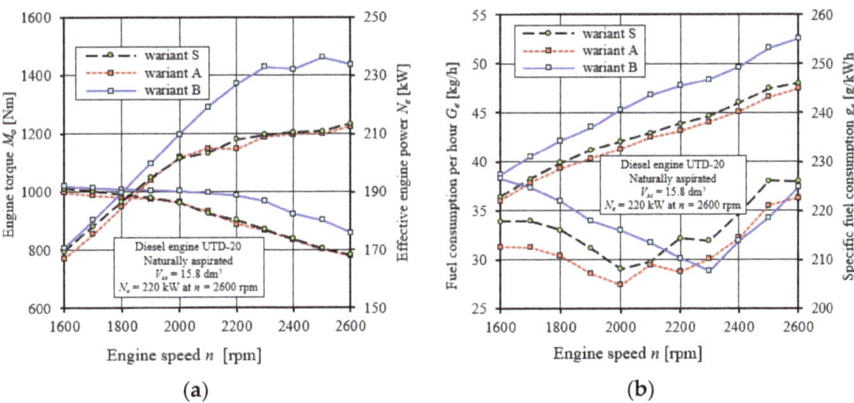

Figure 13. External characteristics of (**a**) effective power N_e and torque M_o, (**b**) hourly G_e and specific fuel consumption g_e of the UTD-20 engine. The figure was made by the authors based on data from the paper [46].

The engine working with air filter (variant A—lower pressure drop) obtained, in the whole speed range, an insignificant increase of effective parameters in comparison with the engine working with filter—variant S. The increase of power and torque is in the range of (0.4–1.2%), with lower values referring to the speed range 2300–2600 rpm. At the same time, lower (on average 1.5%) specific fuel consumption was obtained.

For the engine with an increased fuel dose and with the modernized air filter (variant B), in comparison with variant S, a significant increase in power and torque was obtained: over 2% for 1600 rpm and over 10% for 2200–2600 rpm, with a slight (2%) increase in specific fuel consumption. The increase in engine power occurred as a result of an increase in fuel delivery and increased air mass resulting from a decrease in air filter pressure drop—indicating an increase in engine filling.

According to the authors of the paper [46], an increase in the pressure drop in the intake system by an average of 10 mm Hg (1.36 kPa) causes a decrease in the power of the Jamz-240N engine by 3.68–4.41 kW and an increase in specific fuel consumption by 2.72–4.08 g/kW/h. The author of the work [47] states that an increase in air pressure drop by 1 kPa causes a decrease in power and an increase in specific fuel consumption in engines with diesel engine (0.3–0.4%) and about 1.3% in engines with SI. The authors of the work [49] state that the turbine engine is more sensitive to the decrease in inlet pressure caused by the installation of air cleaning devices than reciprocating engines, and each increase in the airpressure drop at the inlet to the turbine engine by 0.5 kPa causes a decrease in its rated power by 1%.

In work [50], the effect of two types of air filter (variant A—standard filter, variant B—filter with a new structure), on the load and speed characteristics of a single-cylinder engine with a diesel engine was studied (Figures 14 and 15). Air filter B is much larger in size than filter A, thus has a lower pressure drop. The effective power, torque, specific fuel consumption, smoke, oil temperature and engine exhaust temperature were investigated and analyzed and compared with the parameters of engine operation without an air filter—variant W. As expected, the engine operating without air filter obtained the highest power and torque and the lowest fuel consumption. The operation of the engine successively with the filter A and B causes a shift of the characteristics $N_e = f(n)$ i $M_o = f(n)$ almost parallel in the direction of smaller values, and the characteristics $g_e = f(n)$ in the direction of larger values, in the whole range of engine rotational speed—Figure 14. When the air filter is not used, the maximum torque is $M_{omax} = 75.3$ Nm at 1500 rpm. When the engine is operated with air filter type A, the maximum torque decreases to 72.5 Nm and the fuel consumption is $g_e = 234.2$ g/(kWh). With air filter type B, the torque increases to $M_{omax} = 73.6$ Nm at 1500 rpm and the fuel consumption decreases to $g_e = 230.8$ g/(kWh). This is 1.6% more torque at M_{omax} and 1.5% less fuel consumption than the engine with air filter type A.

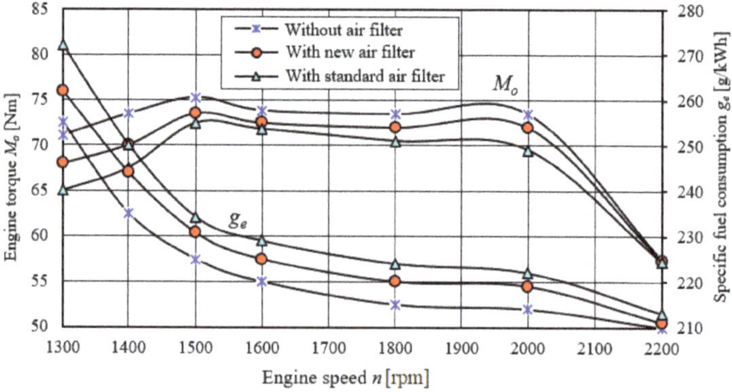

Figure 14. Torque M_o and specific fuel consumption g_e as a function of engine speed n. Figure made by the authors based on data from [50].

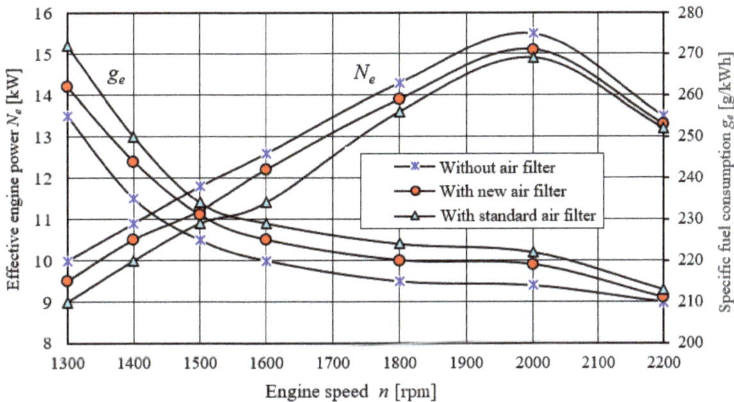

Figure 15. Effective power N_e and specific fuel consumption g_e as a function of engine speed n. Figure made by the authors based on data from [50].

The maximum power increases by 1.1% when a newly designed type B air filter is used when compared to type A air filter. The maximum smoke value (light absorption coefficient) during engine operation without air filter, with air filter type A and with air filter type B is: $k = 1.7$, 2.36 and 2.0 m^{-1} respectively.

The paper [51] experimentally and numerically studied an air filter with a pleated filter cartridge with different parameters to improve the performance of a compression ignition engine. It was investigated what effect the paper cartridge parameters pleat height, pleat spacing, pleat shape (flat pleat, V-shaped pleat, and sinuous shape), filter paper thickness and air flow velocity have on reducing filter pressure drop. A filter media optimized for least pressure drop was designed. An experimental study was conducted on the effects of three air filters: dust contaminated filter, clean standard filter and optimized filter on engine performance. The results showed that the filter with the optimized sinusoidal air cartridge had the lowest pressure drop (Figure 16). The optimized filter also provided the lowest exhaust temperature of 218 °C and the lowest fuel consumption of 290 g/kWh [51].

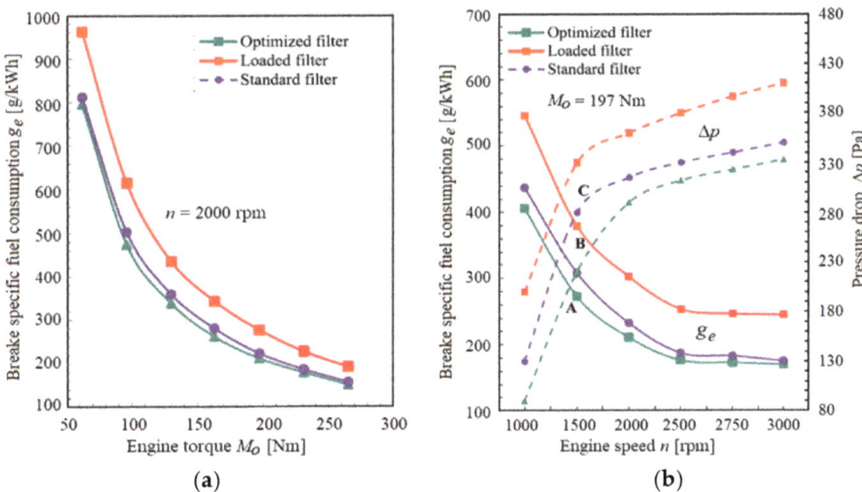

Figure 16. (**a**) variation of engine specific fuel consumption versus engine torque with different types of filters at constant engine speed $n = 2000$ rpm, (**b**) optimization between engine specific fuel consumption and pressure drop versus engine speed with different types of filters at constant engine torque $M_o = 197$ Nm. The figure was made by the authors based on data from the paper [51].

It was found that specific fuel consumption decreased with increasing engine torque for all filter types at constant 2000 rpm (Figure 16a). The optimized filter, due to the lowest pressure drop, provided lower fuel consumption, while the dirty filter was the cause of higher fuel consumption at the same engine torque and speed values. At a specific engine torque value of 163 Nm, the fuel consumption of the optimized filter was 262 g/kWh, while that of the standard and dirty filter was 281 and 344 g/kWh, respectively. Operating the engine with the optimized filter results in a decrease in fuel consumption when compared to the standard and dirty filter of 6.7% and 23.7%, respectively.

The intersection point of the relationship between pressure drop and specific fuel consumption, at a given engine speed and torque, was used as the optimization criterion. For the optimized pleated air filter, the recommended optimization point is located at point A (Figure 16b), where the specific fuel consumption was 290 g/kWh at a pressure drop of 205 Pa and an engine speed of 1450 rpm.

On the other hand, for the standard air filter (with pleated cartridge) and the air filter with dust contaminated cartridge, the optimization criteria were located at the intersection points B and C, for which the values of specific fuel consumption of pressure drop are higher—Figure 16b. The optimized pleated air filter with lower resistance than the others provides minimum fuel consumption at high engine speed and lower pressure loss.

The presented analysis shows that an increase in the air filter pressure drop Δp_f by 1 kPa results, on average, in a decrease in engine power by SI (1–1.5)% and an increase in specific fuel consumption by about 0.7. For compression-ignition engines, these values are respectively a (0.4–0.6)% decrease in power and a (0.3–0.5)% increase in specific fuel consumption.

The author of the paper [52] states that each increase in the pressure drop in the air filter inlet to the compressors by 10 kPa causes an increase in energy consumption by 1%.

2.3. Possibilities for Increasing the Filling and Power of the Internal Combustion Engine

Pleated filter media show great potential for reducing pressure drop at the design stage. Lower filter pressure drop translates directly into increased fill and engine power. The great interest in this problem is manifested by a significant number of both numerical and experimental studies conducted [53–61]. It turned out that each pleated filter element had an optimum number of pleats, for other fixed parameters, at which it showed a minimum pressure drop.

For example, numerical studies performed by the authors of the paper [53] showed that there is an optimum number of pleats for clean filters for which the pressure drop reaches a minimum, regardless of the in-plane orientation of the fibers. At the same time, triangular pleats result in a smaller pressure drop. After accounting for particle deposition, the intensity of the increase in pressure drop decreases as the number of pleats increases. A larger number of pleats results in a higher flow velocity inside the pleat channels, which causes a greater inhomogeneity of dust deposition on the pleats. It was observed that this effect is less pronounced when the pleats are triangular in shape.

The author of this paper [54] analyzed the fluid flow through the pleated medium air filter of a four-cylinder spark ignition engine by performing experimental and computational fluid dynamics (CFD) analysis to reduce the pressure drop. It was found that the filter with triangular pleats achieves higher filtration efficiency and lower pressure drop than the filter with rectangular pleats.

For example, a numerical methodology for predicting the pressure drop in an air filter with a pleated filter element was presented in [55]. It was found that pleat geometry and inlet velocity are key parameters for filter element optimization. The optimum pleat pitch to achieve minimum pressure drop was determined and found to be significantly dependent on pleat height.

The authors of the paper [56] developed a macro-scale CFD simulation to investigate the optimal pleated air filter design under dust loading conditions. It was found that the optimum pleat density in the clean condition can lead to higher pressure drop and energy consumption during the filtration process involving dust. It was observed that the optimum dust deposition resulting from pleat density depends on the pleat height and is lower with higher pleat height.

In [57], a numerical study was presented on the pressure drop and filtration efficiency of an air filter having different numbers of rectangular and triangular pleats in both depth and surface filtration regimes with 1, 5 and 10 μm diameter particles and in the filtration velocity range of 0.5–5.0 m/s. It has been shown that filters with rectangular pleats can potentially provide better performance than their triangular counterparts at high dust loads.

The authors of the paper [58] numerically and experimentally evaluated the effect of the geometric features of pleats on the performance of fiber filters in the deep filtration phase of submicron aerosols. The paper [59] presented simulation results of the effect of the shape of pleats made of different filter materials on the pressure drop of a pleated air filter.

The paper [60] presents the study and optimization of different samples of pleated filter elements by varying the angle between pleat sides, pleat length and number of pleats. The highest filtration efficiency was obtained when the dimensionless pleat coefficient, defined as the quotient of pleat height (vertical distance from the pleat top to the base), and pleat pitch (distance between pleat tops) reached a value of 1.48. Above this value, a systematic increase in filter pressure drop was noted.

On the other hand, the authors of the paper [61] optimized the pleat geometry of automotive filters using the developed dimensionless model. Pressure drop on clean pleated filters was determined taking into account the geometric characteristics of the pleats (distance between pleats 1–3.5 mm, pleat heights 27, 32, 40, 48 mm and filtration velocity in the range 0.01–0.10 m/s). It was found that for a given pleat height and a constant air flow, there is such a width between the pleats for which the pressure drop reaches the smallest value.

The length, diameter and inner surface finish of the inlet pipes directly affect the resistance to flow of the medium. Intake pipes have a complex geometry, in part because of the space available around the engine. The geometry of the intake system affects the mass flow of air, which is hindered by interference from pressure fluctuations generated by moving pistons and valves. In order to improve engine filling, increase engine performance and reduce fuel consumption, research work is carried out including: optimization of the geometry of the intake system components, particularly the length of the intake manifold [62–65], use of a Helmholtz resonator to generate pulsatile flow in the engine intake manifold [66,67], exploitation of airflow dynamics in the form of wave phenomena and pressure wave resonance in the intake system of an internal combustion engine resulting from cyclic opening and closing of valves [19,68,69], generation of pulsatile flow in the exhaust manifold by changing the length of the exhaust pipe [70].

The authors of the paper [71] obtained an increase in engine performance and a decrease in fuel consumption of a naturally aspirated engine by optimizing the airbox geometry. The effects of inlet diameter, airbox volume, throttle diameter and intake manifold length were analyzed. The following optimal intake system component dimensions were determined: intake diameter 81.07 mm, airbox volume, throttle body diameter 44.63 mm and intake manifold length 425 mm. This allowed for maximum engine performance (torque and power) and minimum fuel consumption.

The authors of the work [2] believe that the air flow in the intake system is disturbed by the tapering and diverging of the system components, which can lead to sudden acceleration and deceleration of the flow and cause excessive turbulence and pressure drop in the intake system, which affect the pressure drop and thus reduce the engine performance.

In [63], the effect of changing the length (volume) of the intake manifold on the performance of a ZI engine with multi-point injection system and electronically controlled fuel injectors was experimentally investigated. The results showed that changing the length of the intake manifold results in improved engine performance characteristics, especially fuel consumption at high load and low engine speeds. Therefore, a variable length intake manifold is indicated especially in urban and suburban areas (roads) where there are frequent stopping and accelerating under starting conditions. It was determined that the length of the intake manifold should be increased for low engine speeds and shortened as the engine speed increases.

Similar conclusions were reached by the authors of the paper [64], who studied on a dynamometer bench the effect of the length and diameter of the inlet pipe on the performance of the SI engine. The experiments were carried out in the engine speed range 1500–6500 rpm. Three intake pipe lengths (0.3, 0.6 and 0.9 m) and three intake pipe diameters (0.044, 0.053 and 0.067 m) were investigated on intake air mass flow rate and fill factor, torque, power, thermal efficiency and specific fuel consumption. The results showed that at low engine speeds, the intake pipe with longer length and smaller diameter provided the best engine performance. On the other hand, the intake pipe with shorter length and larger diameter provided the best engine performance at high engine speeds. The length and diameter of the intake pipe have no clear effect on specific fuel consumption and thermal efficiency of the engine.

In order to improve engine filling, the authors of [66] experimentally investigated the pulsatile flow characteristics in the engine intake manifold involving a Helmholtz resonator with variable internal volume. Using an electronic control, the internal volume of the resonator was changed and adjusted to the valve operating frequency and the intake manifold natural frequency. The results showed that the internal volume of the resonator and the frequency tuning affect the mass flow rate of the air in the intake system. For a fixed volume, the average increase in mass flow rate was 17.8%, and when the volume was tuned to the valve frequency, the average increase was 24.7%. The highest intake air mass flow rate was increased by 31.5% when the resonator was tuned to the system frequency.

The authors of the paper [19] presented the effect of variable valve timing and valve lift on internal combustion engine performance and fuel consumption. The valve lift and valve profile, which are the main factors affecting the dynamics of the gas pressure wave in the combustion chamber, were changed at all engine speeds in order to obtain an enhanced pressure wave that increases the filling of the engine cylinders. As a result of changing the above parameters, engine torque and power increased by an average of 6.02% over the entire engine speed range. In the lower speed range of 3000–4000 rpm the improvement is about 18.72%.

The paper [68] presents the performance increase obtained by fine tuning the intake system of a spark ignition engine over its entire speed range. During numerical tests, the diameter of the intake manifold and the timing phases of the intake valves were changed. When only the intake manifold diameter was changed and optimized for each engine speed, an improvement of approximately 8.5% in fill factor was obtained over the performance of the factory engine. Changing the intake valve timing phases at each engine speed resulted in a fill rate improvement of about 3%. However, combining a change in intake manifold diameter with a change in intake valve timing produced an improvement of about 12%.

In [69], the individual and combined effects of varying intake manifold length and intake valve lift on internal combustion engine performance at engine speeds from 3000 to 9000 rpm were studied. The resulting combined effect shows an average power improvement of 7.02% over the entire engine speed range.

In [70], the effect of tuning a pulsating sound wave produced over a wide range of speeds in the exhaust manifold on the power and torque of a single cylinder SI engine was analyzed. The simulation results were compared with standard data obtained during engine testing. An average of 7% increase in torque and 6% increase in horsepower was observed with constant change in tailpipe length. With a constant change in tailpipe diameter, a 6% increase in torque and power was observed. A combination of changes in exhaust pipe length and diameter resulted in 8.5% increase in torque and 9% increase in power, respectively.

A particular way to increase engine cylinder filling is to reduce the air temperature in the intake system by using an insulating layer between the engine and the intake system [71]. Insulating the intake system resulted in an increase in intake air density and an increase in fill factor of 5.88%. The mixture throttle opening with the isolated intake system required to achieve the same engine speed was reduced by about 19.8%.

From the above analysis, it can be seen that there are many feasible technologies and ways to increase engine fill and power, which boil down to reducing the pressure drop of the intake system, and this includes reducing the pressure drop of the air filter in its design phase.

It should be stated that, as of today, in the available literature there are no results of investigations of the influence of the flow resistance in the inlet system on the performance of a modern diesel engine used for driving trucks and truck tractors currently travelling on the roads and constituting the basic means of transporting goods. Therefore, it is purposeful to determine experimentally the relation between parameters of operation of the inlet system of a modern truck engine and its performance in terms of changes in emission of particular components of exhaust gases and traction properties of the vehicle.

3. Experimental Study of the Influence of the Air Filter Pressure Drop on the Performance of the Diesel Engine

3.1. Aim and Subject of Research

The objective of this study was to experimentally evaluate the effect of air filter pressure drop Δp_f on the performance parameters of a modern diesel engine with electronic control of fuel supply and supercharging and air cooling systems.

The subject of the research was a six-cylinder diesel engine of a Volvo D13C460 EURO V EEV truck with a maximum power of 338 kW, which is a power unit of a Volvo FH13 truck tractor with a mileage of 790,500 km. The total operating time of the engine was 11,800 h, and the engine consumed 229,100 dm^3 of diesel fuel from the beginning of operation to the day of testing. The tested engine meets the requirements of EURO V standard according to the homologation documents. The factory external characteristics of the engine are shown in Figure 17 and the basic engine parameters are shown in Table 1 [72].

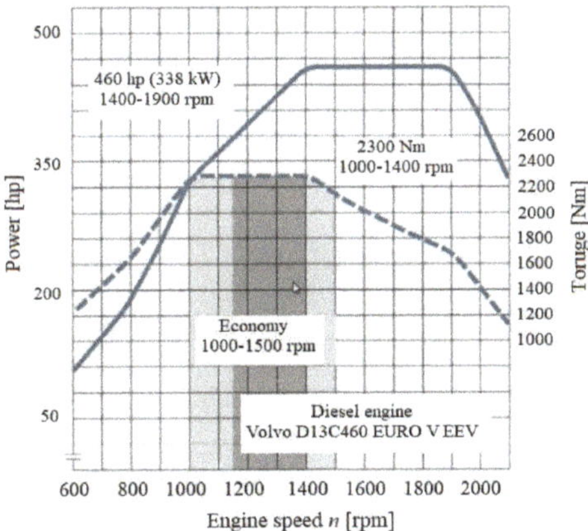

Figure 17. External characteristics of the VOVLO DC13C460 motor (338 kW) according to manufacturer specifications. The figure was made by the authors based on data from the paper [72].

A 6-cylinder, in-line, classic engine with a direct fuel injection—with an electronically controlled pump-injector power system—was studied. The engine has four valves per cylinder, which are controlled using hydraulic tappets driven from a central camshaft located on the head. This shaft also drives the electronically controlled pump injectors using a piezo-quartz valve. The valve controls the timing of the start of a specific stage of fuel injection, as well as its duration, which determines the fuel dose based on information about: intake manifold air pressure and temperature, exhaust gas temperature, ambient conditions and the desired torque resulting from the position of the accelerator pedal. The pump-injector supply system in relation to the sectional pump, shows a number of advantages. The most important include: the absence of high-pressure lines, minimal ignition delay due to the absence of delay caused by the pumping of fuel through the injection lines and high injection pressure even at minimum revolutions per minute. There are no problems with engine start, which is very important in winter conditions. The injection pressure is higher than in other systems (205 MPa), which creates good conditions for mixing fuel with air and its rapid evaporation and combustion. There is no phenomenon of injector leakage, which reduces fuel consumption. All these features make the engine powered by a pump injector have a high efficiency (low specific fuel consumption).

Table 1. Basic parameters of the engine D13C460 EURO V [72].

Name of Device	Range
Engine type	D13C460 EURO V
Maximum power at 1400–1900 rpm	460 hp (338 kW)
Maximum engine speed	2100 rpm
Maximum torque at 1000–1400 rpm	2300 Nm
Number of cylinders	6
Cylinder diameter	131 mm
Piston stroke	158 mm
Displacement	12.8 dm^3
Compression ratio	17.8:1

The air supply system consists of an air intake located on the right side of the cabin at its highest height (Figure 18), an external intake duct of approximately rectangular cross-section located on the rear wall of the cabin, a rubber accordion element connecting the external duct with the air filter intake duct and a single-stage (baffle) air filter. A permissible pressure drop sensor is installed on the air filter outlet line, set at Δp_{fdop} = 4.8–5.0 kPa. Proper filling of engine cylinders is assured by turbocharger and charge air cooler operating in the "air-to-air" system. Prior to testing, the entire vehicle intake system was inspected to check for leaks and was thoroughly cleaned. This action was done to reduce the impact of unidentified damage to the intake system on the test results.

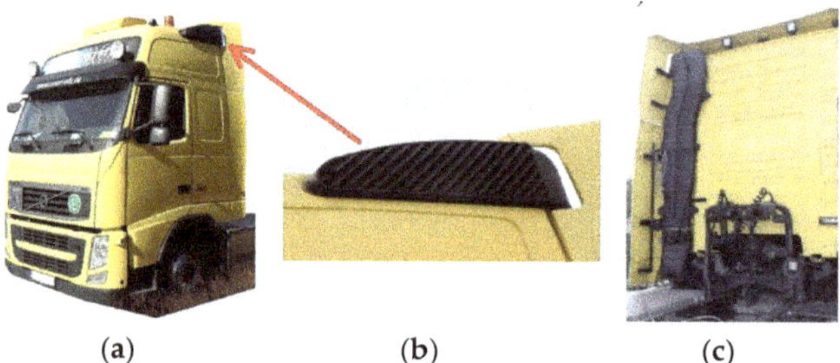

(a)   (b)   (c)

Figure 18. A Volvo FH13 truck tractor: (**a**) general view of the vehicle, (**b**) air intake, (**c**) view of the air supply system.

The filter element of the filter is a cylindrical cartridge made of pleated filter paper (Figure 19) with an active area A_C = 13.72 m^2.

Figure 19. Cylindrical air filter element for Volvo D13C460 EURO V engine.

3.2. Test Methodology and Conditions

Tests were carried out on a standard dynamometer bench. The motor was loaded with a water brake type Zöllner PS1-3812/AE with maximum power of 1250 kW. The torque M_o generated by the motor was measured with a strain gauge transducer connecting the swinging brake housing with the foundation. The engine speed n was recorded with a pulse transducer cooperating with a gear wheel located on the dynamometer flange. Hourly fuel consumption G_e was measured using an AVL fuel balance with a five-second time interval, and the measurement was then averaged over a 60 s interval. The coolant temperature t_{ch} during engine testing was set equal to the operating temperature of t_{ch} = 87–92 °C and was maintained using an external heat exchanger. The opacity of the exhaust gases was determined using an AVL 439 OPACIMETER opacimeter, which works on the principle of measuring the light absorption coefficient. The volumetric air demand Q_S by the engine was recorded with a thermo-areometric flow transducer. The air pressure drop Δp_f was determined as the pressure difference p_1 before and p_2 after the air filter, using a TESTO 400 differential pressure gauge.

Due to the turbocharged, electronically controlled fuel injection air supply system used in the engine, it was determined that the increased air flow resistance through the filter system would have the greatest effect on the maximum power speed characteristics. The research presented in [42,45] shows that the increased flow resistance does not significantly affect the engine performance at part load. The research was conducted using the passive experiment method, by stepping the engine speed at its full load—100% accelerator pedal position.

The following methodology was used during engine testing. After establishing the thermal equilibrium conditions of the engine, the accelerator pedal was set to the position corresponding to the full engine load, and then the lowest (n = 1000 rpm) possible rotational speed, at which the engine was still running steadily, was set using the brake control system. After the operating conditions of the engine had stabilized, its operating parameters were recorded. The engine load was then reduced to obtain the next higher speed (in 100 rpm increments), and after the engine operating conditions stabilized, the engine operating parameters were again recorded. In the above manner, the engine operating parameters were recorded for rotational speeds in the range of 1000–2100 rpm. Due to the characteristics of the water brake used, it was not possible to perform measurements below a speed of 1000 rpm, however this did not affect the test results obtained, as speeds below 1000 rpm are not used during normal operation of vehicles with this type of engine.

During the tests, the following engine operating parameters and air parameters in the intake system were directly measured for each speed:

- engine torque, M_o [Nm],
- engine rotational speed, n [rpm],
- hourly fuel consumption, G_e [kg/h],
- engine air demand, Q_s [m^3/h],
- exhaust gas temperature, t [°C],
- Smoke opacity of exhaust gases—light absorption coefficient, k [m^{-1}],
- air pressure before p_1 and after air filter p_2, [kPa],
- charge air pressure, p_d, [kPa].

On the basis of directly measured values of engine working parameters the following were determined:

- effective engine power, N_e [kW],
- specific fuel consumption, g_e [g/(kWh)],
- air filter pressure drop Δp_f [kPa].

All tests were repeated in duplicate to eliminate coarse errors that could lead to incorrect inferences.

The control of engine load, engine speed and brake load was performed from the control and measurement cabin, where there are also indicators for measuring engine operating parameters.

During the tests, the engine operation was continuously controlled by using the diagnostic interface NAVIGATOR TXTs with the software IDC 5 TRUCK. The schematic diagram of the test stand is shown in Figure 20. The applied measurement equipment and its accuracy are presented in Table 2. Before and after the tests, all measurement systems important for the measurement results were checked using external standards: the exhaust gas analyzer—with standard gases, the opacimeter—with optical filters of known light absorption coefficient, the torque measuring system—with standard mass standards, the speed measuring system—with a standard tachometer, the pressure measuring systems—with standard sensors, the fuel consumption measuring system—with a dedicated mass standard.

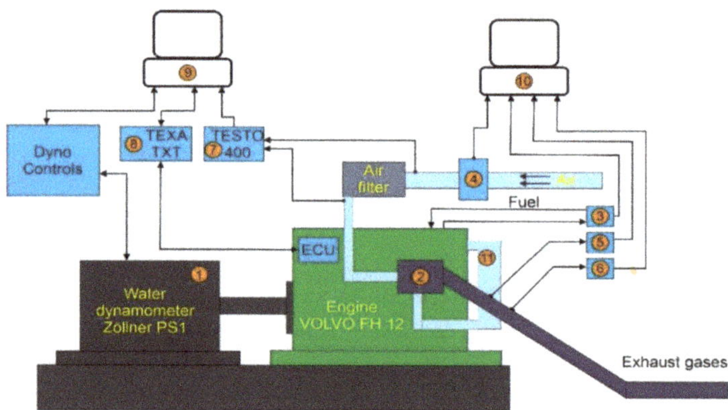

Figure 20. Diagram of the dynamometric test stand with a Volvo D13C460 EURO V engine: 1—water brake, 2—turbocharger, 3—fuel consumption measuring system, 4—air consumption measuring system, 5—smoke opacity measuring system, 6—exhaust gas temperature measuring system, 7—pressure measuring system in the engine intake system, 8—TEXA TX diagnoscope, 9—computer controlling the operation of the dynamometric brake and recording engine operating parameters, 10—computer controlling the operation of measuring systems, 11—charge air cooler.

Table 2. List of investigation equipment used during investigation.

No.	Name of Device/ Measured Quantity	Type	Range	Accuracy
1.	Water dynamometer • torque—M_o • rotated speed—n	Zöllner PS1-3812/AE	$M_o = (0–7000)$ Nm, $n = (0–3000)$ rpm $Ne = (0–1250)$ kW	± 1 Nm ± 1 rpm ± 1 kW
2.	Fuel weight-meter (diesel)—G_e	AVL 733S Fuel Balance	$(0–200)$ kg/h	± 0.005 kg/h
3.	Smoke concentration extinction coefficient of light radiation—k	AVL Opacimeter 4390	$(0.001–10.0)$ m^{-1}	± 0.002 m^{-1}
4.	Thermocouple-measuring of exhaust temperature—t_s	NiCr-NiAl (K)	$(-50–1100)$ °C	± 1 °C
5.	Mass air consumption—Q_s	SensyMaster FMT430 Thermal Mass Flowmeter	$(100–6000)$ m^3/h	± 1.0 m^3/h
6.	Vacuum in the intake system	TESTO 400	$(-100–200)$ hPa	0.3 Pa + 1% measured quantity

The results of engine operating parameters: power and hourly fuel consumption, obtained during the tests, were reduced to normal conditions in accordance with the PN-ISO 15550:2009 standard [73].

3.3. Analysis of Research Results

During the experimental research the influence of four (New, A-33, B-66, C-90), differing in pressure drop, technical states of the same air filter on the external characteristics of the Volvo D13C460 EURO V engine was determined. In each case, the same parameters characterizing the engine operation were measured. After each test, the engine was inspected to detect any damage to the intake system that could adversely affect the test results.

Increase of the value of the filter pressure drop Δp_f was modeled by means of obscuring a part of the active filtration surface of the cylindrical cartridge. As a result, four technical states were obtained, differing in the value of the pressure drop of the same air filter.

- condition New—air filter with clean, brand new, paper air filter insert, $\Delta p_f = 0.58$ kPa
- condition A-33—air filter with an air filter insert that has had approx. 33% of its active filtration surface obscured, $\Delta p_f = 0.604$ kPa
- condition B-66—air filter with a filter insert, which is obscured by approx. 66% of the active filtering surface, $\Delta p_f = 0.757$ kPa
- condition C-90—air filter with a filter insert, which has approx. 90% of its active filtering surface obscured, $\Delta p_f = 2.024$ kPa.

The results of the engine tests with different air filter conditions are presented as characteristics $N_e = f(n)$, $g_e = f(n)$, $G_e = f(n)$, $p_d = f(n)$, $Q_s = f(n)$, $\Delta p_f = f(n)$, $t_s = f(n)$, $k = f(n)$ in Figures 21–26.

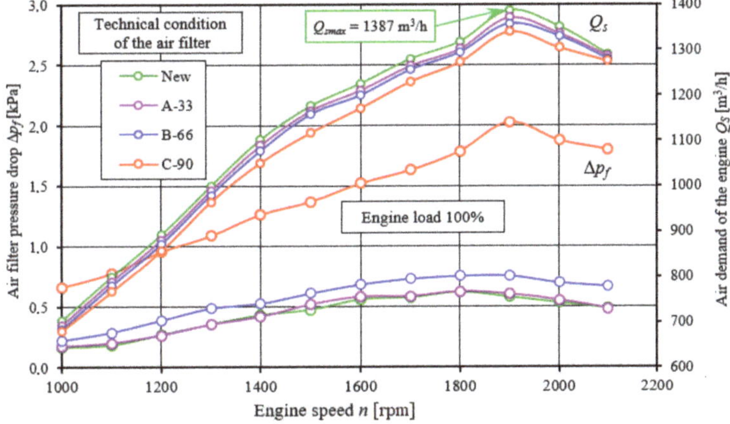

Figure 21. Pressure drop of different air filter states (New, A-33, B-66, C-90) and air demand of Volvo D13C460 EURO V engine as a function of speed.

Figure 21 shows the pressure drop of four technical states (New, A-33, B-66, C-90) of the same air filter as a function of engine speed n of a Volvo D13C460 EURO V engine. The analysis of the test results was performed over the entire range of speeds that are used in the operation of a truck tractor equipped with the type of engine that was tested. As the engine speed increases in the range of $n = 1000–2100$ rpm, the pressure drop of the air filter, regardless of the percentage obscuration of the active area of the cartridge, increases its value until the engine speed reaches $n = 1900$ rpm, which is associated with the achievement of the maximum air demand by the engine (Figure 20) and the maximum power. A further increase in the engine speed causes a decrease in the filter pressure drop, which results from a decrease in the air flow rate Q_s. When 33% of the active surface of the cartridge (A-33) was obscured, no significant differences in the pressure drop values were found when compared to the new filter (New). The recorded differences of the pressure drop are within the limits of the measurement errors and do not significantly affect the other parameters of engine operation presented in the following figures.

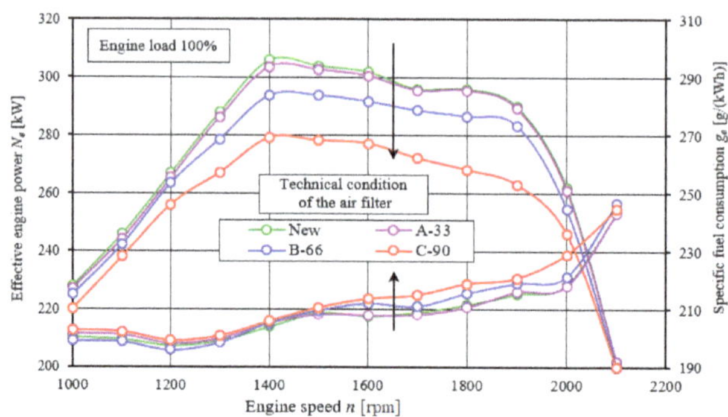

Figure 22. Effective engine power N_e and specific fuel consumption N_e of the VOVLO DC13C460 engine as a function of speed n for different air filter states New, A-33, B-66, C-90.

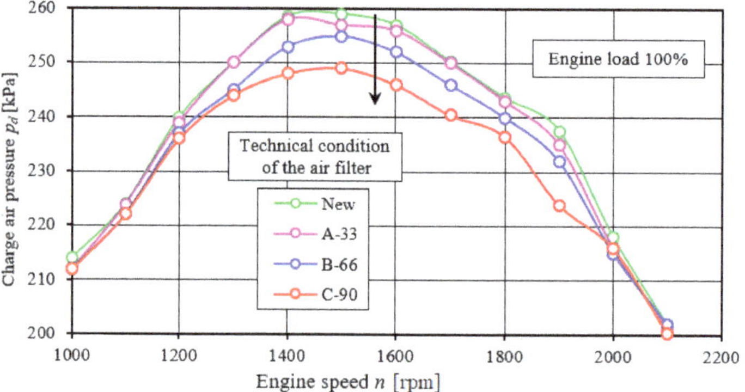

Figure 23. Charge air pressure p_d in the intake manifold of VOVLO DC13C460 engine as a function of engine speed n for different air filter states: New, A-33, B-66, C-90.

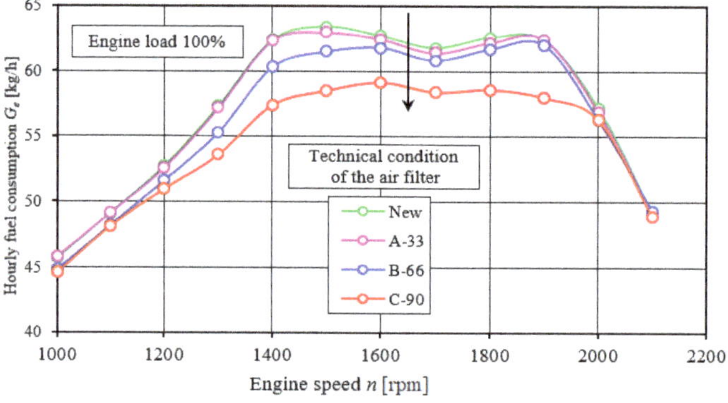

Figure 24. Hourly G_e fuel consumption of VOVLO DC13C460 engine as a function of engine speed n for different air filter conditions: New, A-33, B-66, C-90.

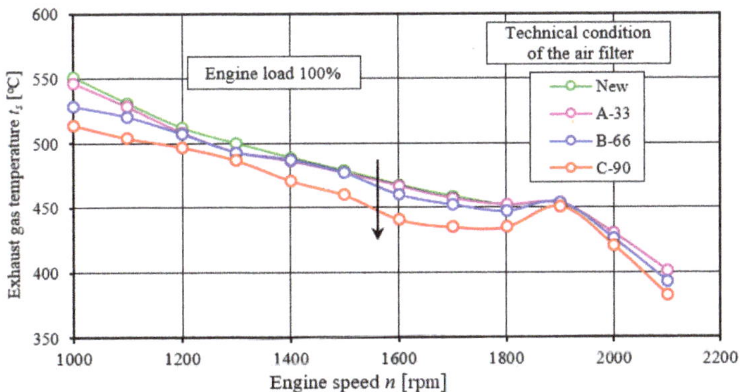

Figure 25. Exhaust gas temperature t_s at the turbine exit of the turbocharger unit of the VOVLO DC13C460 engine as a function of engine speed n for different air filter states: New, A-33, B-66, C-90.

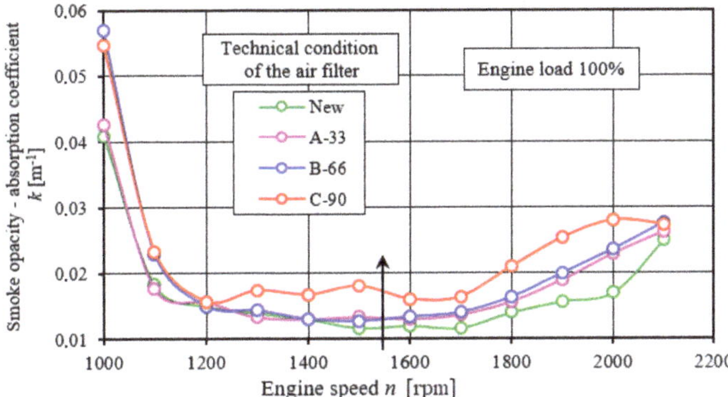

Figure 26. VOVLO DC13C460 engine smoke opacity—light absorption coefficient k (absorption) as a function of engine speed n for different air filter states: New, A-33, B-66, C-90.

Obscuring 66% of the filter cartridge surface caused an increase in the filter pressure drop in the range of 10–15%, which at $n = 1900$ rpm reached a maximum value of 0.757 kPa. A significant (more than threefold) increase of the filter pressure drop in comparison to the filter with "New" cartridge was obtained only after covering about 90% of the cartridge active area. At $n = 1900$ rpm the filter reaches a maximum value of 2.024 kPa. For this air filter, the manufacturer has determined the permissible resistance of $\Delta p_f = 6$ kPa measured at $n = 1900$ rpm. A sensor installed in the duct after the air filter is used to record this value. Obtaining this value by the air filter means the necessity of performing maintenance consisting in filter cartridge replacement. The authors obscured 90% of the surface area and did not obtain an acceptable value, suggesting that only mineral dust retained on paper can provide an effective barrier to airflow. The low (0.604 kPa) value of the pressure drop of the new air filter ("New") results from the large ($A_c = 13.72$ m^2) filter area of the paper, which guarantees a low velocity of air flow through the paper surface, called in the literature the filtration velocity v_F calculated from the relation

$$v_F = \frac{Q_s}{3600 \cdot A_c} \ [\text{m/s}], \tag{7}$$

where: Q_s—engine air demand [m^3/h], A_c—filter cartridge active surface area [m^2].

In the case of the tested engine, the air demand Q_s, irrespective of the technical condition of the air filter, increases rapidly until the engine reaches the rotational speed

from its dust loading during operation. The value of the permissible pressure drop Δp_{fdop} shall be selected individually for each engine depending on the demand for air and the anticipated operating conditions, and in particular on the concentration of dust in the air.

5. The use of air filter after exceeding the value of admissible resistance Δp_{fdop} (not replacing the filter cartridge) is possible as the filter still retains high filtration efficiency. However, there is a sharp increase in the pressure drop, which causes additional loss of engine power and deterioration of the vehicle's traction properties, in particular its climbing ability.

6. From the analysis of the available test results of the engines found in the literature, it results that the increase in the resistance to the flow of the air filter Δp_f by 1 kPa causes on average a decrease in the power of the SI engine (carburetor) of (1.0–1.5)% and an increase in specific fuel consumption of about 0.7. For diesel engines with a classical injection system, these values are respectively (0.4–0.6)% decrease of power and (0.3–0.5)% increase of specific fuel consumption.

7. The research carried out on an engine dynamometer on the influence of the air filter pressure drop on the parameters of the modern, supercharged, with an inlet air cooler diesel engine, being a driving unit of a truck tractor, only partly confirms the previous research results presented in the literature. This is due to different injection systems and fuel delivery strategies, which in modern engines is aimed at minimizing smoke and reducing emissions of toxic exhaust components, rather than maximizing power.

8. The increase of 2 kPa in the air filter pressure drop above that of the "New" one results in a 9.31% decrease in the effective power of the tested engine, which is 4.66% per 1 kPa increase in the pressure drop and is 10 times higher than in the case of diesel engines with classic injection systems. The decrease in power is correlated with a decrease in engine inlet flow (3.39%) and boost pressure (4.28%), a decrease in hourly fuel consumption (7.87%) and an increase in specific fuel consumption (2.52%).

9. The decrease of the engine power caused by the increase of the flow resistance in the intake system (air filter contamination) will have a very negative effect on the traction properties of the vehicle (truck tractor), from which the tested engine originated. As the engine power decreases, the traction properties of the vehicle deteriorate, in particular: the ability to overcome a hill in individual gears, the change of the maximum speed value depending on the road inclination angle and the time and distance of acceleration of the vehicle–tractor–trailer combination.

10. Increase of the flow resistance in the inlet system of the tested engine did not significantly affect the increase of smoke, which is a very beneficial phenomenon, since there is no increase in the emission of particulate matter, which is currently one of the most significant components polluting the atmosphere.

11. Comparing the test results obtained with the results presented in the literature for older engine types, it should be stated that the modern truck engines with electronically controlled fuel injection and turbocharging systems are less susceptible to the negative influence of contamination of the intake system in comparison with engines with mechanical supply systems—injection pump, carburetor. This is due to the fact that the ECU of the modern engine on the basis of input signals—boost pressure, mass of intake air, vacuum in the intake system—corrects the fuel dose in such a way as to offset the negative effects of increased resistance to flow. The dose of fuel and the method of its application is selected in such a way as to limit power loss with no increase in smoke.

12. The obtained results showed the influence of the flow resistance in the inlet system of the modern truck engine on its performance. It is advisable to continue the work, the final effect of which should be the determination of the maximum/admissible flow resistance, the exceeding of which should eliminate the vehicle from further operation due to deterioration of the economic, energetic, ecological and traction properties of the vehicle.

13. It is advisable to extend the presented research by an analysis of the influence of increased pressure drop in the air filtration system on the emission of toxic components of exhaust gases and a change in the traction characteristics of the vehicle equipped with this type of inlet system.

Author Contributions: Conceptualization, T.D. and M.K.; methodology, M.K.; software, T.D.; validation, T.D. and M.K.; formal analysis, T.D. and M.K.; investigation, T.D. and M.K.; data curation, T.D. and M.K.; writing—original draft preparation, T.D. and M.K.; writing—review and editing, T.D.; visualization, T.D. and M.K.; supervision, T.D. and M.K.; project administration, M.K.; funding acquisition, T.D. and M.K. All authors have read and agreed to the published version of the manuscript.

Funding: The article was written as part of the implementation of the university research grant No. UGB 759/2022.

Institutional Review Board Statement: Not applicable.

Informed Consent Statement: Not applicable.

Data Availability Statement: Data is contained within the article.

Conflicts of Interest: The authors declare no conflict of interest.

Abbreviations

ε	compression ratio
ζ	dimensionless pressure drop coefficient of the inlet system
G_e	hourly fuel consumption
g_e	specific fuel consumption
i	number of cylinders
k	smoke opacity of exhaust gases—light absorption coefficient
k	stroke number factor
L_t	theoretical air demand
λ	excess air factor
M_o	engine torque
N_e	effective engine power
$\dot{m}_{rz}$	mean real flow rate of air supplied to engine cylinders
$\dot{m}_t$	theoretical mass flow rate of air supplied to engine cylinders
n	engine rotational speed
η_i	indexed efficiency
η_m	mechanical efficiency
η_v	filling factor
p_d	charge air pressure
p_H	ambient pressure
p_N	pressure at the end of the filling stroke
p_r	pressure of the rest of the exhaust gas
p_1	pressure before
Δp_f	air filter pressure drop
T_H	ambient temperature
T_N	temperature at the end of the filling stroke
T_r	temperature of the rest of the exhaust gas
Q_s	engine air demand
V_s	cylinder displacement
$v_Ł$	average charge velocity in the inlet system
W_d	fuel calorific value
ρ_{pow}	air density
$\rho_Ł$	charge density

References

1. Yun, J.-E. Optimal Design of Off-Road Utility Terrain Vehicle Air Filter Intake. *Energies* **2021**, *14*, 2269. [CrossRef]
2. Mustafa, N.S.; Ngadiman, N.H.A.; Abas, M.A.; Noordin, M.Y. Application of Box-Behnken Analysis on the Optimisation of Air Intake System for a Naturally Aspirated Engine. *Int. J. Automot. Mech. Eng. (IJAME)* **2020**, *17*, 7607–7617. [CrossRef]

3. Szczepankowski, A.; Szymczak, J.; Przysowa, R. The Effect of a Dusty Environment Upon Performance and Operating Parameters of Aircraft Gas Turbine Engines. In Proceedings of the Specialists' Meeting-Impact of Volcanic Ash Clouds on Military Operations NATO AVT-272-RSM-047, Vilnius, Lithuania, 17 May 2017. [CrossRef]

4. Tang, R.-J.; Hu, B.-F.; Zhang, M.; Lu, Z.-J. Study on Separation Characteristics of Dust and Droplet on Air Intake Pre-filtration Systems of CV Based on CFD Simulation and Test. In Proceedings of the International Conference on Artificial Intelligence and Computing Science (ICAICS 2019), Hangzhou, China, 24–25 March 2019.

5. Cardozo, J.I.H.; Sánchez, D.F.P. An experimental and numerical study of air pollution near unpaved Roads. *Air Qual. Atmos. Health* **2019**, *12*, 471–489. [CrossRef]

6. Rienda, I.C.; Alves, C.A. Road dust resuspension: A review. *Atmos. Res.* **2021**, *261*, 105740. [CrossRef]

7. Smialek, J.L.; Archer, F.A.; Garlick, R.G. Turbine Airfoil Degradation in the Persian Gulf War. *JOM* **1994**, *46*, 39–40. [CrossRef]

8. Bojdo, N.; Filippone, A. Effect of desert particulate composition on helicopter engine degradation rate. In Proceedings of the 40th European Rotorcraft Forum, Southampton, UK, 2–5 September 2014.

9. Dziubak, T.; Dziubak, S.D. A Study on the Effect of Inlet Air Pollution on the Engine Component Wear and Operation. *Energies* **2022**, *15*, 1182. [CrossRef]

10. Wróblewski, P.; Rogólski, R. Experimental Analysis of the Influence of the Application of TiN, TiAlN, CrN and DLC1 Coatings on the Friction Losses in an Aviation Internal Combustion Engine Intended for the Propulsion of Ultralight Aircraft. *Materials* **2021**, *14*, 6839. [CrossRef]

11. Wróblewski, P. Analysis of Torque Waveforms in Two-Cylinder Engines for Ultralight Aircraft Propulsion Operating on 0W-8 and 0W-16 Oils at High Thermal Loads Using the Diamond-Like Carbon Composite Coating. *SAE Int. J. Engines* **2022**, *15*, 129–146. [CrossRef]

12. Wróblewski, P.; Koszalka, G. An Experimental Study on Frictional Losses of Coated Piston Rings with Symmetric and Asymmetric Geometry. *SAE Int. J. Engines* **2021**, *14*, 853–866. [CrossRef]

13. Dziubak, T.; Dziubak, S.D. Experimental Study of Filtration Materials Used in the Car Air Intake. *Materials* **2020**, *13*, 3498. [CrossRef]

14. Maddineni, A.K.; Das, D.; Damodaran, R.M. Numerical investigation of pressure and flow characteristics of pleated air filter system for automotive engine intake application. *Sep. Purif. Technol.* **2019**, *212*, 126–134. [CrossRef]

15. Fleck, S.; Heim, M.; Beck, A.; Moser, N.; Durst, M. Realitätsnahe Prüfung von Motoransaugluftfiltern. *MTZ-Mot. Zeitschrift.* **2009**, *70*, 414–418. [CrossRef]

16. Dziubak, T.; Bąkała, L. Problems of selecting filter partition in passenger car engine intake air filters. *Combust. Eng.* **2021**, *185*, 44–59. [CrossRef]

17. Rieger, M.; Hettkamp, P.; Löhl, T.; Madeira, P.M.P. Effcient Engine Air Filter for Tight Installation Spaces. *ATZ Heavy Duty Worldw.* **2019**, *12*, 56–59. [CrossRef]

18. Dahlan, A.A.; Muhammad Said, M.F.; Abdul Latiff, Z.; Mohd Perang, M.R.; Abu Bakar, S.A. Acoustic Study of an Air Intake System of SI Engine using 1-Dimensional Approach. *Int. J. Automot. Mech. Eng. (IJAME)* **2019**, *16*, 6281–6300. [CrossRef]

19. Sawant, P.; Bari, S. *Effects of Variable Intake Valve Timings and Valve Lift on the Performance and Fuel Efficiency of an Internal Combustion Engine*; SAE Technical Paper 2018-01-0376; SAE International: Warrendale, PA, USA, 2018.

20. Poon, W.S.; Liu, B.Y. *Dust Loading Behavior of Engine and General Purpose Air Cleaning Filters*; SAE Technical Paper 1997; SAE International: Warrendale, PA, USA, 2018.

21. Tian, X.; Ou, Q.; Liu, J.; Liang, Y.; Pui, D.Y.H. Particle Loading Characteristics of a Two-Stage Filtration System. *Sep. Purif. Technol.* **2019**, *215*, 351–359. [CrossRef]

22. Payen, J.; Vroman, P.; Lewandowski, M.; Perwuelz, A.; Callé-Chazelet, S.; Thomas, D. Influence of fiber diameter, fiber combinations and solid volume fraction on air filtration properties in nonwovens. *Text. Res. J.* **2012**, *82*, 1948–1959. [CrossRef]

23. Dziubak, T.; Boruta, G. Experimental and Theoretical Research on Pressure Drop Changes in a Two-Stage Air Filter Used in Tracked Vehicle Engine. *Separations* **2021**, *8*, 71. [CrossRef]

24. Sun, Z.; Liang, Y.; He, W.; Jiang, F.; Song, Q.; Tang, M.; Wang, J. Filtration performance and loading capacity of nano-structured composite filter media for applications with high soot concentrations. *Sep. Purif. Technol.* **2019**, *221*, 175–182. [CrossRef]

25. Toma, M.; Bobâlc, C. Research on Drivers' Perception on the Maintenance of Air Filters for Internal Combustion Engines. *Procedia Technol.* **2016**, *22*, 961–968. [CrossRef]

26. Bugli, N.; Scott, D.; Scott, F. *Investigating Cleaning Procedures for OEM Engine Air Intake Filters*; SAE Technical Paper 2007-01-1431; SAE International: Warrendale, PA, USA, 2018.

27. Bugli, N. *Automotive Engine Air Cleaners-Performance Trends*; SAE Technical Paper 2001-01-1356; SAE International: Warrendale, PA, USA, 2018.

28. Jaroszczyk, J.; Wake, J.; Connor, M.J. Factors Affecting the Performance of Engine Air Filters Factors Affecting the Performance of Engine Air Filters. *J. Eng. Gas Turbines Power* **1993**, *115*, 693–700. [CrossRef]

29. Toma, M. A Study of the Air Filters' Maintenance for Automotive Internal Combustion Engines. *Sci. Bul. Ser. D Mech. Eng.* **2014**, *76*, 101–110.

30. Toma, M.; Andreescu, C.; Negurescu, N. Some Aspects Concerning the Period of Automotive Air Filters Replacement on the Basis of their Technical Conditions. In Proceedings of the Innovative Manufacturing Engineering International Conference, Iasi, Romania, 21–22 May 2015.

31. Barris, M.A. *Total Filtration^{TM}. The Influence of Filter Selection on Engine Wear. Emissions and Performance*; SAE Technical Paper: 952557; SAE International: Warrendale, PA, USA, 2018.

32. Bugli, N.J.; Green, G.S. Performance and Benefits of Zero Maintenance Air Induction Systems. In Proceedings of the 2005 SAE World Congress, Detroit, MI, USA, 11–14 April 2005.

33. Norman, K.; Huff, S.; West, B. *Effect of Intake Air Filter Condition on Vehicle Fuel Economy*; Oak Ridge National Lab. (ORNL): Oak Ridge, TN, USA; U.S. Department of Energy (DOE) Information Bridge: Washington, DC, USA, 2009.

34. Oszczypała, M.; Ziółkowski, J.; Małachowski, J. Reliability Analysis of Military Vehicles Based on Censored Failures Data. *Appl. Sci.* **2022**, *12*, 2622. [CrossRef]

35. Żurek, J.; Zieja, M.; Ziółkowski, J. Reliability of supplies in a manufacturing enterprise. In *Safety and Reliability-Safe Societies in a Changing World—Proceedings of the 28th International European Safety and Reliability Conference (ESREL 2018), Trondheim, Norway, 17–21 June 2018*; CRC Press: Boca Raton, FL, USA, 2018; pp. 3143–3148. ISBN 978-1-138-62937-0.

36. Bernhard, M.; Dobrzyński, S.; Loth, E. *Silniki Samochodowe*; WKŁ: Warszawa, Poland, 1988. (In Polish)

37. Mysłowski, J. *Doładowanie Silników*; WKŁ: Warszawa, Poland, 2011. (In Polish)

38. Dziubak, T.; Piętak, A. Braking investigations on influence of pressure drop of air filter on operation parameters of combustion piston engine. *Bull. Mil. Univ. Technol.* **1994**, *1*, 113–121.

39. Wajand, J.A. *Silniki o Zapłonie Samoczynnym*; WNT: Warszawa, Poland, 1988. (In Polish)

40. Karczewski, M.; Szczęch, L. Influence of the F-34 unified battlefield fuel with bio components on usable parameters of the IC 1126 engine-Wpływ mieszanin jednolitego paliwa pola walki F-34 z biokomponentami na parametry użyteczne silnika. *Eksploat. Niezawodn. Maint. Reliab.* **2016**, *18*, 358–366. [CrossRef]

41. Dziubak, T. Analiza wpływu oporu przepływu filtru powietrza na napełnienie tłokowego silnika spalinowego. *Zesz. Nauk. Politech. Szczecińskiej* **1992**, *19*, 33–40.

42. Dziubak, T. Diagnostics of military vehicle engine air filters. *Bull. Mil. Univ. Technol.* **1998**, *10*, 73–89.

43. Dziubak, T.; Trawiński, G. The experimental Research of the air filter flow drag influence on the T359E engine operation parameters. *Bull. Mil. Univ. Technol.* **2001**, *4*, 135–149.

44. Piętak, A.; Trawiński, G. Prędkość obrotowa silnika jako parametr diagnostyczny układu T-P-C. *Arch. Motoryzacji* **2004**, *7*, 513–518.

45. Abdullah, N.R.; Shahruddin, N.S.; Mamat, A.M.I.; Kasolang, S.; Zulkifli, A.; Mamat, R. Effects of Air Intake Pressure to the Fuel Economy and Exhaust Emissions on a Small SI Engine. *Procedia Eng.* **2013**, *68*, 278–284. [CrossRef]

46. Dziubak, T.; Trawiński, G. The experimental assessment of air supply system modification on inlet air filtration efficiency and military vehicle engine effectiveness improvement. *J. KONES Powertrain* **2010**, *17*, 79–86.

47. Slabov, E.; Antropov, B.; Nilov, V. Influence of Air Purification on Engine Life. *Automob. Transp.* **1978**, *4*, 41–42.

48. Jaroszczyk, T. Jakość filtrów powietrza tłokowych silników spalinowych. *Siln. Spalinowe* **1978**, *4*, 26–31.

49. Szczeciński, S.; Otyś, J.; Dzierżanowski, P.; Wiatrek, R. *Turbinowe Napędy Samochodów*; WKŁ: Warszawa, Poland, 1974. (In Polish)

50. Yang, A.-J.; Song, Z.-J.; Yang, J.-W.; Zhao, C.-J. Effect of Air Filter on Performance of S1110 Diesel Engine. IOP Conference Series. *Mater. Sci. Eng.* **2019**, *688*, 022050. [CrossRef]

51. Allam, S.; Mimi Elsaid, A. Parametric Study on Vehicle Fuel Economy and Optimization Criteria of the Pleated Air Filter Designs to Improve the Performance of an I.C Diesel Engine: Experimental and CFD Approaches. *Sep. Purif. Technol.* **2020**, *241*, 116680. [CrossRef]

52. Bulejko, P. An analysis on energy demands in airborne particulate matter filtration using hollow-fiber membranes. *Energy Rep.* **2021**, *7*, 2727–2736. [CrossRef]

53. Fotovati, S.; Tafreshi, H.V.; Pourdeyhimi, B. A macroscopic model for simulating pressure drop and collection efficiency of pleated filters over time. *Sep. Purif. Technol.* **2012**, *98*, 344–355. [CrossRef]

54. Mahesh, J. Parametric study and CFD analysis of air filter. *Asian J. Converg. Technol. (AJCT)* **2019**, *5*, 1–9.

55. Maddineni, A.K.; Das, D.; Damodaran, R.M. Air-borne particle by fibrous filter media under collision effect: A CFD-based approach. *Sep. Purif. Technol.* **2018**, *193*, 1–10. [CrossRef]

56. Feng, Z.; Long, Z. Modeling unsteady filtration performance of pleated filter. *Aerosol Sci. Technol.* **2016**, *50*, 626–637. [CrossRef]

57. Saleh, A.; Tafreshi, H. A simple semi-numerical model for designing pleated air filters under dust loading. *Sep. Purif. Technol.* **2014**, *137*, 94–108. [CrossRef]

58. Théron, F.; Joubert, A.; Coq, L. Numerical and experimental investigations of the influence of the pleat geometry on the pressure drop and velocity field of a pleated fibrous filter. *Sep. Purif. Technol.* **2017**, *182*, 69–77. [CrossRef]

59. Wiegmann, A.; Rief, S.; Kehrwald, D. Computational study of pressure drop dependence on pleat shape and filter media. In Proceedings of the Filtech 2007 International Conference for Filtration and Separation Technology, Wiesbaden, Germany, 27 February–1 March 2007; Volume 1, pp. 79–86.

60. Park, B.; Lee, M.; Jo, Y.; Kim, S. Influence of pleat geometry on filter cleaning in a PTFE/glass composite filter. *J. Air Waste Manage. Assoc.* **2012**, *62*, 1257–1263. [CrossRef] [PubMed]

61. Del Fabbro, L.; Laborde, J.C.; Merlin, P.; Ricciardi, L. Air Flows and Pressure Drop Modelling for Different Pleated Industrial Filters. *Filtr. Sep.* **2002**, *39*, 35–40. [CrossRef]

62. Abd Halim, M.A.; Nik Mohd, N.A.R.; Mohd Nasir, M.N.; Dahalan, M.N. Experimental and Numerical Analysis of a Motorcycle Air Intake System Aerodynamics and Performance. *Int. J. Automot. Mech. Eng. (IJAME)* **2020**, *17*, 7607–7617. [CrossRef]

63. Ceviz, M.A.; Akın, M. Design of a new SI engine intake manifold with variable length plenum. *Energy Convers. Manag.* **2010**, *51*, 2239–2244. [CrossRef]

64. Costa, R.C.; Hanriot, S.M.; Sodré, J.R. Influence of intake pipe length and diameter on the performance of a spark ignition engine. *J. Braz. Soc. Mech. Sci. Eng.* **2013**, *36*, 29–35. [CrossRef]

65. Alves, L.O.F.T.; Santos, M.G.D.; Urquiza, A.B.; Guerrero, J.H.; Lira, J.C.; Abramchuk, V. *Design of a New Intake Manifold of a Single Cylinder Engine with Three Stages*; SAE Technical Paper Series 2017; SAE International: Warrendale, PA, USA, 2018. [CrossRef]

66. Hanriot, S.M.; Queiroz, J.M.; Maia, C.B. Efects of variable-volume Helmholtz resonator on air mass fow rate of intake manifold. *J. Braz. Soc. Mech. Sci. Eng.* **2019**, *41*, 79. [CrossRef]

67. Song, H.S.; Cho, H.M. A Study on Duct Integrated Resonator of Automobile Intake System. *Int. J. Appl. Eng. Res.* **2017**, *12*, 7696–7699.

68. Ghodke, S.; Bari, S. Effect of Integrating Variable Intake Runner Diameter and Variable Intake Valve Timing on an SI Engine's Performance. *SAE Tech. Pap. Series* **2018**, *1*, 380. [CrossRef]

69. Sawant, P.; Bari, S. Combined effects of variable intake manifold length, variable valve timing and duration on the performance of an internal combustion engine. In Proceedings of the ASME 2017 International Mechanical Engineering Congress and Exposition, Tampa, FL, USA, 3–9 November 2017.

70. Aradhye, O.; Bari, S. Continuously Varying Exhaust Pipe Length and Diameter to Improve the Performance of a Naturally Aspirated SI Engine. In Proceedings of the International Mechanical Engineering Congress and Exposition IMECE2017, Tampa, FL, USA, 3–9 November 2017.

71. Chhalotre, S.; Baredar, P.; Soni, S. Experimental investigation to analyze the effect of variation in intake air temperature on the throttle response of the s.i. engine. *Int. J. Mech. Eng. Technol. (IJMET)* **2018**, *9*, 972–981.

72. Opis Techniczny FH, FM Silnik D13C460, EU5SCR-M, Informator Techniczny Firmy VOLVO. 2021. Available online: https://stpi.it.volvo.com/STPIFiles/Volvo/FactSheet/D13C460,%20EU5SCR-M_Pol_02_1144722.pdf (accessed on 27 March 2022).

73. *PN-ISO 15550:2009*; Silniki Spalinowe Tłokowe-Określanie i Metoda Pomiaru Mocy Silnika-Wymagania Ogólne. PKN: Warszawa, Poland, 2009.

74. Karczewski, M.; Wieczorek, M. Assessment of the Impact of Applying a Non-Factory Dual-Fuel (Diesel/Natural Gas) Installation on the Traction Properties and Emissions of Selected Exhaust Components of a Road Semi-Trailer Truck Unit. *Energies* **2021**, *14*, 8001. [CrossRef]

75. Rozporządzenie Ministra Infrastruktury z Dnia 31 Grudnia 2002 r. w Sprawie Warunków Technicznych Pojazdów Oraz Zakresu ich Niezbędnego Wyposażenia (Dz. U. z 2016 r. poz. 2022, z 2017 r. poz. 2338 oraz z 2018 r. poz. 855). Available online: https://isap.sejm.gov.pl/isap.nsf/download.xsp/WDU20220000122/O/D20220122.pdf (accessed on 27 March 2022). (In Polish)

Article

Experimental Assessment of the Impact of Replacing Diesel Fuel with CNG on the Concentration of Harmful Substances in Exhaust Gases in a Dual Fuel Diesel Engine

Mirosław Karczewski [1,*], Grzegorz Szamrej [2] and Janusz Chojnowski [1]

[1] Faculty of Mechanical Engineering, Military University of Technology, 2 Gen, Sylwestra Kaliskiego Street, 00-908 Warsaw, Poland; janusz.chojnowski@wat.edu.pl
[2] Military University of Technology in Warsaw, Gen. Sylwestra Kaliskiego 2, 00-908 Warsaw, Poland; grzegorz.szamrej@wat.edu.pl
* Correspondence: miroslaw.karczewski@wat.edu.pl

Abstract: The problem of global warming and related climate change, as well as rising oil prices, is driving the implementation of ideas that not only reduce the consumption of liquid fuels, but also reduce greenhouse gas emissions. One of them is the use of natural gas as an energy source. It is a hydrocarbon fuel with properties allowing the reduction of CO_2 emissions during its combustion. Therefore, solutions are being implemented that allow natural gas to be supplied to means of transport, which are trucks of various categories and purposes. This article presents the results of tests of an engine from a used semi-truck, to which an innovative compressed natural gas (CNG) supply system was installed. This installation (both hardware and software), depending on the engine operating conditions, enables mass replacement by natural gas (up to 90%) of the basic fuel—diesel oil. During the tests, on the basis of the obtained results, the influence of the diesel fuel/CNG exchange ratio under various engine operating conditions on the concentration of toxic CO_2, CO, NO, NO_2, CH_4, C_2H_6, NMHC, NH_3 and exhaust smoke was assessed. The test results confirm that, compared to conventional fueling, the diesel/CNG-fueled engine allows for a significant reduction in CO_2 concentration even in a car operated for several years with diesel fuel and with high mileage. The use of a non-factory installation significantly increased the concentration of methane CH_4, nitrogen dioxide NO_2 and carbon monoxide CO in the exhaust gas. It was found that the smoke content and the temperature of exhaust gases did not decrease with increasing ratio of fuel replacement. The concentration of CO, NO_X, CH_4 and NMHC was increased, while the concentration of CO_2, C_2H_6, NH_3 and the consumption of diesel fuel by the engine, decreased significantly. The innovation of the research is based on the use of a modern and unique engine gas fuel system control system where the original fuel supply system with unit pumps is able to reduce diesel oil consumption by up to 90%.

Keywords: dual fuel; diesel-CNG; CNG

Citation: Karczewski, M.; Szamrej, G.; Chojnowski, J. Experimental Assessment of the Impact of Replacing Diesel Fuel with CNG on the Concentration of Harmful Substances in Exhaust Gases in a Dual Fuel Diesel Engine. *Energies* **2022**, *15*, 4563. https://doi.org/10.3390/en15134563

Academic Editor: George Kosmadakis

Received: 1 June 2022
Accepted: 20 June 2022
Published: 22 June 2022

Publisher's Note: MDPI stays neutral with regard to jurisdictional claims in published maps and institutional affiliations.

1. Introduction

In a dual fuel engine, fuel with a high octane number, which in the case of our tests was natural gas (NG), is fed into the combustion chamber of the engine first, along with the air supplied to the cylinder. After reaching a sufficiently high pressure and temperature, resulting from the compression of the gas-air mixture, a second fuel is injected into the combustion chamber which has good self-ignition properties allowing it to initiate self-ignition of the injected fuel. In the presented research, the high cetane number autoignition initiating fuel was classic diesel fuel commonly used in single-fuel diesel engines. Such a combination is the most popular way of powering "dual-fuel" compression ignition (CI) engines due to the possibility of achieving a very high degree of replacement of diesel fuel by natural gas, the use of which is supported by the cost and ecological aspects and the uncomplicated modification of the engine installation [1].

The degree of substitution in dual-fuel engines is defined as the ratio of the amount of high-octane fuel consumed to the total amount of fuel consumed by the engine. The diesel/CNG compressed natural gas power supply system used in the research determined the degree of mass replacement in this way. The ratio is expressed as the ratio of the amount of energy supplied to the engine in the form of high-octane fuel, in our case, CNG, to the total amount of energy contained in the fuel supplied to the single-fueled engine. In simple terms, it is also assumed to be the ratio of the amount of energy supplied to the engine in the form of high-octane fuel to the total amount of energy that is supplied at a given point of operation in the form of two fuels—the results in the case of both methods of calculation will be identical when the efficiency of the engine in both cases (mono-fuel and dual fuel power) does not change. This relationship can easily be calculated by relying only on the hourly fuel consumption based on the change in the consumption of diesel (or other high-cetane fuel) alone at the operating point where the substitution is measured and calculated. This can be done on the basis of Formula 1 below, without knowing the calorific value of the fuels used [2]:

$$\frac{diesel}{CNG} = \frac{Ge_{diesel} - Ge_{diesel\_CNG}}{Ge_{diesel}} * 100\% = 1 - \frac{Ge_{diesel\_CNG}}{Ge_{diesel}} * 100\% \tag{1}$$

Ge_{diesel}—hourly consumption of diesel fuel when running only on diesel oil,
$Ge_{diesel\ on\ CNG}$—hourly diesel fuel consumption when running on CNG/diesel.

In the case of the controller that was used in our research, the degree of replacement was calculated from a relation based on the mass fuel consumption of the engine, in which the calorific ratio of one fuel to another was expressed by the coefficient set by the operator at the beginning of the tests. The value of the adopted ratio was diesel/CNG 1:1.1 [2].

The degree of replacement of diesel fuel by natural gas (CNG) in a dual fuel engine depends on a number of factors, but the most important limitation that does not allow a high degree of replacement of diesel fuel by substitute fuel is the appearance of knock combustion and the expenditure of the CNG power supply installation. On the basis of research by many authors [3–5], an approximate diagram based on [6] has been developed, and is shown in Figure 1 schematically showing the possibilities of substitution of diesel by natural gas in dual fuel engines according to [7].

As part of the dual fuel supply, various fuels can be used, but the greatest number of advantages is characterized by compressed or liquefied natural gas. It is possible to co-burn diesel fuel in diesel autoignition engines together with a mixture of propane and butane gases, high-octane biofuels, or with gasoline. However, at the current level of technology in widespread use, such installations only achieve a partial, relatively small replacement of diesel fuel by these fuels. Diesel oil in classic installations of this type constitutes not less than half of the energy value of the fuel supplied for combustion. Depending on the source of information, it is possible to achieve a different replacement ratio with liquefied petroleum gas (LPG) fuel, but most often it does not exceed 30% [8]. However, in the case of using natural gas combined with diesel oil, which is the main research goal of this work, it already allows for a significant reduction in the operating costs of the engine and ecological improvement. The exchange ratio of diesel with natural gas is usually much higher than in the case of gasoline, biofuels, LPG or alcohols, mainly due to the much higher resistance of methane to knock combustion and its higher calorific value [9]. The same applies to the use of natural gas in liquid form, which has some additional advantages over its use in compressed form. These advantages relate to a large extent to the operation of the vehicle in which the installation of this type was placed, but also the possibility of direct injection of this fuel straight into the combustion chamber under high pressure. In the case of the gas combustion process, it may be more efficient if an appropriate fuel installation is used in which high gas pressure [10] or the physical state in which it is located [11] is used. The first installations of this type already exist in the world and are used both in motor vehicles [12] and in marine engines [13]. Although they are not yet widely available solutions, intensive work is being carried out on their dissemination and further development. The biggest advantage of dual fuel engines is the combination of many advantages of CI and SI engines. There are a few ways to supply fuel to those engines and the two most important differ in the place of delivery of gaseous fuel [14]. The specificity of dual fuel power supply with direct injection of both fuels by one injector is shown in the schematic drawing of the injector in Figure 2.

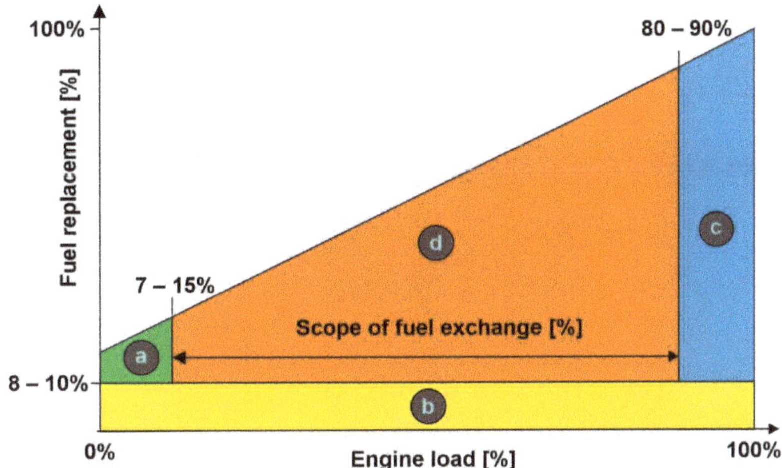

Figure 1. The various operating conditions known as "modes" for dual fuel engines, where horizontal is the engine load between 0–100%, and vertical is the rate of fuel replacement from 0% to 100%. The individual areas above result from the characteristics of the engine, the diesel fuel system and the CNG fueling system. Area (**a**) results from the properties of the CNG supply system—e.g., minimum CNG injector opening time, (**b**) minimum fuel dose necessary for proper ignition of the air-fuel mixture (its minimum value is also limited by the minimum opening time of the unit injector), (**c**) maximum load engine (**d**) range of proper dual fuel engine operation on diesel/CNG fuels. Authors source based on the chart from [6].

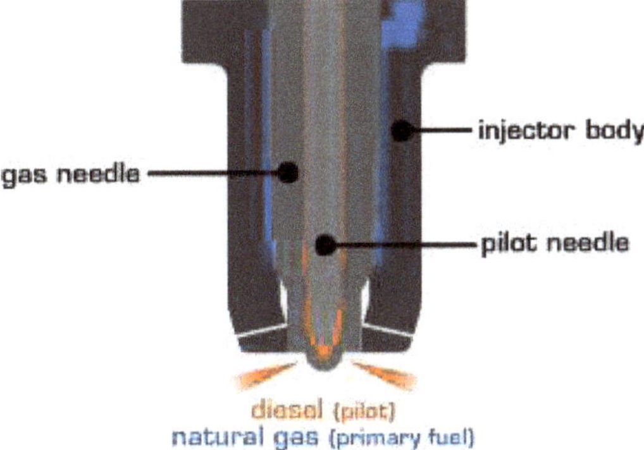

Figure 2. Schematic drawing of a direct dual fuel injector in use at the moment of diesel fuel injection [15].

This type of injector is developed especially for dual fuel engines working in diesel/CNG mode. This type of fuel supply is still very rare. A much more popular type of fuel supply is indirect injection of CNG and direct injection of diesel fuel. That was the type of supply researchers used in their project. A schematic drawing of an engine at work with this type of fueling is shown in Figure 3.

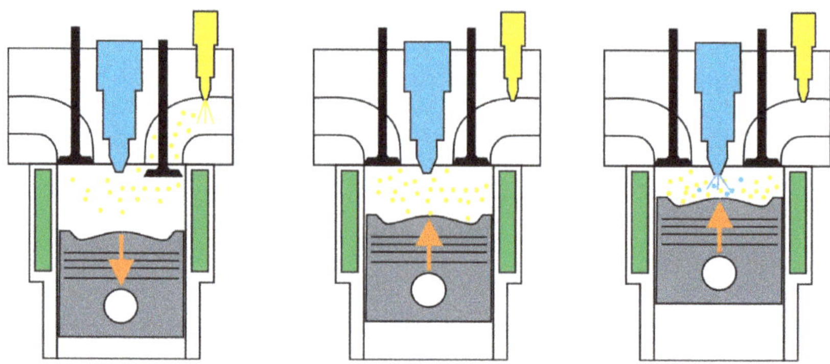

Figure 3. Schematic drawing of fueling of a working dual fuel engine with indirect injection of CNG fuel (yellow drops) and direct injection of diesel fuel (blue drops) [Authors' source].

The type of fueling shown in the Figure 3 is much more common, simpler and well-researched, and is a much more popular way to make a dual fuel engine from a diesel CI engine. In dual fuel engines, during operation the injected liquid fuel is dosed into the combustion chamber in the form of a pilot dose, initiating the ignition of the fuel already supplied to the combustion chamber under the effect of a steady increase in pressure resulting from self-ignition of the injected high-cetane fuel. Figure 4 below shows a schematic view of the method of supply and ignition in a dual fuel engine powered by a mixture of CNG and diesel oil.

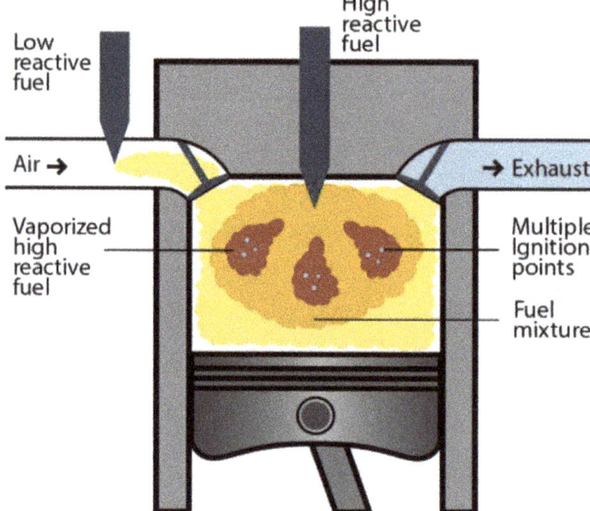

Figure 4. Autoignition and combustion in a Dual Fuel CI engine, own artwork based on [16].

An engine fueled by dual fuel CNG and diesel oil requires, for standard work mode, only the already mentioned, pilot dose of diesel fuel to initiate the ignition. When using LPG as an additional fuel, it is necessary to use a greater amount of injected diesel fuel than in the case of natural gas, due to the insufficient resistance of the additional gaseous fuel to knocking combustion, which is dangerous due to the possibility of engine damage. Its presence determines the possibility of using fuel in the engine and the degree of replacement of the basic fuel, which is diesel fuel in a dual fuel CI engine. Highly loaded dual fuel engines operate under conditions that make it impossible to replace as much chemical energy contained in high-cetane number fuel, as is possible at low loads There is a lower ratio of replacement at such operating points due to the need to use more high-cetane number fuel with a higher energy density than gaseous fuels.

The sense of using natural gas in CI engines results not only from the physicochemical properties allowing its effective use in such engines, but also the ecological and economic impact that this fuel could have on the industry based on the traditional piston internal combustion (IC) engine. Along with the development of civilization and the increase in consumerism, the demand for the transport of goods using wheeled vehicles will increase, especially in increasingly crowded cities where people have to deal with the negative health effects caused by pollution from the growing number of trucks [17,18]. It is estimated that in Poland alone, about 50,000 people die every year due to poor air quality, or even twice as many taking into account the percentage of deaths in Europe caused by exposure to fine particulate matter (PM 1.0 and PM 2.5) resulting from the combustion of fossil fuels for populations aged >14 years [19,20]. The use of a low-emission alternative fuel instead of the coal and diesel fuel commonly used in Poland could contribute to avoiding these deaths, which is why research is currently placing so much emphasis on the use of natural gas to power thermal machines.

Public transport can offer a more rational use of energy [21] as it reduces fuel consumption, exhaust emissions and vehicle noise [22–24]. The concept of sustainable transport combines high-quality transport with care for the natural environment [25]. Therefore, these vehicles should have a power source other than diesel or gasoline and reducing carbon dioxide (CO_2) emissions can help limit the development of global warming in the world.

An example of such activities may be the increasing restrictiveness of requirements for products created during the combustion of a fuel mixture. Their implementation means that, for example, even relatively new semi-trucks (5 years old) are classified as polluting and their use is charged with additional fees. This forces vehicle manufacturers to look for technical solutions that reduce the emission of harmful exhaust components. The obvious path of development seems to be the construction of trucks with hybrid or only electric drive. However, in this case, the range remains a limitation, which for an electric car of average payload ("N2") is currently about 200 km, while for a tractor unit with a semi-trailer ("N3") it is, according to available data, only 100 km [26]. In addition, the price of an electric tractor unit in 2021 was five times the price of the same conventionally powered tractor unit (based on the example of DAF and Volvo cars). In addition, increasing the weight of the semi-truck as a result of the usage of batteries that are still very heavy, reduces the weight of the load that can be transported, and the charging process is time-consuming. Both of these factors have a very negative impact on the logistical costs of using such a vehicle [27]. Therefore, a good direction of change is still the introduction of more environmentally friendly solutions into transport, mainly through the use of alternative and ecological fuels in relation to diesel, including natural gas in various forms: compressed natural gas and Liquid Natural Gas (LNG) and its ecologically produced varieties Renewable Natural Gas—RNG, BioCNG, or eCNG and natural gas enriched with hydrogen fuel—HCNG, or H2CNG. Many fuels can be used as high-cetane fuel, the properties of which do not differ significantly from those of diesel oil. Very promising are the possibilities of using, together with ecological gaseous fuels, also ecological, alternative fuels replacing diesel oil in this case, such as oil made of algae [28], rice [29], pongamia, palm, peanuts, mushrooms, jatrophy, coffee grounds, soybean, coconut, linseed, olives or even animal fat, which the authors described in detail in [30]. Also, a gas used as a fuel corresponding to the properties of natural gas can be produced in many ways and has great potential as a clean alternative fuel for additional reduction of GHG emissions in the process of its combustion or production [29]. Such a gas can be produced under the same conditions as bio-oil from many plant and animal products [30] or waste [31] and individual final products will differ mainly in the concentration of methane and other components of this fuel.

Combustion of natural gas, due to its different composition compared to diesel oil, with a lower proportion of carbon and a greater proportion of hydrogen, results in lower emissions of carbon dioxide, which is one of the main greenhouse gases emitted by human activity. In the short and medium term, natural gas may become the most important alternative fuel [32]. Its large global resources increase the potential of this fuel as a green fuel used in transport. Many European countries have a well-developed infrastructure adapted for the reception and distribution of CNG and LNG. Natural gas can also be a relatively cheap fuel compared to diesel oil. In 2020, the average price of natural gas accounted for approx. 50% of the diesel oil price. The average cost of using natural gas to transport heavy goods vehicles is approx. 40% lower than that of diesel fuel [33].

The properties of natural gas (in particular in the form of compressed CNG) in terms of the use of IC piston engines in transport or in utility vehicles, as well as the growing number of supply sources in individual European countries justify undertaking scientific, research and development works on the use of CNG to supply the IC engines of heavy-duty vehicles. This is in line with the energy policy of the European Union—Directive 2009/33/EC of 23 April 2009 on the promotion of ecologically clean and energy-efficient road transport vehicles. The presented work is an example of such activities and describes the effects of the application of an innovative concept of effective adaptation of an engine of a utility vehicle to a CNG supply.

2. Materials and Methods

The aim of the study was to experimentally determine the impact of the exchange ratio (diesel/CNG), defined by Equation (1), of the replacement fuel, which was CNG, to the base fuel, which was diesel oil, for an innovative non-factory two-fuel CNG/diesel fuel supply installation in a used tractor unit engine on the emissions of individual exhaust components: carbon dioxide (CO_2), carbon monoxide (CO), nitric oxide (NO), nitrogen dioxide (NO_2), methane (CH_4), ethane (C_2H_6), non-methane hydrocarbons(NMHC), ammonia (NH_3) and smoke in fixed operating states.

"The tests and calculations were carried out for the D13C460 EU5EEV (338 kW) engine, used in Volvo FH13 series tractor units. The mileage of the car, the unit of which was tested on an engine dynamometer and on which a non-factory diesel/CNG installation was adapted, was 790,500 km. The total operating time of the engine is 11,800 h, and the engine has used 229,100 L of diesel fuel since the beginning of its operation. The tested engine, according to the approval documents, met the requirements of the EURO V standard when it left the factory" wrote the authors in [33]. As a part of our research, the engine operation points were determined on the basis of good engineering judgment and as equivalents of the operating points in which ISO 8178 [34] tests are carried out. In many cases found in the literature, the research program is prepared with the use of certain algorithms [28,31,35] however, the authors' experience shows that this type of methodology is not fully effective in the areas in which the authors of this work operate. The research presented in the paper is an investigation of an innovative, high-pressure dual fFuel installation. The planned tests were divided into the following stages:

- Determination of external and load characteristics of the D13C460 engine when supplying the basic fuel, which was diesel fuel, in order to assess its technical condition.
- Installing a dual fuel supply system on the engine.
- Carrying out research into the impact of the diesel/CNG replacement ratio in the range from 10% to the knocking limit at selected engine operating points—the condition for "tuning the CNG supply system" was to obtain the same useful power during supply (diesel/CNG) as during supply of the basic fuel, which was diesel oil.
- On the basis of the obtained test results, the impact of the replacement factor (diesel/CNG) on the concentration of toxic exhaust gas components at selected test points was assessed.

The tests were carried out on a standard dynamometric stand. The engine was loaded with a Zöllner PS1-3812/AE water brake with a maximum braking power of 1250 kW. The torque generated by the motor was measured using a strain gauge. The rotational speed was measured by an impulse transducer in cooperation with a gear which was located on the collar of the dynamometer. Measurement of liquid fuel consumption was carried out with the use of a fuel scale by AVL, measuring at a frequency of 5 s, and the temperature of the cooling liquid was maintained in the range of 87–92 °C using an external heat exchanger. The smoke opacity was tested with an AVL 439 OPACIMETER, which operated on the principle of light absorption—measuring the extinction coefficient of the absorbed radiation. The composition of the exhaust gases was measured with the use of an AtmosFIRt analyzer, working on a hot sample of 180 °C, using the method of infrared spectroscopy using the Fourier transform (FTIR) method with an oxygen analyzer installed in the measuring chamber using the zirconium cell method. In the configuration used, the device allowed the measurement of the following components: CO_2, CO, NO, NO_2, CH_4, C_2H_6, NMHC and NH_3. The measurement results were automatically converted to normal conditions (101,325 Pa, 273.15 K = 0 °C). The spectral range of the analyzer was 485–8500 cm^{-1}. The ambient conditions were determined using the ABB SensyMaster FMT430 air flow meter, which was also used to determine the amount of air sucked in by the engine.

The temperature of the exhaust gases was measured using a NiCr-NiAl thermocouple installed in the exhaust manifold with a measuring range from −200 °C to 1200 °C. Selected engine parameters, measured by standard sensors mounted on the engine, were read from the engine controller using a Texa Navigator diagnoscope. These were defined as:

- boost pressure,
- coolant temperature,
- engine oil temperature,
- air temperature in the intake manifold.

The tests were carried out while keeping the thermal state of the engine approximately constant, controlled with the use of an air heat exchanger (cooling tower) and a cooling liquid circulation system. During the tests, the temperature of the cooling liquid and engine oil was within 81–96 °C.

Measurements of individual quantities were carried out under the specified engine operating conditions when the force on the brake and the engine rotational speed were constant for at least 1 min. During the measurements, a constant engine speed was automatically maintained (range n = constant water brake operation).

Measurements of fuel consumption (from which the value of hourly fuel consumption was obtained) were carried out for each measuring point during approx. 1 min. Measurement of other quantities (brake force, ambient temperature, ambient pressure, ambient relative humidity, exhaust temperature and smoke (light absorption coefficient), air consumption by the engine and exhaust gas analysis was carried out simultaneously after establishing the test conditions.

The following parameters of the engine operation and the engine's intake system were measured directly:

- engine torque, Mo [N·m]
- engine speed, n [rpm],
- hourly fuel consumption diesel oil, GeON [kg/h],
- hourly fuel consumption CNG, GeCNG [kg/h],
- hourly air consumption, Gp [kg/h],
- temperature in individual engine systems T [°C],
- smoke D [m^{-1}],
- Concentration of gaseous components in exhaust gases: CO_2, CO, NO, NO_2, CH_4, C_2H_6, NMHC and NH_3,

On the basis of the quantities measured in a direct way, the quantities measured indirectly were determined:

- Knock combustion signal, [–],

On the basis of the directly measured quantities, the values measured indirectly were determined:

- actual diesel/CNG exchange rate

The value of the useful power of the engine was determined on the basis of the value of the loading torque realized by the brake and was determined from the dependencies:

$$N_e = \frac{n M_o}{9550} K_{d1} \; [\text{kW}] \tag{2}$$

where:

n is rotational speed in rpm,

Ne is engine useful power in kW,

Mo is engine loading torque (rotational) read from the brake controller in Nm,

K_{d1} is correction factor for normal conditions determined in accordance with the PN-ISO 15550: 2009 standard.

Determination of the engine power value and knowledge of the concentration of harmful exhaust components allowed us to determine the emissions of a given exhaust component at a specific point of engine operation. For all results, the uncertainties of the measurement results were calculated, taking into account the possibility of both systematic and random errors. Systematic errors of direct measurements were determined on the basis of the accuracy class of the measuring instruments or on the basis of an elementary division on a scale. Errors of the value determined indirectly (from mathematical dependencies) were calculated after determining the size of systematic error Δx_i (where $i = 1, 2, 3, \ldots , k$) values of measurements of quantities measured directly as the maximum error Δy from the general dependence in the form [36]:

$$y = \sum_{i=1}^{k} \left| \frac{f(x_1, x_2, \ldots, x_3)}{x_1} \right| x_i \tag{1}$$

The values of systematic errors of the measured quantities are presented in Table 1.

Table 1. List of investigation equipment used during investigation.

No.	Name of Device/Measured Quantity	Type	Range	Accuracy
1.	Water dynamometer Torque—Mo, rotated speed—n	Customs officer PS1-3812/AE	$Mo = (0 \div 7000)$ N·m, $n = (0 \div 3000)$ rpm $Ne = 0 \div 1250$ kW	± 1 N·m, ± 1 rpm, ± 1 kW,
2.	Fuel weight-meter (diesel)	AVL 733S Fuel Balance	$(0 \div 200)$ kg/h	± 0.005 kg/h
3.	Fuel weight-meter (CNG) 1.2–104.6 kg/h $\pm$ 0.6% measured quantity	SwirlMaster FSS450 Intelligent Swirl Flowmeter	1.2–104.6 kg/h	$\pm 0.6\%$ measured quantity
4.	Exhaust analyser—measuring of toxic elements concentration in exhaust gases - carbon dioxide (CO_2), - carbon monoxide (CO), - nitrogen oxides (NO), - nitrogen diooxide (NO_2), - methane (CH_4) - ethane (C_2H_6) - ammonia (NH_3)	Atmosphere FIR, emissions monitoring FTIR systems	CO_2 (0.01 $\div$ 23)% CO (1.0 $\div$ 11,000) ppm, NO (1.0 $\div$ 6000) ppm NO_2 (1.0 $\div$ 6000) CH_4 (1 $\div$ 5000) ppm C_2H_6 (1 $\div$ 1000) ppm NH_3 (1 $\div$ 5000) ppm)	$\pm 0.1\%$ measured quantity
5.	Smoke concentration—extinction coefficient of light radiation—k.	AVL Opacimeter 4390	$(0.001 \div 10.0)$ m^{-1}	± 0.002 m^{-1}
6.	Thermocouple—measuring of exhaust temperature—T	NiCr—NiAl (K)	$(-50 \div 1100)$ °C	± 1 °C
7	Mass air consumption	SensyMaster FMT430 Thermal Mass Flowmeter	$120 \div 7000$ kg/h	± 1.2 kg/h

Control of the engine load, as well as its rotational speed, was carried out from the control and measurement cabin. In this cabin there is a cabinet on which instruments are mounted to control the brake load and to control the servo. In this cabinet there are also engine speed indicators, as well as an indicator for measuring the temperature of individual systems.

During the tests, the operation of the engine was constantly controlled by the use of the NAVIGATOR TXTs diagnostic interface with the IDC 5 TRUCK software. A diagram of the test bench is shown on a Figure 5.

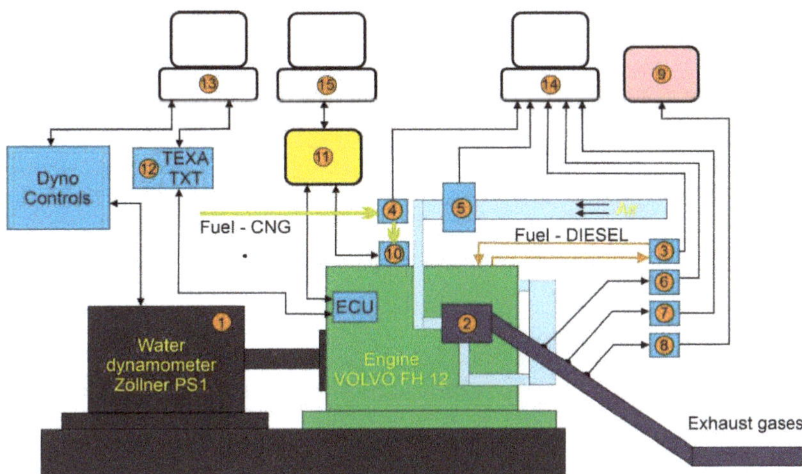

Figure 5. Diagram of the dynamometer with Volvo D13C460 EURO V engine: 1—water brake, 2—turbocharger, 3—basic diesel fuel consumption measurement system, 4—CNG gas consumption measurement system, 5—air consumption measurement system, 6—exhaust temperature measurement system, 7—smoke measurement system, 8—gas component consumption system for analysis, 9—FTIR analyzer, 10—CNG injection system for the engine intake manifold, 11—CNG power supply system with software, 12—TEXA TXT diagnoscope, 13—computer controlling the operation of the torque brake and recording engine operating parameters, 14—computer controlling the operation of measuring systems, 15—computer for programming the CNG power supply system controller.

Tests and results of useful measurements of the engine have been reduced to normal conditions in accordance with PN-ISO 15550:2009—Reciprocating IC engines—Determination and method of measuring engine power—General requirements [36], which recommends correcting the engine power according to ambient conditions by multiplying the value determined directly by the calculated correction factors.

Methodology for Determining the Effect of the CNG/ON Substitution Factor on Engine Performance

The beginning of the test was to determine the external characteristics and load characteristics of the D13C460 engine. The tests were carried out on a dynamometer station located in the Department of Engines and Operation Engineering of the Military University of Technology in Warsaw. For supplying the engine commercial diesel fuel was used. This diesel fuel was from one production batch and was used during all tests. As a result of these activities, measuring points were designated to carry out further work.

Then, an original gas supply system was installed on the engine, allowing us to bring two types of fuel to the combustion chambers—diesel and CNG. The installation consists of a gas supply rail mounted between the engine head and the intake manifold. Gas injectors were installed in the rail, 2 per cylinder and the working pressure of the CNG gas used was 0.8 MPa. The installation works on the principle of shortening the time of injection of the basic fuel, which was diesel oil in proportion to the set coefficient of replacement of diesel fuel by natural gas.

In the course of adaptation activities, the installation was subjected to an optimization process, in which the main criterion was to determine the optimal angle of advancement of CNG gas fuel injection for different rotational speeds and engine loads relative to the position of the crankshaft at which the piston TDC (top dead center) occurs.

Subsequently, for such an optimized installation, studies were carried out on the impact of the coefficient of replacement of mass diesel fuel by natural gas on smoke and exhaust gas composition. The test was carried out in selected engine operating states corresponding to the most commonly used engine operating conditions when performing a road transport task. In the individual operating states selected by our team, the ratio of diesel to natural gas exchange was changed in leaps and bounds, with a jump of 10%, starting from 10% of replacement, up to the knock combustion. At each point of the engine's operation, the amount of gas supplied was corrected using the CNG controller in such a way that the power developed by the engine was the same as the engine power obtained when running on the basic fuel, which was diesel oil. The assumed conversion rate of diesel to natural gas was set using a computer controlling the operation of the CNG installation, and the obtained exchange rate was determined on the basis of changes in the hourly consumption of the basic fuel. The research was carried out until the exchange ratio was achieved at which point the signal level from the knock combustion measurement system exceeded the level of 2.

During the tests, commercial diesel from one production batch in accordance with en-590 was used, due to the fact that the normative document [36] allows the use of a fuel similar to the reference fuel in a situation where the appropriate reference fuel is not available. The advantage of choosing this method of proceeding is to determine the actual values of the engine's operating characteristics with fuel of a random composition (random petrol station). It was remembered to reduce the impact of fuel properties on the result of comparative tests as much as possible, so for all tests, diesel fuel from one delivery was used.

The measurements were made using the prepared research version of the CNG dosing controller in which the optimal individual control functions were determined and control maps were developed.

The results of the research were illustrated by comparing them in subsequent drawings to changes in the concentration of the components of individual gaseous components, i.e., smoke as a coefficient of extinction of absorbed radiation (Figure 10), CO and CO_2 (Figures 11 and 12), NO and NO_2 (Figures 13 and 14), and CH_4 (Figure 15). To facilitate the analysis process and to quantify the changes taking place, the results are also included in a relative way, comparing those obtained for diesel/CNG power supply with only On power supply (right side of drawings). The concentration of individual gaseous components and smoke opacity were determined directly at the exit of the engine, before the exhaust after-treatment systems, so as to eliminate the result of the exhaust gas neutralization systems. As part of the verification of the technical condition of the engine, the external characteristics of the Volvo D13C460 engine used in the Volvo FH13 car were measured, determined when it was powered by diesel, and presented and compared with the original characteristics as shown in Figure 6.

Then, on the basis of the analysis of the course of useful (effective) power and torque of this engine, research engine operating states were developed in which the impact of the degree of diesel/CNG replacement on the concentration of exhaust gas components was assessed. The designated external characteristics indicate that the engine develops a maximum torque of about 2300 Nm in the speed range of 1000–1500 rpm, and a maximum value of useful power of about 338 kW (460 hp) in the speed range of 1400–1900 rpm. Above an engine speed of 1500 rpm, the torque decreases sharply and reaches a value of about 1700 Nm at a speed of about 1900 rpm, above which the useful power of the engine decreases. Analysis of the course of useful power and torque and the operating conditions of the vehicle from which the engine came showed that the range of economical, operational engine speed is approx. 1100–1500 rpm. In this range, the engine develops a constant (approximately) maximum torque, and at the same time increases (as the rotational speed increases) its useful power.

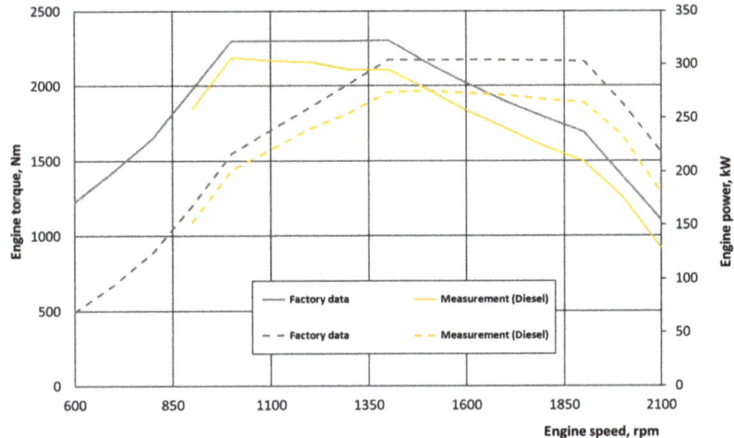

Figure 6. External characteristics of the VolvoD13C46 engine.

In order to influence the supply of the engine with different proportions of diesel/CNG to determine the optimal proportion of fuel supply with the CNG/diesel mixture (percentage replacement of diesel oil with compressed natural gas), a test program was developed taking into account the above-mentioned results of the analysis of the external characteristics of the engine powered by diesel fuel. The scope of research was limited to determining the effective parameters of the engine in the most economical operating speed range, 1100–1500 rpm. In this respect, the basic engine operating parameters were measured at the dynamometer for three speed values (1100 rpm, 1300 rpm, 1500 rpm) at a constant value of loading torque.

The tests were carried out:

- at a constant engine speed of 1100 rpm for two load torque values—approx. 520 Nm (61 kW) and 1150 Nm (139 kW);
- at a constant engine speed of 1300 rpm for two load torque values—approx. 480 Nm (67 kW) and 980 Nm (134 kW);
- at a constant approximately variable engine speed of 1500 rpm for two load torque values—approx. 500 Nm (79 kW) and 1040 Nm (163 kW).

For each engine speed and the value of (loading) torque, tests were performed for different proportions of the diesel/CNG fuel supply—from diesel only (replacement rate 0%) to the size of the replacement, at which, under the given measurement conditions, the occurrence of detonation combustion was recorded by means of a sensor attached to the engine block.

3. Results

The graphs showing the results of the research were divided according to the measured parameter. Each of the graphs presents the results for all six operating states in which the engine was tested. In each of the engine operating states, the engine speed and load were strictly determined, while the degree of replacement of diesel fuel with natural gas was variable. The test results are presented in graphs. The value of the degree of replacement of diesel by CNG is expressed as the real value of replacement calculated on the basis of fuel consumption, which is visible in the diagrams in which the marked points are usually near the assumed value of the replacement ratio.

Figures 7–18 show the effect of the diesel/CNG replacement factor for six engine operating states, differing in speed and load, on the engine operating parameters or the concentration of individual exhaust gas components.

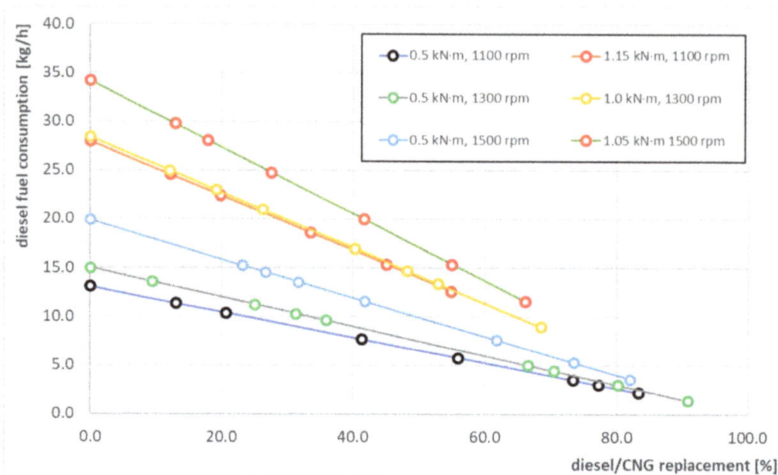

Figure 7. Hourly diesel fuel consumption for individual engine operating points (rotation speed, torque) in the diesel/CNG replacement function.

Figure 7 shows the effect of the diesel/CNG replacement factor for the six engine operating states, differing in speed and load, on the hourly consumption of the diesel base fuel. A linear decrease in hourly diesel fuel consumption is visible as a function of the replacement coefficient with Ge diesel = f(diesel/CNG) for the tested CNG power supply system of the Volvo D13C460 EURO V engine. The nature of the changes in fuel consumption are illustrated by the linear relationship between the consumption of the basic fuel with the designated replacement coefficient, which due to the compliance of the results with the theoretical assumptions indicates the high accuracy of the measurements made. A similar correlation can be observed for changes in hourly CNG consumption GeCNG = f(diesel/CNG) as shown in Figure 8.

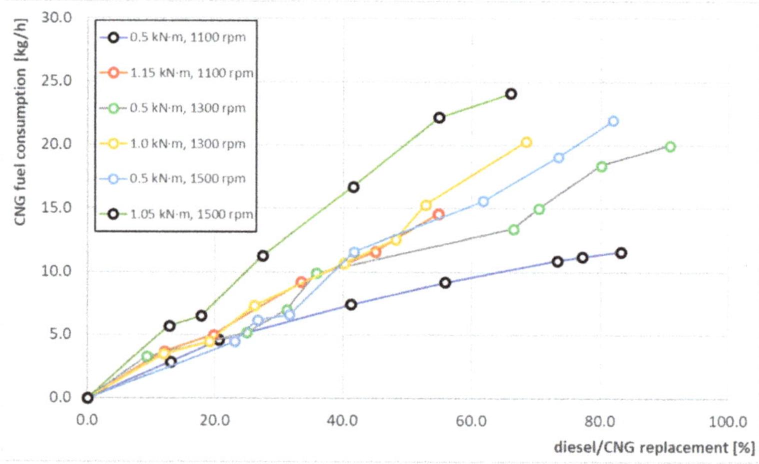

Figure 8. Hourly CNG fuel consumption for individual engine operating points (rotation speed, torque) in the diesel/CNG replacement function.

Differences in hourly CNG consumption for individual engine operating points are, of course, due to different values of power generated by the engine, but deviations from the linear nature of the graphs may have different backgrounds. A shape more parabolic than the linear graph showing the hourly fuel consumption may indicate, in the case of the tested engine loaded at 1100 rpm with a torque of ~0.5 kN·m, a systematic increase in the efficiency of the tested engine. The change in the efficiency with which the energy supplied to the engine was used also potentially had an impact on the results of fuel consumption in other measurements in which some nonlinear changes are clearly visible at individual stages of fuel exchange. The smallest increases in gaseous fuel consumption can be seen in the lower load points, where higher degrees of replacement have also been achieved than with a higher engine load. This is also correlated with the lower achieved temperature of the exhaust gases, which can be seen in the graphs shown in Figure 9.

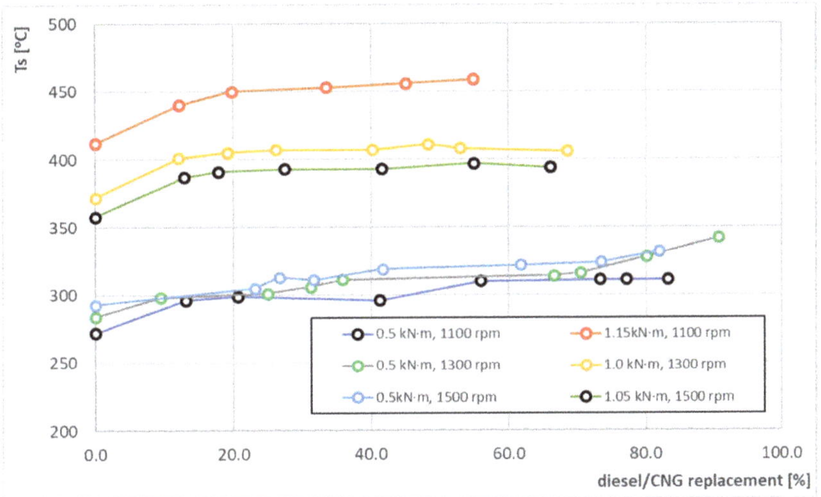

Figure 9. Exhaust gas temperature for individual engine operating points (rotation speed, torque) as a diesel/CNG replacement function.

Figure 9 shows the effect of the diesel/CNG replacement factor for six engine operating states, differing in speed and load, on the exhaust gas temperature at the turbocharger output Tsp = f(diesel/CNG). A clear increase in exhaust gas temperature is visible with an increase in the diesel/CNG replacement coefficient of about 50–60 °C. This phenomenon is caused by a change in the way the fuel is burned and with a decrease in the stoichiometric lambda. Natural gas, due to its density, occupies a much larger space in the combustion chamber than diesel fuel, which leads to a decrease in the amount of energy ballast that is mainly nitrogen, which leads to a noticeable increase in the temperature of exhaust gases.

In the case of our application, in which the geometry of the combustion chamber was not modified for a dual fuel power supply, and the mixture of natural gas and air was not homogeneous, some of the unburned natural gas molecules could burn only in the exhaust system, which in some cases may result in a significant increase in the temperature of the exhaust gases. This is a very unfavorable phenomenon due to the possibility of increasing the emission of nitrogen oxides and increasing the heat load of the engine. It should be emphasized that these changes are the strongest in the low range, where differences in higher loaded engine working states between the 1100 rpm and 1500 rpm of speed causing an increase of about 60 °C between this curves. For the exchange rates between 20% 70% and higher engine loads at medium speeds, an increase in engine speed above 1300 rpm causes a decrease in exhaust gas temperature to 300 °C and a maximum temperature in this engine operation state increase of about 50 °C. A higher exhaust gas temperature at higher loads and a slightly lower one at higher speeds is natural due to the engine using more fuel at high loads and significantly higher flow through the engine at high speed, which is clearly visible in the diagram.

Seeing the values of the exhaust gas temperature achieved at higher engine loads, one can immediately see the association of these values with the achieved smoke in those engine operating states where these values (visible in Figure 10) are clearly higher than in states in which the engine was loaded with less braking torque. This figure shows the dependence of smoke on the degree of replacement in six defined operating states.

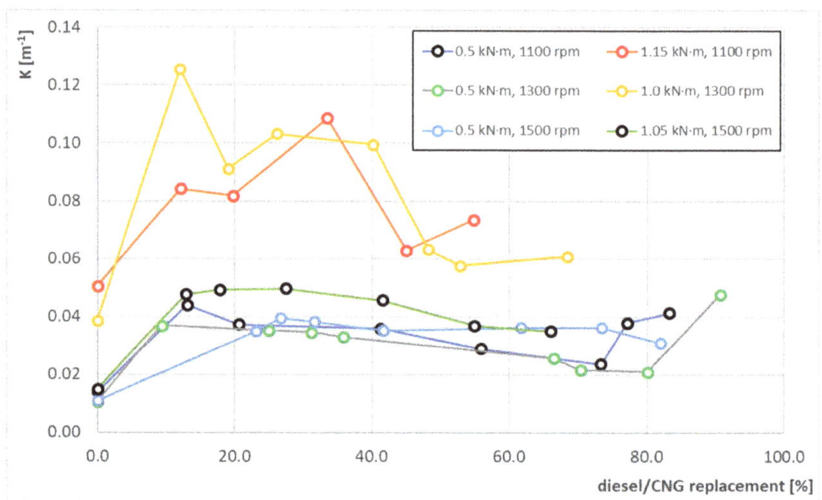

Figure 10. Smoke opacity for individual engine operating points (rotation speed, torque) in the diesel/CNG replacement function.

Figure 10 shows a clear increase in smoke with an increase in the diesel/CNG replacement ratio of about 200–300% at different points, depending on the operating conditions. Smoke opacity at low engine load regardless of replacement is slightly higher than for smoke obtained for an engine sown exclusively with diesel fuel. In the case of engine operation at a higher load, the smoke increases many times and at some degrees of replacement significantly deviates from the trend, which is a critical case, and which will be additionally verified during subsequent tests of this engine. Significantly higher concentrations recorded for two engine operation states may result from high engine load at lower rpm, but comparing the absolute values of concentrations for individual engine operation states is not advisable without checking the emissions at these operating points. For low loads, only at small and high replacement values, did smoke increase, which will require more in-depth diagnostics in the future.

Unlike smoke and exhaust gas temperature, the dependence of carbon dioxide concentration as a replacement function is laid. Theoretically, carbon dioxide emissions should decrease as the rate of replacement of diesel by natural gas increases. The diagram shown in Figure 11 shows a decrease in the value of carbon dioxide concentrations in the exhaust gas is visible and in most cases it takes on a close to linear character.

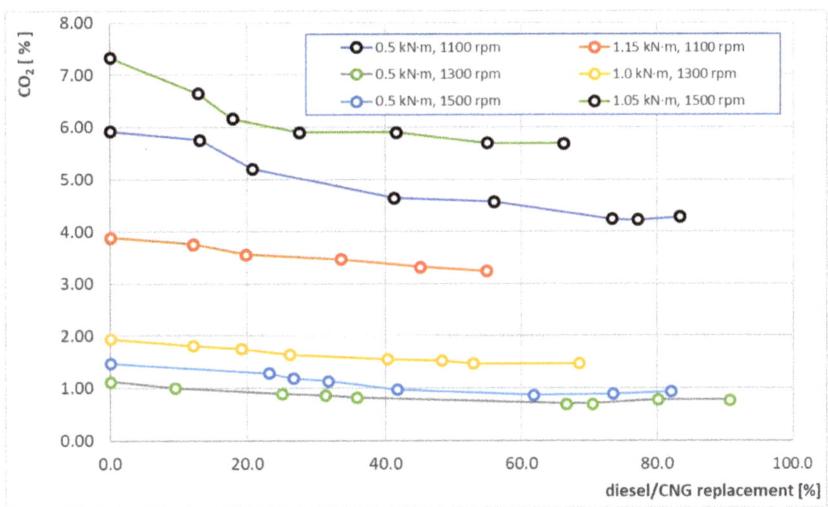

Figure 11. Concentration of carbon dioxide in the exhaust gas for individual engine operating points (rotation speed, torque) as a function of replacing diesel/CNG.

The strongest concentration drops occur at lower replacement values, then the level of carbon dioxide concentration in the exhaust gases begins to stabilize. There is a small visible decrease in carbon dioxide concentration with an increase in the diesel/CNG ratio, which reaches a maximum decrease of about 20–30% depending on the operating conditions. The reduction in the concentration of carbon dioxide is associated with a change in the elementary H/C ratio (hydrogen/carbon) as a result of replacing diesel fuel with a high carbon content, with a gaseous fuel, methane, which has a high hydrogen content. This phenomenon is accompanied by an increase in the water content of the exhaust gas [37].

The linear relationship is no longer visible in the diagrams in Figure 12 which shows the effect of the diesel/CNG replacement factor on the concentration of carbon monoxide in the exhaust gases.

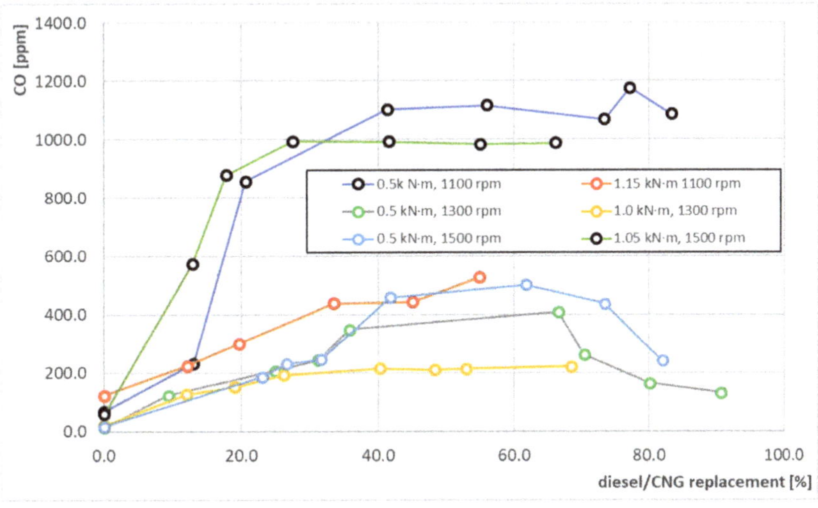

Figure 12. Concentration of carbon monoxide in the exhaust gas for individual engine operating points (rotation speed, torque) as a diesel/CNG replacement function.

Figure 12 shows a clear increase in the concentration of carbon monoxide with an increase in the degree of diesel/CNG replacement. The course of curves is difficult to associate between individual engine operating states, the mileage of the graphs clearly differ from each other. The CO content of diesel/CNG was up to 20 times higher in some dual fuel engine operating states than when powered solely by diesel and generally increases with an increase in the degree of diesel-CNG replacement. The most intense increase in CO concentration can be observed with small (up to 30%) replacement values. The CO content at an engine load of ~0.5 kN·m and at 1300 crankshaft revolutions per minute increases with an increase in CNG conversion to 60%. Above this level, it decreases to a value of about 150 ppm, but it is still 10 times higher than when the engine is powered solely by diesel. Possible decreases in the concentration of CO in the exhaust gases occur at the replacement level of 40–70%, but they are many times higher than when powered only by diesel.

The increasing share of CO in the exhaust gases, which increases with the increase in the degree of replacement, is correlated inversely with the NO content in the exhaust gases, which is consistent with theoretical assumptions [38]. The concentration of nitric oxide in the exhaust gas is shown in Figure 13.

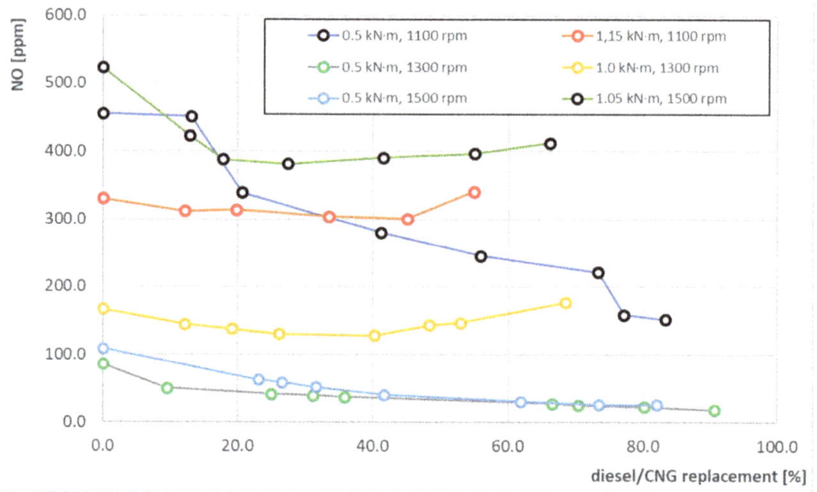

Figure 13. Concentration of nitric oxide in the exhaust gas for individual engine operating points (rotation speed, torque) as a diesel/CNG replacement function.

The graph shows a decrease in the concentration of NO in the exhaust gases with an increase in the degree of replacement to a value of about 30% of the degree of replacement, while with a further increase in the degree of replacement, from a level of about 40%, the concentration of NO in the exhaust gas in some operating states increased. There is no clear correlation with the temperature of the exhaust gases, so the increases may result from the formation of local combustion areas with elevated temperatures. This is due to the failure to adapt the geometry of the combustion chamber to the needs resulting from the supply of natural gas to this engine. Decreases in NO concentration, on the other hand, most likely result from reduced availability of methane used by the methane supplied to the engine, because along with an increase in the degree of exchange, the value of the excess air coefficient decreases.

The reverse situation can be seen in the diagram shown in Figure 14, showing the effect of the diesel/CNG replacement factor on the concentration of nitrogen dioxide in the exhaust gas.

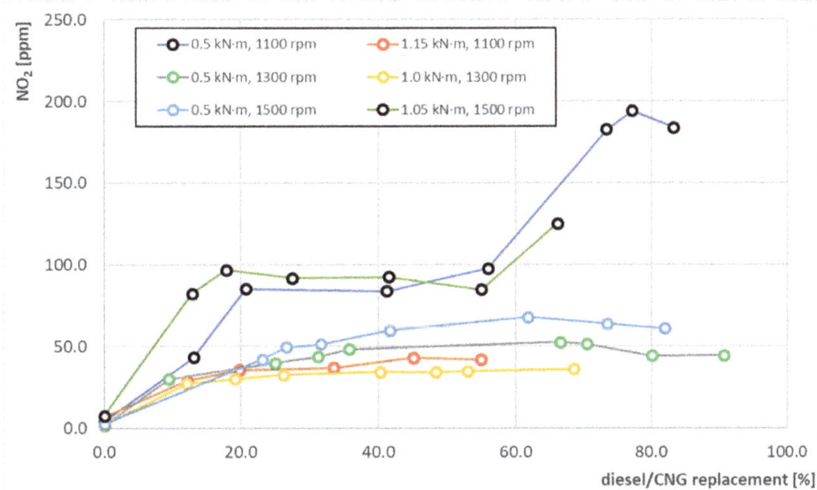

Figure 14. Concentration of nitrogen dioxide in the exhaust gases for individual engine operating points (rotation speed, torque) as a function of replacing diesel/CNG.

Intensive growth with an initial small degree of replacement may indicate the formation of a larger frontal surface of the flame, which results in an intensification of the phenomenon of frontal (fast) NO$_X$ formation, including NO$_2$. There is no correlation of such a clear increase in NO$_2$ concentration with the increased temperature of the exhaust gases, which, given the lack of this correlation also in the case of NO concentration, may indicate the widespread occurrence of a reduction reaction (2) in flame areas with low temperatures, the OH component is stable and can react with NO formed in areas with a higher temperature [39].

$$NO_2 + HNO + OH \qquad\qquad (4)$$

It should be remembered, however, that the combustion temperatures that could occur locally are uneven, and at low temperatures a small concentration of hydrocarbons can accelerate the process of transition of nitric oxide to nitrogen dioxide. When the exhaust gas components forming the so-called NO$_x$ are in chemical equilibrium with each other, then the amount of NO$_2$ reaches very low concentrations compared to NO [40]. However, there is a clear increase in the concentration of NO$_2$ in the exhaust gas in the case of two engine operating states—at a load of ~0.5 kN·m and 1100 crankshaft revolutions per minute, and at a load of ~1.05 kN·m and 1500 revolutions per minute. The concentration increases most strongly from the initial low replacement to a value of about 20%, then for subsequent replacement values it stabilizes so that at values exceeding 60% of the replacement, it increases again. This trend does not occur in other engine operating states, in which the NO$_2$ concentration values after the initial increase stabilize and remain approximately constant in the remaining replacement values exceeding 15–20%. The linear course of these graphs indicates the stability of combustion for different degrees of replacement in these engine operating states, while the states in which a clear increase in NO$_2$ concentration is observed are also correlated with the increase in CO concentration visible in the previously presented Figure 12 and with the increase in CH$_4$ shown in Figure 15, illustrating the influence of the diesel/CNG on the methane concentration in the exhaust gases as shown below.

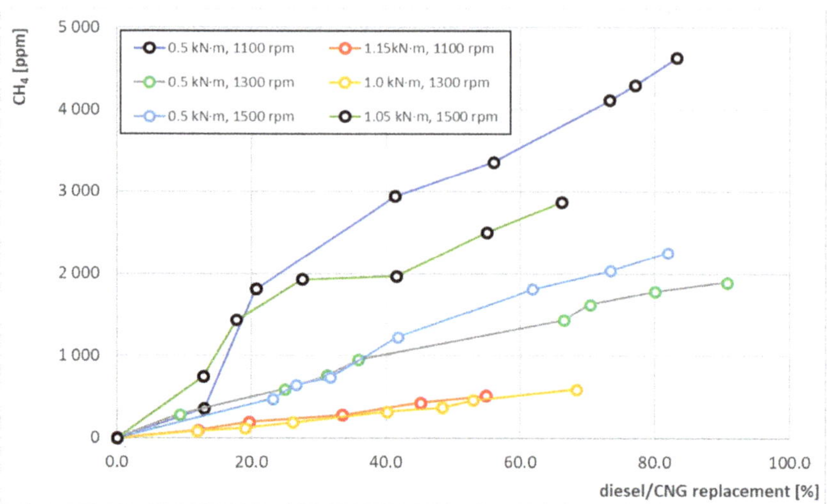

Figure 15. Methane concentration in the exhaust gas for individual engine operating points (rotation speed, torque) as a function of replacing diesel/CNG.

The graph shows a clearly higher concentration of CH$_4$ for those engine operating states for which higher concentrations of CO and NO$_2$ have also been recorded. This clearly indicates an increase in the phenomenon of incomplete combustion of hydrocarbons supplied to the engine, which is most likely due to the previously described incompatibility of the combustion chamber geometry to supply the gaseous fuel supplied to the engine. Incomplete combustion can occur in various engine operating states, because it depends to a large extent on the composition of the natural gas-air mixture in the combustion chamber. The formation of poorer and richer fuel zones results in part, from the degree of turbulence of the [41] mixture, which, with a combustion chamber not adapted to such a mode of operation, may differ significantly in different engine operating conditions. Hence, in individual engine operating states, there may be intense increases in emissions of unburnt CH$_4$ which is difficult to explain by other reasons. The increasing emission of CH$_4$ results directly from the increase in the exchange ratio, i.e., the appearance in the combustion chamber of more methane that will not be fully burned [42]. Its appearance in a high concentration has an impact on the formation of NO$_2$, which, according to the Fenimore mechanism, converts the nitrogen contained in the mixture into its oxides as a result of hydrocarbons present in the fuel [43], which in the case of the correlation of their high concentration with high concentration of unburnt methane suggests an intense occurrence of this phenomenon. Another reason for the presence of methane in the exhaust gases is the failure to adjust the geometry of the camshaft to the change of method of fuel supply. A common rule is the use of "cylinder flushing" in turbocharged engines, i.e., part of the supercharged air in the final phase of the exhaust process gets, along with the exhaust gases, into the engine exhaust system. In this case, the inlet air to the engine contains methane, which causes an increase in its concentration in the exhaust gases.

In the case of a working state in which the engine was loaded with a braking torque of ~1.05 kN·m at a crankshaft speed of 1500 rpm, there was also a high concentration of ethane, as can be seen in Figure 16, showing the influence of the diesel/CNG replacement coefficient on ethane concentration in the exhaust gases.

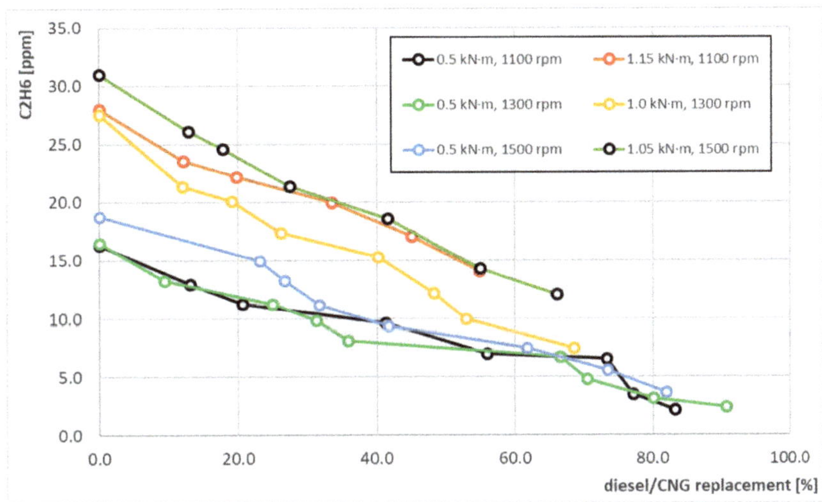

Figure 16. Ethane concentration in the exhaust gases for individual engine operating points (rotation speed, torque) as a function of diesel/CNG replacement.

A clear decrease in ethane concentration is visible with an increase in the diesel/CNG replacement coefficient in all engine operating states. The values of decreases exceed 75%, which is mainly due to the reduction in the share of hydrocarbons from injected diesel fuel, whose energy share with the increase in fuel replacement is systematically decreasing.

However, the upward trend can be seen in the diagrams in Figure 17, which shows the concentration of all hydrocarbon components excluding methane—NMHC (Non Methane Hydro Carbons)—depending on the degree of replacement, in which the points which are extremely different from the adjacent operating points clearly indicate incomplete combustion resulting from insufficient turbulence of the mixture in a given load state of the IC engine, which, as previously demonstrated, results from the inadequacy of the design of this engines combustion chamber for combustion of natural gas.

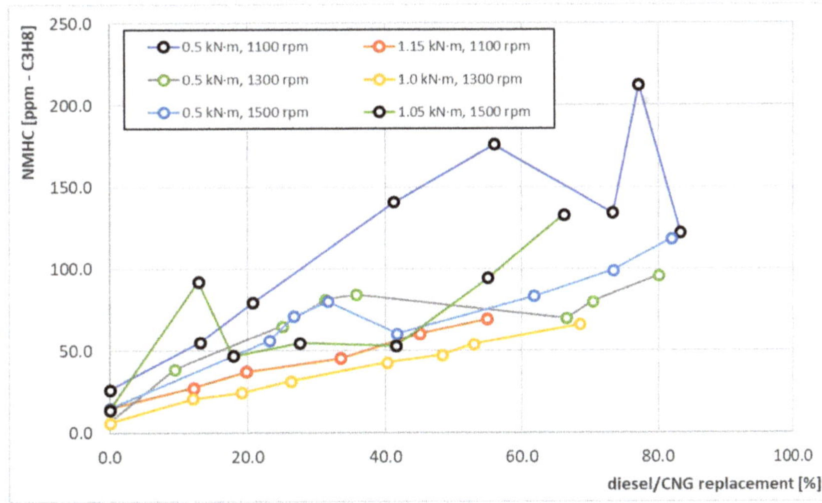

Figure 17. NMHC concentration in the exhaust gases for individual engine operating points (rotation speed, torque) as a function of diesel/CNG replacement.

In general, the charts show an upward trend of NMHC concentrations with an increase in the degree of replacement of diesel by CNG. Despite the decrease in the amount of injected diesel fuel, the number of unburned hydrocarbons heavier than CH_4 is increasing, which indicates an increase in the share of unburnt fuel in exhaust gases. This phenomenon suggests that the combustion chamber of the tested engine should be rebuilt in order to improve the possibility of using natural gas as an additional fuel in this engine.

On the other hand, a clear decrease in all concentration values with an increase in the degree of replacement is also visible for NH_3, which is visible in Figure 18 showing the influence of the diesel/CNG replacement coefficient on the concentration of NH_3 in the exhaust gas at the output from the engine.

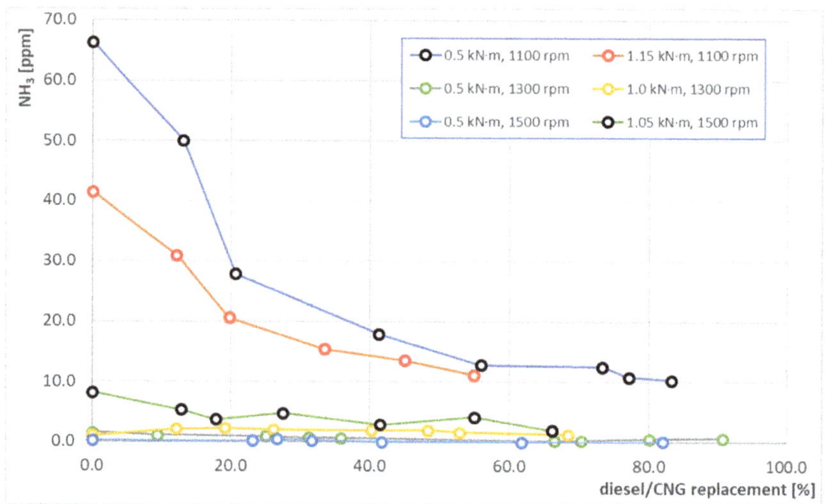

Figure 18. Detonation status for individual engine operating points (rotation speed, torque) in the diesel/CNG replacement function.

Clear decreases in the concentration of ammonia in the exhaust gases are an interesting phenomenon, because they are associated with the emissions of other harmful exhaust components already presented, and the phenomenon of direct NH_3 emissions from the IC engine most strongly affects engines powered by LPG or CNG. In the case of a dual fuel engine, it is therefore interesting to look at the phenomenon of the formation of this harmful substance in the process of co-combustion of diesel fuel and natural gas.

NH_3 can be emitted from a piston engine every time it is operating on stoichiometric or close to stochiometric parameters of combustion. NH_3 emissions are mainly created by the engine NO emissions by reactions (5) and (6).

$$2\,NO + 4\,CO + 2\,H_2O + H_2 \rightarrow 2\,NH_3 + 4\,CO_2 \tag{5}$$

$$2\,NO + 5\,H_2 \rightarrow 2\,NH_3 + 2\,H_2O \tag{6}$$

As the concentration of CO, NO and H_2 in the exhaust gas increases, the amount of NH_3 increases simultaneously. The mentioned substances are natural components of the exhaust gases, which determine the amount of NH3 produced. NO is formed when nitrogen-containing substances are involved in the combustion process (it can come from both air and fuel). CO is most strongly produced in rich fuel-air mixtures, where there is a small amount of oxygen for the fuel to be burned. H2 can be produced when fuel is burned under high pressure and in high temperatures, because under these conditions the chemical bonds that hold the hydrogen contained in the fuel break down. Other possibilities of obtaining hydrogen are water gas shift (7) or steam reforming (8) [44].

$$CO + H_2O \rightarrow CO_2 + H_2 \tag{7}$$

$$CnHm + n\,H_2O \rightarrow n\,CO + (m/2 + n)\,H_2 \tag{8}$$

Higher engine NO, CO, HC and H_2 emissions could bring about higher NH_3 emissions, but not all of these substances were measured in our research. It can be seen that the type of fuel used may have an influence on the amount of H_2 produced in the combustion chamber according to reaction (8), so the level of replacement can influence it too.

Summing up, along with the enrichment of the air-fuel mixture, the conditions for the formation of NH_3 improve in general, but in our case, the high content of methane led to a reduction in its concentrations due to the reduction of the concentrations of substances generated mainly in the process of combustion of fuels with longer hydrocarbon chains [45].

The authors of the article says that the main way to reduce NH_3 emissions in SI engines is to reduce CO and NO emissions in the exhaust gas. A similar situation can be found in dual fuel engines using a large amount of gaseous fuel. The authors state that "in particular, this may be a significant problem in gaseous engines". Variability in gaseous SI or CI dual fuel engines should be lower than that for gasoline engines.

In the case of our research, the concentration of NH_3 in the case of four engine operating states did not change significantly, while in the case of two others it decreased significantly. No concentrations in the exhaust gases were fairly stable, while CO increases were more intense, but in operating states where an intense decrease in NH_3 concentration was observed, it is difficult to find a correlation between their concentrations. On the basis of the theoretical assumptions cited, for a more thorough analysis it will be necessary to know the concentration of water and hydrogen in future tests. Reactions (9) and (10) between ammonia and nitrogen oxides inside the (inactive) SCR reactor may have affected the final emission of both of these substances, which also makes it impossible to clearly indicate the cause of such intense changes in NH3 concentration in some engine operating states [39].

$$4\,NO + 4\,NH_3\;th\;O_2 \rightarrow 4\,N_2 + 6\,H_2O \tag{9}$$

$$NO + 2\,NO_2 + 4\,NH_3 \rightarrow 4\,N_2 + 6\,H_2O \tag{10}$$

Summary of the Results

Summarizing the obtained test results and dividing them into individual engine operating states, the following analysis of the obtained results can be carried out:

- Test results obtained for an engine loaded with approximately 520 Nm at 1100 rpm

Smoke opacity, regardless of the CNG content, is similar to the smoke obtained for an engine sewn exclusively with diesel fuel. Only with a 20% replacement, the smoke is approximately 3027 times greater than the average smoke value for the remaining CNG substitution factor, which will require more in-depth diagnostics in the future. The CO content of diesel/CNG is up to 20 times higher than that of diesel-only and generally increases as the degree of replacement of diesel oil by CNG increases. The content of NO, NOx and ammonia NH_3 decreases with increasing degree of substitution, despite the increase in exhaust gas temperature. When increasing the CNG content, the share of methane CH_4 and ethane C_2H_6 in the exhaust gases increases.

- Test results obtained for an engine loaded with a torque of approx. 1150 Nm at a speed of 1100 rpm.

For these test conditions, it was not possible to achieve stability, without detonating engine operation with diesel replacement by CNG above 50%. Smoke opacity, regardless of the CNG content, is similar to the smoke obtained for an engine supplied exclusively with diesel fuel. Only with a replacement of 15% is the smoke opacity 10 times greater than the average smoke value for the remaining CNG diesel replacement values, which is a critical case, and which will be further verified in subsequent tests of this engine. As the CNG content increases, the share of CO increases. The content of NO decreases and NO_2 increases as the conversion rate increases. With a replacement rate of 50%, the NO_2 content increases significantly compared to the nitrogen oxide content when powered solely by diesel. This is despite the decrease in exhaust gas temperature in this range of engine operation. When increasing the CNG content, the share of methane CH_4 (up to five times higher than when powered only by diesel) and ethane C_2H_6 (three times compared to diesel-only) increases in the exhaust gases.

- Test results obtained for an engine loaded with approximately 480 Nm at 1300 rpm

For these test conditions, it was possible to obtain stable, stemless operation of the motor with the replacement of diesel by CNG even at the level of 85%. Smoke opacity regardless of the CNG content is very small and similar. It ranges from 0.01 to 0.05 m^{-1}. The CO content increases as CNG conversion increases up to 60%. Above this level, it decreases (to a value of approx. 150 ppm), but it is still 10 times higher than when the engine is powered only by diesel. An increase in the degree of conversion causes a decrease in both the amount of NO and NOx. The proportion of ammonia NH_3 decreases with increasing degree of conversion (up to 60%), and then increases slightly. With an increase in the share of CNG, the content of CH_4 is systematically increasing. A similar relationship was observed for ethane (C_2H_6) to 50%, and then above 80%. The share of CH_4 increases from 10 ppm (when powered only by diesel) to 130 ppm for 85% CNG conversions.

- Test results obtained for an engine loaded with a torque of approximately 980 Nm at a speed of 1300 rpm

For these test conditions, it was possible to obtain stable, stemless operation of the motor with the replacement of diesel by CNG up to 70%. The smoke with diesel is 0.04 m^{-1}. With a 15% share of CNG, it increases to a value of about 0.13 m^{-1}, and then generally decreases. The CO content increases as CNG conversion increases. At 40–70% share, CNG is 9 times higher than when powered solely by diesel. An increase in the conversion rate to 40% causes a decrease in the amount of NO. At higher conversion rates, NO increases. With a conversion of 70%, the proportion of nitrogen oxides is greater than when powered solely by diesel. With an increase in the share of CNG, the content of both methane CH_4 and ethane C_2H_6 is systematically increasing. A similar relationship was observed for C_2H_6.

- Test results obtained for an engine loaded with approximately 500 Nm at 1500 rpm

For these test conditions, it was possible to obtain stable, smooth operation of the motor with the replacement of diesel by CNG up to 80%. The smoke is very small and is 0.01 to 0.04 m^{-1}. As the CNG content increases, the CO_2 content decreases, while the share of CO increases. The share of both NO and NOx decreases with increasing CNG conversion rate. When increasing the share of CNG, the share of methane CH_4 and propane C_2H_6 increases.

- Test results obtained for an engine loaded with approximately 1040 Nm at 1500 rpm

For these test conditions, it was possible to obtain stable, stemless operation of the motor with the replacement of diesel by CNG only up to 60%. Smoke opacity increases in the conversion rate range of 0–30% from 0.015 m^{-1} (for diesel only) to 0.05 m^{-1} for a conversion rate of 15–30%. With a further increase in the share of CNG, the smoke decreases to 0.035 m^{-1} with the largest share of CNG. The share of CO increases with the increase in the degree of CNG conversion (up to 30%). Above this value, the share of CO is approximately constant. In this respect, it is 20 times larger than when the engine is powered solely by diesel. As the CNG content increases, there is a tendency to reduce the proportion of both NO, NO_X and NH_3. At the same time, for the degree of CNG conversion above 50%, the share of nitrogen oxides NO_X increases rapidly).

The aggregate graphs clearly show the twin trends occurring for the same degrees of diesel fuel replacement by CNG at individual engine operating points. On the XY diagrams, a clear deviation from the trend of the NO_2 value can be seen at an engine load of 0.5 at 1500 rpm. This is also visible for CH_4 and CO emissions, where also for a rotational speed of 1300 rpm the CO concentration value was higher than at other operating points. At this engine speed, regardless of the load, significantly higher smoke was recorded than at other operating points.

Based on the analysis of the obtained test results, the conversion values (share) of CNG in the fuel dose were determined, which allowed the minimum lowest smoke k min, emissions CO, NO, NO_2, CH_4, NH_3 and C_2H_6 for the considered operating parameters of the Volvo FH13 engine to be obtained—these values are summarized in Table 2.

Table 2. The replacement of diesel/CNG setting with the lowest concentration of harmful components.

Rotational Speed [RPM]	Load [Nm]	CNG Replacement Setting [%] When There Is a Minimum Share of:						
		CO	NO	NO_X	NH_3	C_2H_6	CH_4	k_{min}
1100	520	20	70	50	80	30	20	60
	1150	15	40	30	50	15	15	40
1300	480	20	85	85	60	20	20	80
	980	15	40	40	70	15	15	60
1500	500	20	70	80	70	50	20	80
	1040	15	30	30	60	20	15	60

4. Conclusions

1. It was noticed that the smoke density and exhaust gas temperature did not decrease with increasing degree of diesel oil replacement with natural gas. The concentration of CO, NO_X, CH_4 and NMHC increased, while the concentration of CO_2, C_2H_6, NH_3 and the consumption of diesel fuel by the engine decreased significantly. The conducted research has shown the possibility of obtaining high degrees of diesel/CNG exchange. An increase in the methane content in the air-fuel mixture will match the increase in the proportion of toxic exhaust gas components (before the exhaust after-treatment system).

2. The biggest amount of fuel replacement that could be carried out without knocking the combustion effect was about 90% of CNG fuel. This is a good result in comparison to other researchers' test results.

3. Unambiguous indication of the optimal settings for adjusting the diesel/CNG ratio due to the different impact of the same CNG proportions on the parameters under consideration is difficult. The performed research allowed us to quantify the impact of the CNG share on the parameters under consideration, and the obtained dependencies, shown in the figures, are often not unambiguous.

4. The tests showed that the use of CNG fuel causes an increase in the share of toxic compounds in the exhaust gases, to the greatest extent CO and HC hydrocarbons. The content of these components at many measuring points is several times to several dozen times higher than for an engine powered solely by diesel. The dependencies shown in the drawings are difficult to explain unequivocally, comparing them with the results available in other studies [46–49] it is possible to clearly state the inadequacy of the engine to supply a dual fuel using natural gas and diesel, and in order to use the full potential of this type of power supply thorough structural changes to the engine are required. The increase in hydrocarbon emissions may result from the course of fuel combustion (however, it can be said with certainty that unchanged design solutions of an engine originally adapted to burn only diesel fuel contribute to a large extent to the emission of unburnt methane, which is the largest share of hydrocarbon compounds emitted by the engine). In turn, the increase in the share of carbon monoxide is difficult to explain; reducing the coefficient of excess air leads to

an increase in the concentration of CO in the exhaust gases. This is due to the much lower availability of oxygen in the combustion chamber than in the case of mono-fuel diesel fuel.

5. A decrease in CO_2 content was found with the diesel/CNG mixture, which is associated with a change in the elementary H/C ratio in the outgoing exhaust gases. This is beneficial from the point of view of toxic and greenhouse gas emissions. Changes in CO_2 content are proportional to changes in the diesel/CNG substitution factor.

6. A dozen or so percent increase in the content of nitrogen oxides—mainly nitrogen dioxide in the exhaust gases—was found. This is related to the change in the organization of the process—from qualitative to quantitative control. The introduction of CNG gas into the intake manifold reduces the excess air coefficient and increases the combustion temperature of the fuel-air mixture. Increasing the combustion temperature results in an increase in the NO_2 content, which affects the total NOx emissions. The vehicle from which the engine comes is equipped with an SCR system, which should reduce NOx emissions to the required level.

7. A several-fold increase in the methane content of the exhaust gases was found. This is related to the contraction of the intake and exhaust system—flushing the combustion chamber. This phenomenon may also cause an increase in carbon monoxide emissions during type-approval tests and, as a result, exceeding the limits resulting from the requirements of EURO standards. This conclusion is confirmed by a several-fold increase in the content of carbon monoxide in the exhaust gases. This is due to the introduced changes in the organization of the combustion process, i.e., the transition from qualitative to quantitative control, as in the case of NOx concentration. The introduction of CNG gas into the intake collect causes the fuel to be mixed with air beforehand, resulting in a mixture close to homogeneous, which is then supplied to the cylinder. After compression, diesel is injected into this gas mixture in the liquid phase. This causes local oxygen deficiencies which promotes incomplete combustion and the formation of carbon monoxide and an increase in smoke when feeding diesel/CNG in the range of 50–120% depending on the operating conditions. This is a very unfavorable phenomenon from the point of view of environmental protection due to the increase in the emission of particulate matter into the atmosphere.

8. The reasons for the increase in smoke can be several:
   - reduction of the excess air coefficient caused by the supply of CNG fuel in gaseous form,
   - change in the method of combustion of the fuel-air mixture, resulting from the shift of the flammability limit and homogenization of the mixture,
   - local oxygen deficiencies, resulting from co-combustion of CNG and diesel, which leads to a local reduction in the coefficient of excess air, especially in the area of diesel combustion.

9. In order to optimize the engine due to the emissions of individual exhaust components, tests must be carried out to select the timing phases and to optimize the CNG fuel injection process.

10. In the opinion of the authors, it is necessary to further evaluate the energy state of the tested engine and to evaluate the emissions of individual exhaust gas components. According to the researchers, some issues, such as drops in NH_3 concentration, increased NMHC concentrations or the lack of noticeable smoke reduction, require further research. Moreover, the concentrations of other exhaust gas components and the achieved overall efficiency of the engine are also interesting and require further analysis.

Author Contributions: Conceptualization, M.K. and G.S.; methodology, M.K.; validation, M.K., J.C. and G.S.; formal analysis, M.K. and G.S.; investigation, M.K. and J.C.; resources, M.K. and G.S.; data curation, G.S.; writing—original draft preparation, G.S.; writing—review and editing, M.K., J.C. and

G.S.; visualization, M.K.; supervision, M.K.; project administration, M.K.; funding acquisition, M.K. All authors have read and agreed to the published version of the manuscript.

Funding: This work was financed by Military University of Technology under research project UGB 758/2022.

Institutional Review Board Statement: Not applicable.

Informed Consent Statement: Not applicable.

Data Availability Statement: Data is contained within the article.

Conflicts of Interest: The authors declare no conflict of interest.

References

1. Reitz, R.D.; Duraisamy, G. Review of high efficiency and clean reactivity controlled compression ignition (RCCI) combustion in internal combustion engines. *Prog. Energy Combust. Sci.* **2015**, *46*, 12–71. [CrossRef]
2. Karim, G.A. *Dual-Fuel Diesel Engines*; CRC Press: New York, NY, USA, 2015.
3. Chojnowski, J.; Karczewski, M.; Szamrej, G. The phenomenon of knocking combustion and the impact on the fuel exchange and the output parameters of the diesel engine operating in the dual-fuel mode (Diesel-CNG). In Proceedings of the NAŠE MORE 2021, 2nd International Conference of Maritime Science & Technology, Dubrovnik, Croatia, 17–18 September 2021; pp. 30–41.
4. Jeż, M. *Silniki Spalinowe Zasada działania, Zastosowania*; Biblioteka Naukowa Instytutu Lotnictwa: Warszawa, Poland, 2008; pp. 36–38.
5. Stelmasiak, Z.; Larisch, J.; Pietras, D. Wpływ dodatku gazu ziemnego na wybrane parametry pracy silnika Fiat 1.3 MultiJet zasilanego dwupaliwowo. *Combust. Engines* **2015**, *54*, 672–682.
6. Kuiken, K. *Gas- and Dual-Fuel Engines for Ship Propulsion, Power Plants and Cogeneration*; Book II: Engine Systems and Environment; Target Global Energy Training: Onnen, The Netherlands, 2016; pp. 1–488.
7. Ryan, T.W.; Callahan, T.J.; King, S.R. Engine Knock Rating of Natural Gases-Methane Number. *ASME J. Eng. Gas Turbines Power* **1993**, *115*, 922–930. [CrossRef]
8. Tira, H.S.; Herreros, J.M.; Tsolakis, A.; Wyszynski, M.L. Characteristics of LPG-diesel dual fueled engine operated with rapeseed methyl ester and and gas-to-liquid diesel fuels. *Energy* **2012**, *47*, 620–629. [CrossRef]
9. Wei, L.; Geng, P. A review on natural gas/diesel dual fuel combustion, emissions and performance. *Fuel Process. Technol.* **2016**, *142*, 264–278. [CrossRef]
10. Westport Fuel Systems. Available online: https://wfsinc.com/our-solutions/hpdi-2.0 (accessed on 15 January 2021).
11. Djermouni, M.; Ouadha, A. Comparative assessment of LNG and LPG in HCCI engines. *Energy Procedia* **2017**, *139*, 254–259. [CrossRef]
12. WEICHAI. *Weichai Westport Secures Chinese Certification for WP12 Natural Gas Engine Powered by HPDI 2.0*; Green Car Congress; WEICHAI: Weifang, China, 2020.
13. Kuiken, K. *Gas- and Dual-Fuel Engines for Ship Propulsion, Power Plants and Cogeneration*; Book I: Principies; Target Global Energy Training: Onnen, The Netherlands, 2016; pp. 1–488.
14. Karczewski, M.; Szamrej, G.; Chojnowski, J. Solutions of CNG and LNG supply systems in modern land and marine CI engines working in dual-fuel (NG—diesel) mode. In Proceedings of the NAŠE MORE 2021, 2nd International Conference of Maritime Science & Technology, Dubrovnik, Croatia, 17–18 September 2021; pp. 177–193.
15. Kheirkhah, P. *CFD Modeling of Injection Strategies in a High-Pressure Direct-Injection (HPDI) Natural Gas Engine*; University of British Columbia: Vancouver, BC, Canada, 2015. [CrossRef]
16. Engine Management Systems for CPBC and RCCI Engines. Available online: https://www.arenared.nl/cpbc+~{}+rcci (accessed on 21 March 2021).
17. Proost, S.; Van Dender, K. Energy and environment challenges in the transport sector. *Econ. Transp.* **2012**, *1*, 77–87. [CrossRef]
18. Pietrzak, K.; Pietrzak, O. Environmental effects of electromobility in a sustainable urban public transport. *Sustainability* **2020**, *12*, 1052. [CrossRef]
19. Dziubak, T. The effects of dust extraction on multi-cyclone and non-woven fabric panel filter performance in the air filters used in special vehicles. *Eksploat. Niezawodn. Maint. Reliab.* **2016**, *18*, 348–357. [CrossRef]
20. Dziubak, T. Experimental Studies of Dust Suction Irregularity from Multi-Cyclone Dust Collector of Two-Stage Air Filter. *Energies* **2021**, *14*, 3577. [CrossRef]
21. Wołek, M.; Wolański, M.; Bartłomiejczyk, M.; Wyszomirski, O.; Grzelec, K.; Hebel, K. Ensuring sustainable development of urban public transport: A case study of the trolleybus system in Gdynia and Sopot (Poland). *J. Clean. Prod.* **2021**, *279*, 123807. [CrossRef]
22. Grijalva, E.R.; López Martínez, J.M. Analysis of the Reduction of CO_2 Emissions in Ur-ban Environments by Replacing Conventional City Buses by Electric Bus Fleets: Spain Case Study. *Energies* **2019**, *12*, 525. [CrossRef]
23. Junga, R.; Pospolita, J.; Niemiec, P.; Dudek, M.; Szleper, R. Improvement of coal boiler's efficiency after application of liquid fuel additive. *Appl. Therm. Eng.* **2020**, *179*, 115663. [CrossRef]

24. Feiock, R.C.; Stream, C. Environmental protection versus economic development: A false trade-off? *Public Adm. Rev.* **2001**, *61*, 313–321. [CrossRef]
25. Wallander, J.L.; Schmitt, M.; Koot, H.M. Quality of life measurement in children and adolescents: Issues, instruments, and applications. *J. Clin. Psychol.* **2001**, 571–585. [CrossRef] [PubMed]
26. Karczewski, M.; Szczęch, L. Influence of the F-34 unified battlefield fuel with bio components on usable parameters of the IC engine. *Maint. Reliab.* **2016**, *18*, 358–366. [CrossRef]
27. Karczewski, M.; Wieczorek, M. Assessment of the Impact of Applying a Non-Factory Dual-Fuel (Diesel/Natural Gas) Installation on the Traction Properties and Emissions of Selected Exhaust Components of a Road Semi-Trailer Truck Unit. *Energies* **2021**, *14*, 8001. [CrossRef]
28. Sharma, P.; Chhillar, A.; Said, Z.; Memon, S. Exploring the Exhaust Emission and Efficiency of Algal Biodiesel Powered Compression Ignition Engine: Application of Box–Behnken and Desirability Based Multi-Objective Response Surface Methodology. *Energies* **2021**, *14*, 5968. [CrossRef]
29. Bhaskor, J.B.; Ujjwal, K.S. Emission Reduction Operating Parameters for a Dual-Fuel Diesel Engine Run on Biogas and Rice-Bran Biodiesel, Technical Papers. *J. Energy Eng.* **2017**, *143*, 04016064. [CrossRef]
30. Karczewski, M.; Chojnowski, J.; Szamrej, G. A Review of Low-CO2 Emission Fuels for a Dual-Fuel RCCI Engine. *Energies* **2021**, *14*, 5067. [CrossRef]
31. Sharma, P.; Sahoo, B.B. Precise prediction of performance and emission of a waste derived Biogas–Biodiesel powered Dual–Fuel engine using modern ensemble Boosted regression Tree: A critique to Artificial neural network. *Fuel* **2022**, *321*, 124131. [CrossRef]
32. Szamrej, G.; Karczewski, M.; Chojnowski, J. A review of technical solutions for RCCI engines. *Combust. Engines* **2021**, *61*, 36–46. [CrossRef]
33. Karczewski, M.; Chojnowski, J. Analysis of the market structure of long-distance transport vehicles in the context of retrofitting diesel engines with modern dual-fuel systems. *Combust. Engines* **2022**, *188*, 13–17.
34. Emission Test Cycles. ISO 8178. Available online: https://dieselnet.com/standards/cycles/iso8178.php (accessed on 20 May 2022).
35. Samar, D.; Debangsu, K.; Bhaskor, J.B.; Pankaj, K.; Vinayak, K. Thermo-economic optimization of a biogas-diesel dual fuel engine as remote power generating unit using response surface methodology. *Therm. Sci. Eng. Prog.* **2021**, *24*, 100935. [CrossRef]
36. Wajand, J.A. *Silniki o Zapłonie Samoczynnym*; WNT: Warsaw, Poland, 1988; pp. 123–130.
37. Kokjohn, S.; Hanson, R.M.; Splitter, D.A.; Reitz, R.D. Fuel reactivity-controlled compression ignition (RCCI): A pathway to controlled high-efficiency clean combustion. *Int. J. Engine Res.* **2011**, *12*, 209–226. [CrossRef]
38. Kuiken, K. *Gas- and Dual-Fuel Engines for Ship Propulsion, Power Plants and Cogeneration*; Book III: Operation and maintenance; Target Global Energy Training: Onnen, The Netherlands, 2016; pp. 326–327.
39. Nova, I.; Tronconi, E. *UREA SCR Technology for deNOx after Treatment of Diesel Exhausts*; Springer Science + Buisness Media: New York, NY, USA, 2014.
40. Kowalewicz, A. *Podstawy Procesów Spalania*; WNT: Warsaw, Poland, 2000; pp. 248–268.
41. Dahodwala, M.; Joshi, S.; Koehler, E.; Franke, M.; Tomazic, D.; Naber, J. *Investigation of Diesel-CNG RCCI Combustion at Multiple Engine Operating Conditions*; SAE International: Warrendale, PA, USA, 2020.
42. Paykani, A.; Amirhasan, K.; Pourya, R.; Rolf, R. Progress and recent trends in reactivity-controlled compression ignition engines. *Int. J. Engine Res.* **2016**, *17*, 481–524. [CrossRef]
43. Radsak, D. Redukcja emisji tlenków azotu w kotłach energetycznych jako konieczność spełnienia europejskich standardów emisyjnych. *Poznań Univ. Technol. Acad. J.* **2017**, *90*, 333–345. [CrossRef]
44. Żółtowski, A.; Gis, W. Ammonia Emissions in SI Engines Fueled with LPG. *Energies* **2021**, *14*, 691. [CrossRef]
45. Muscala, W. Tworzenie i Destrukcja Tlenków Azotu w Procesach Energetycznego Spalania Paliw. Available online: https://docplayer.pl/5623726-Tworzenie-i-destrukcja-tlenkow-azotu-w-procesach-energetycznego-spalania-paliw.html (accessed on 19 May 2022).
46. Stelmasiak, Z. *Wybrane Problemy Stosowania Gazu Ziemnego do Zasilania Silników o Zapłonie Samoczynnym*; *Archiwum Motoryzacji*; Akademia Techniczno-Humanistyczna w Bielsku-Białej: Bielsko-Biała, Poland, 2006.
47. Pachiannan, T.; Zhong, W.; Rajkumar, S.; He, Z.; Leng, X.; Wang, Q. A literature review of fuel effects on performance and emission characteristics of low-temperature combustion strategies. *Appl. Energy* **2019**, *251*, 113380. [CrossRef]
48. Heywood, J.B. *Internal Combustion Engine Fundamentals*, 2nd ed.; McGraw-Hill Education: New York, NY, USA, 2018.
49. Stoumpos, S. Marine dual fuel engine modelling and parametric investigation of engine settings effect on performance-emissions trade-offs. *Ocean Eng.* **2018**, *157*, 376–386. [CrossRef]

 *energies*

Article

Experimental Studies of the Effect of Air Filter Pressure Drop on the Composition and Emission Changes of a Compression Ignition Internal Combustion Engine

Tadeusz Dziubak * and Mirosław Karczewski

Faculty of Mechanical Engineering, Military University of Technology, 2 Gen, Sylwestra Kaliskiego St., 00-908 Warsaw, Poland; miroslaw.karczewski@wat.edu.pl
* Correspondence: tadeusz.dziubak@wat.edu.pl; Tel.: +48-(26)-1837121

Abstract: This paper presents an experimental evaluation of the effect of air filter pressure drop on the composition of exhaust gases and the operating parameters of a modern internal combustion Diesel engine. A literature analysis of the methods of reducing the emission of toxic components of exhaust gases from SI engines was conducted. It has been shown that the air filter pressure drop, increasing during the engine operation, causes a significant decrease in power output and an increase in fuel consumption, as well as smoke emission of Diesel engines with the classical injection system with a piston (sectional) in-line injection pump. It has also been shown, on the basis of a few literature studies, that the increase in the resistance of air filter flow causes a change in the composition of car combustion engines, with the effect of the air filter pressure drop on turbocharged engines being insignificant. A programme, and conditions of tests, on a dynamometer of a modern six-cylinder engine with displacement V_{ss} = 15.8 dm^3 and power rating 226 kW were prepared, regarding the influence of air filter pressure drop on the composition of exhaust gases and the parameters of its operation. For each technical state of the air filter, in the range of rotational speed n = 1000–2100 rpm, measurements of exhaust gas composition and emission were carried out, as well as measurements and calculations of engine-operating parameters, namely that of effective power. An increase in the pressure drop in the inlet system of a modern Diesel truck engine has no significant effect on the emissions of CO, CO$_2$, HC and NO$_x$ to the atmosphere, nor does it cause significant changes in the degree of smoke opacity of exhaust gases in relation to its permissible value. An increase in air filter pressure drop from value Δp_f = 0.580 kPa to Δp_f = 2.024 kPa (by 1.66 kPa) causes a decrease in the maximum filling factor value from η_v = 2.5 to η_v = 2.39, that is by 4.5%, and a decrease in maximum power by 8.8%.

Keywords: internal combustion engines; air filter pressure drop; engine power; exhaust components; emission of toxic exhaust components

Citation: Dziubak, T.; Karczewski, M. Experimental Studies of the Effect of Air Filter Pressure Drop on the Composition and Emission Changes of a Compression Ignition Internal Combustion Engine. *Energies* 2022, *15*, 4815. https://doi.org/10.3390/en15134815

Academic Editor: Evangelos G. Giakoumis

Received: 30 May 2022
Accepted: 28 June 2022
Published: 30 June 2022

Publisher's Note: MDPI stays neutral with regard to jurisdictional claims in published maps and institutional affiliations.

1. Introduction

There are over one and a half billion passenger cars and commercial vehicles in use in the world, and forecasts indicate that these numbers will continue to increase [1]. Road transport is the most common means of transport, and the most popular form of vehicle propulsion are internal combustion engines, among which compression ignition engines play a decisive role [2]. These engines are widely used in automotive, agricultural and industrial applications, as well as being used as stationary and military equipment, due to their high energy conversion efficiency and durability. Currently used compression ignition engines are fueled by mineral diesel derived from refined petroleum with minor additions of bio-components.

The main disadvantage of using Diesel engines is the emission of harmful substances, especially nitrogen oxides (NO$_x$), particulate matter (PM), carbon monoxide (CO) and unburned hydrocarbons (HC). The combustion of the fuel–air mixture is a process whose end results are often different from the desired ones. Although modern diesel vehicles provide lower fuel consumption, lower emissions and good driveability, NO_x and PM emitted by diesel engines are considered a serious public health problem and contribute to respiratory and cardiovascular diseases.

According to the authors of the paper [3], road transport is largely responsible for emissions of nitrogen oxides (30%), carbon monoxide (20%) and, to a lesser extent, for particulate matter emissions—a few percent. The authors of paper [4] analysed the actual driving conditions of passenger cars on five different routes in Delhi and showed that the average emission rates of CO, HC and NO_x were 3.99, 0.34 and 0.54 g/km for diesel vehicles, respectively. For petrol vehicles, the emission factors were higher and were: 7.26, 0.17 and 0.62 g/km, respectively. Emissions were shown to increase with increasing vehicle speed and acceleration. Moreover, the emissions were minimal at the speed of 40–60 km/h and acceleration, below 0.5 m/s^2. On the other hand, according to the authors of works [5,6], almost 30% of the world's greenhouse gas emissions come from the transport sector, which leads to global warming, and compression ignition engines are the main source of airborne particulate matter.

A car engine exhaust is a mixture of many substances, molecules, compounds and groups of chemical compounds. For the most part, they are non-toxic gases normally contained in the air that humans breathe. Only a relatively small part of the exhaust gases is a burden on the environment (harmful substances), and an even smaller part has poisonous properties (toxic substances)—Figure 1. It is generally accepted that the toxic substances in exhaust gases are carbon monoxide CO, nitrogen oxides NO_x, hydrocarbons HC and particulate matter PM. The toxic components in the exhaust of an SI engine are about 1% and in Diesel engines only 0.3% [7].

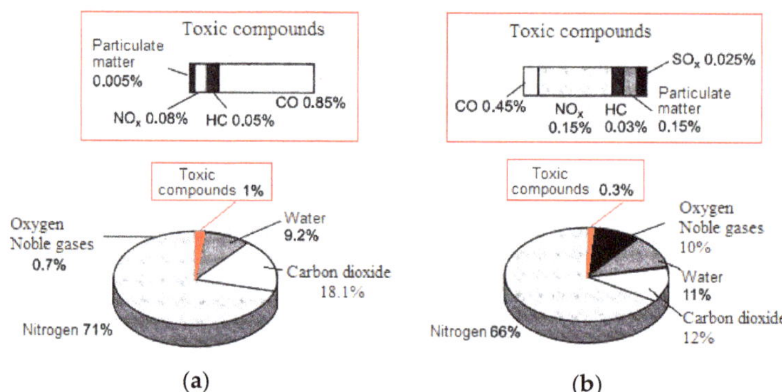

Figure 1. Exhaust gas composition: (**a**) SI engine, (**b**) Diesel engine. The figure was made by the authors based on information from the paper [7].

PM (Particulate Matter) is soot that is a byproduct of complete and incomplete combustion of hydrocarbons caused by local oxygen deficiency. Exhaust particulate matter is not only concentrations of carbon atoms (soot), but also unburned hydrocarbons from fuel and lubricating oil, water vapor, abrasive wear products from metals, sulfur compounds and ash. Soot itself, as a chemically pure carbon, is not dangerous to the human body. It is the compounds on its surface that are dangerous.

The harmful effect of particulate matter on the environment and living organisms is due to the fact that: they persist for a long time in the atmosphere due to their small size (0.01–30 μm), so they are easily absorbed by the respiratory system. They enable heavy metals (lead), sulphur compounds, nitrogen compounds and other hydrocarbons to enter the body.

Nitrogen oxides NO_x are formed during combustion in the fuel chamber at high temperatures. They are the most toxic components of exhaust gases. Of the five different nitrogen oxides (N_2O, NO, N_2O_3, NO_2 and N_2O_4) in the atmosphere, NO nitric oxide and

NO_2 nitrogen dioxide are most abundant. Nitric oxide NO, when absorbed into the human body, reacts rapidly with hemoglobin to form NO-hemoglobin (HBNO). Its affinity for hemoglobin is 1500 times higher than that of carbon dioxide (CO_2) [8]. The formation of NO_x is favoured by a high compression ratio, high engine load, high combustion temperature, early ignition or injection advance angle and a lack of air–fuel mixture turbulence.

Carbon monoxide (CO) is an odorless gas produced during unstable combustion processes caused by local oxygen deficiency. This compound binds easily with hemoglobin (200–300 times faster than oxygen). The formation of CO is favoured by a too-rich mixture in SI engines or a low excess air ratio in compression-ignition engines, low engine load, low temperature of cylinder walls (underheated engine, idling), late fuel injection advance angle or late ignition advance angle, small load turbulence and the use of exhaust gas recirculation [8].

Hydrocarbons (HC) emitted into the atmosphere by motor vehicles are primarily formed by complete and incomplete combustion of rich mixtures, or during local oxygen deficiency and by emissions from the fuel system. Large amounts of hydrocarbons are formed in the case of combustion of too-lean mixtures due to ignition loss and a prolonged combustion process resulting from a low combustion speed of such mixtures. The formation of hydrocarbons is favoured by the low temperature of the combustion chamber walls (underheated engine), late fuel injection or late ignition, non-uniform mixture and low charge turbulence.

In addition to fuel combustion products, motor vehicles emit particulate matter resulting from abrasive wear processes on tires and road surfaces [9–14], friction linings and brake discs [15–17], and clutch disc linings.

In order to minimise the problem of environmental pollution from car engine exhausts and to meet increasingly stringent emission standards, intensive research is being conducted that focuses on two directions: first, the development of technology to provide fuel economy; and second, the reduction in exhaust emissions. Research to reduce exhaust emissions is being conducted intensively in many areas. One of them is combustion management, mainly related to fuel injection control [18]. Higher emissions of particulate matter (PM) and nitrogen oxides (NO_x), and the resulting trade-off between them, are the main drawbacks of conventional diesel engines. Many researchers have shown that PM emissions from compression ignition engines are generally 10–100 times higher than those from gasoline-fueled spark ignition engines [19,20].

Combustion, exhaust emissions and performance characteristics of a diesel engine are directly influenced by several factors, including fuel injection pressure, time of injection start, amount of fuel injected, injection pattern, number of nozzles, spray pattern, etc. However, some of these parameters also have an indirect effect on engine power, and one such important parameter is heat transfer.

The parameters of the injection system and the injection characteristics have a very large influence on the combustion, emissions and performance of a compression ignition engine. The primary technique used in modern common rail injection diesel engines is the multiple injection method, which is used depending on the purpose to be served [21–24]. Depending on the engine speed and load, modern diesel engines have pilot injection, primary injection and secondary injection.

Typically, most of the fuel for the engine duty cycle is injected in the main injection cycle. Pilot injection reduces NO_x and PM emissions, reduces combustion noise and peak cylinder pressure and exhaust temperature. Final injection controls particulate emissions and exhaust gas temperature, the adequate value of which is necessary for the proper operation of exhaust after-treatment systems or turbochargers [25]. The authors of paper [26] found that final injection provides additional energy that improves mixing and accelerates soot oxidation at the end of the combustion process. They also found that end injection increases the temperature in the combustion chamber and the rate of soot oxidation. On the other hand, Kumar et al. [27] studied and compared the effect of injection start angle and intake air temperature on the combustion process and exhaust emissions in a compression ignition engine running on a mixture of ethanol and biodiesel (cottonseed oil product). An increased injection starting angle results in an earlier start of combustion relative to TDC. As the piston moves toward TDC, the charge in the cylinder is compressed, causing an increase in pressure and temperature. At the same time, there is an increase in pressure and temperature in the cylinder resulting from the progressive combustion of the mixture. As a result, there is a significant increase in the rate of heat release, which results in reduced HC emissions and increased NO_x emissions, due to the high combustion temperature. On the other hand, a delayed injection starting angle results in the opposite trend. For an extended injection timing and a higher intake of air temperature, soot and CO emissions show a decreasing trend due to the better reaction of fuel with oxygen. With shortened injection timing, soot and CO emissions show a reverse trend. Increased intake air temperature when using fuel blend containing ethanol and biodiesel resulted in an increased peak of in-cylinder temperatures. The increased charge temperature compensates for the higher heat of vaporisation of the ethanol fuel, resulting in reduced ignition delay. This results in increased NO_x emissions and decreased HC. As a result of improved in-cylinder evaporation and combustion, it can be concluded that preheating the intake air can potentially reduce CO emissions and smoke opacity.

Exhaust after-treatment technology is another area of research in reducing exhaust emissions from automobile engines [28–30]. There is also research work related to exhaust gas recirculation [31,32]. Mossa et al. [32] investigated the effect of hot exhaust gas recirculation (EGR) system on engine power and torque, average cylinder pressure, fuel consumption and exhaust emissions of a direct injection (DI) diesel engine. The test object was a single-cylinder, four-stroke engine with an air-cooled system with a rated speed of 3600 rpm and a displacement of 0.219 dm³. The tests were conducted in a speed range from 1600 to 3600 rpm with different percentages of EGR exhaust contributions (5%, 7%, 10% and 15%). The results showed that increasing the proportion of EGR exhaust decreased engine power and torque while increasing fuel consumption. The effect of EGR exhaust gas recirculation reduced NO_x from 800 to 240 ppm and CO_2 from 9% to 4%, while increasing CO from 2% to 4% and HC from 10 to 100 ppm.

Another method used to reduce exhaust emissions is the use of various alternative fuels to power engines. Currently, the most studied alternative fuels for compression ignition engines are biodiesel, vegetable oils, alcohols, dimethyl ether, dimethyl carbonate, natural gas, liquefied petroleum gas, methane, methane, propane, hydrogen and waste materials [33–50]. Some of these alternative fuels can be used in pure form while others are used as dual fuels, fuel blends and emulsions. For example, in [42], changes in exhaust emissions and effective parameters of an indirect injection compression ignition (IDI) engine were investigated when running on diesel fuel and in dual-fuel mode by adding compressed natural gas (CNG). The tests were carried out at three engine load steps of 100%, 75% and 50% and with different fixed CNG percentages of 20, 30 and 40%, and no gas. It was found that the addition of CNG has a positive effect on the thermal efficiency of the combustion process and a negative effect on incomplete combustion products, such as hydrocarbons (HC), which increase on average from 0 ppm to 320 ppm. Carbon monoxide (CO) increases on average by about 70% at high loads and by about 90% at medium and low loads compared to diesel fuel. Carbon dioxide (CO_2) emissions decrease by an average of 7%. Karczewski et al. [43] made an extensive analysis of the exhaust gas emission problem in the context of possibilities of its reduction by using fuels with increased H/C ratio. Such fuels may be a solution to the problem of increasing limitations of exhaust gas emission.

Kukharonak et al. [44] studied the use of n-butanol (biobutanol) in blends with gasoline in SI internal combustion engines. An experimental study of the effect of *n*-butanol (0%, 20% and 40% by volume) and gasoline blends on the parameters (M_o, N_e and g_e) of an SI engine was conducted. The results show that increasing the concentration of *n*-butanol in the mixture causes changes in engine performance. Power and torque decrease by 1.6 kW (12.7%) and 6.8 Nm (13.8%), respectively, and specific fuel consumption increases by 6.3 g/(kWh) (2%) when the n-butanol concentration is increased to 40%. The results show that, without changing the ignition advance angle αz, the gasoline mixture with 10% n-butanol has essentially no negative effect on engine power N_e, torque M_o and specific fuel consumption g_e.

Freitas et al. [45] evaluated the performance and NO_x, CO, and CO_2 content of a compression-ignition engine fueled with three blends of diesel, biodiesel, and ethanol: pure B7, B7E3 (B7 with 3% ethanol), and B7E10 (B7 with 10% ethanol). The experimental studies were performed for fixed engine speeds in the range of n = 1000–1750 rpm and for fixed loads in the range of 10–100% of maximum torque. B7 was taken as the base for the study and then ethanol additions (3% and 10% by weight) were evaluated. An increase in ethanol content in the diesel/biodiesel blend resulted in a 1.4-fold increase in exhaust emissions. An increase in engine speed resulted in a five- to seven-fold increase in exhaust emissions. As the ethanol content in the fuel blend increased, NO_x emissions were higher, with NO values being much higher than NO_2. CO and CO_2 concentrations increased with increasing ethanol content.

Yang et al. [46] believe that methanol in diesel impairs power performance but improves fuel economy and emissions in compression ignition engines. The methanol ratio should be kept at 10–15% to balance the power and emission performance.

Electric vehicle technologies are used to solve the problem of environmental pollution and to meet increasingly stringent emission standards. Pielecha et al. [48] compared exhaust emissions of two hybrid cars of the same manufacturer made in plug-in version and HEV version (petrol engine + electric motor). The comparative exhaust emission tests were carried out on a chassis dynamometer and in real conditions of use—in road traffic. The obtained values of emission of exhaust components: CO2, CO, NOx and HC were lower for the plug-in hybrid vehicle by 3%, 2%, 25% and 13%, respectively, compared to the HEV. Fuel consumption was 3% lower for the plug-in hybrid vehicle and particulate matter was 10% lower compared to the HEV. In real driving conditions, the differences were more pronounced in favour of the plug-in hybrid vehicle: CO_2 emissions in the road test were 30% lower, NO_x emissions were 50% lower, and particle counts were 10% lower.

According to the authors of the paper [50], dimethyl ether (DME) is a promising alternative to diesel in compression ignition (CI) engines used in various industrial applications. However, the high emission of nitrogen oxides (NO_x) during DME combustion limits its application. The main reason for high NO_x emission is high combustion temperature. In this study, high exhaust gas recirculation (EGR) was used while testing a CI engine with common rail direct injection, suitable (with minor modifications) for a passenger car. The modified fuel supply system provided high injection pressure during the combustion performance evaluation.

The use of alternative fuels in a compression ignition engine can have several undesirable consequences, such as increased fuel consumption and NO_x emissions, reduced engine power, piston ring jamming and sometimes cold engine starting problems [51–53]. These drawbacks can often be overcome by proper management of the combustion process and by using appropriate fuel additives. In recent years, nanomaterials are becoming very promising additives for diesel engine fuels. Their aim is to reduce harmful emissions from diesel engines and improve their performance [54–63]. Nanomaterials exhibit excellent properties so that they can be used as fuel additives to improve the performance of diesel engines.

For example, in [61], the effect of adding cerium oxide (CeO_2) nanoparticles, which were added to rapeseed oil (C30D) and diesel fuel at two concentrations (50 and 100 mg/L), on engine performance, NO_x and PM emissions was investigated. The addition of CeO_2 nanoparticles to C30D oil and diesel fuel was tested under a secondary fuel injection (PI) strategy. The results showed that when the engine was operated on a C30D + CeO_2 blend, the engine power increased by 9.42% compared to a diesel + CeO_2 operation with and without PI strategy. In addition, specific fuel consumption decreased by 16.46% for C30D + CeO_2 blends compared to CeO_2 nanoparticles in diesel. The results showed that the addition of CeO_2 nanoparticles with the same concentration (100 mg/L) to C30D and diesel fuel resulted in a decrease in NO_x emissions by 10.64% and 8.73%, respectively. The addition of CeO_2 nanoparticles at concentrations of 50 and 100 mg/L to C30D reduced the PM concentration in the engine exhaust by 28.62% and 42.72%, respectively, compared to engine operation on fuel without additives. In contrast, adding both concentrations of CeO_2 to diesel fuel reduced the PM concentration by 16.74 and 24.83%, respectively, compared to diesel fuel without additives. The data from this experiment showed that the introduction of the PI strategy and the addition of CeO_2 nanoparticles at 100 mg/L to the fuel has a positive effect of increasing engine performance and reducing NO_x emissions and PM concentration.

In work [62], an evaluation of specific fuel consumption and emission of selected components of exhaust gases from a six-cylinder compression-ignition engine with direct injection and a displacement of 11,051 dm^3 was carried out. The engine was fed with fuel to which a catalyst containing 5% ferric chloride was added. The results obtained show that the component added to the fuel has a positive effect on exhaust emissions. There was an average decrease of 16.7% in particulate matter (PM), 10.1% in CO and 7.9% in hydrocarbons (HC). In contrast, there was a 1.2% increase in NO_x. The difference between the specific consumption of fuel to which ferric chloride was added and pure diesel fuel does not exceed the level of measurement error and amounts to 0.5%.

Adzmi et al. in [63] conducted an experimental evaluation of combustion characteristics, engine performance and exhaust emissions after addition and mixing of nanoparticles with palm oil methyl ester (POME) on a single-cylinder compression-ignition engine. Aluminum oxide (Al_2O_3) and silicon dioxide (SiO_2) nanoparticles at 50 ppm and 100 ppm were used. SiO_2 and Al_2O_3 were blended with POME and designated as PS50, PS100 and PA50, PA100, respectively. The test results for PS and PA fuels were compared with those for the POME test fuel. The tests were conducted at engine loads of: 7 Nm, 14 Nm, 21 Nm and 28 Nm and no load, at a constant engine speed $n = 1800$ rpm. The results show that the highest maximum in-cylinder combustion pressure obtained for the fuel containing nanoparticles is 16.3% higher compared to the POME fuel. Moreover, the peak engine torque and engine power show a significant increase of 43% and 44%, respectively, recorded in the test with PS50 fuel. For the fuel containing nanoparticles, NO_x emissions decreased by 10%, CO_2 emissions decreased by 6.3% and CO emissions decreased by 0.02%.

An important area of efforts to reduce carbon dioxide (CO_2) emissions into the atmosphere is the use of hydrogen fuel cell vehicles [64].

From the above analysis, there is a wide range of possibilities to reduce exhaust emissions from engine exhaust systems, but in many cases, this involves a decrease in engine performance.

One of the factors that determine the changes in exhaust emissions from internal combustion engines is the pressure drop of the intake system, and mainly the air filter pressure drop defined as the static pressure drop downstream of the filter. The air filter is an important component of an internal combustion engine that is the power unit of a motor vehicle, as its performance determines the reliability and life of the engine. Suitable purity of the air intake to the combustion engines of commercial vehicles is ensured by baffle air filters, where the filtering element is a pleated paper insert. A characteristic feature of baffle filters is that during operation, as a result of the deposition and accumulation of dust particles in the filter bed, its capacity decreases and the air filter pressure drop Δp_f increases steadily. The higher the value of the dust concentration in the air sucked into the engine, the faster the filter reaches the permissible value—Δp_{fdop}. The pressure drop of the air filter hinders the air flow to the engine cylinders, which results in a decrease in the engine filling and power, and an increase in specific fuel consumption. In the available literature, it is very difficult to find a sufficiently complete picture of the effect of air filter pressure drop on engine performance metrics, especially on changes in exhaust emissions. In the available literature, there is a full description of the study of the effect of air filter pressure drop on the performance of naturally aspirated carbureted and diesel internal combustion engines with a classic injection system with an in-line piston (sectional) injection pump [65–68]. The engines of modern passenger cars are equipped with gasoline direct injection systems with an electronic control system, while the engines of trucks are equipped with high-pressure diesel injection systems, such as common rail or systems with electronically controlled pump injectors. The ECU (Engine Control Unit) of the engine controls all the quantities which affect the value of the generated torque produced by the engine, while at the same time meeting the requirements in the area of exhaust emissions and fuel consumption throughout the life of the vehicle. In the available literature, the results of empirical investigations determining the influence of the flow resistance of the air supply system, including the filter, on the performance of a modern car engine and, in particular, on the composition of exhaust gases, are not encountered very often. Empirical tests are expensive and labour intensive, which explains the scarce number of available results. It is considered to be the research method that produces the most reliable results.

Dziubak and Karczewski in [69] carried out extensive experimental research on an engine dynamometer on the influence of air-filter flow resistance on the basic parameters of operation (filling factor, power and specific fuel consumption) of a modern Diesel engine equipped with the common rail fuel supply system. The investigated engine is a driving unit of a truck tractor. The present work is a continuation of the experimental research, the results of which will allow to fill the gap in the scope of the influence of the flow resistance of the air supply system on the change of the quantitative and qualitative composition of the exhaust gases from the modern Diesel engine.

2. Literature Analysis on the Effect of Intake System Pressure Drop on Engine Performance and Exhaust Emission Change

As dust settles on the surface of the filter material, the air flow decreases due to increasing resistance to air flow through the filtration system. The operation of an engine with an air filter of increased pressure drop causes a decrease in the cylinder filling with fresh charge and torque, resulting in a decrease in maximum engine power. The result is a reduction in the dynamic properties of the vehicle. A high air filter pressure drop (a significant vacuum created behind the filter) can cause the breakaway forces to exceed the grain adhesion forces to the filter substrate. The result of this phenomenon can be an avalanche of dust detachment (so-called secondary emission), and sucking it back together with the air into the engine cylinders will cause accelerated wear of P-PR-C (piston-piston ring-cylinder) components. In extreme cases the high pressure drop of the air filter can cause mechanical damage (breakage, rupture) of the filter insert, which will result in increased wear of engine components. For this reason, when the filter reaches a certain resistance value, it is necessary to service the air filter by replacing the filter cartridge. In passenger cars, due to low values of dust concentration in the air, and thus small increments of the filter pressure drop, this operation is performed depending on the mileage of the vehicle or its operation time.

Trucks, off-road vehicles, and special vehicles are operated with high and variable dust concentrations in the air. For this reason, the increase in pressure drop is significant (by 5–8 kPa), and occurs with varying intensity. If the systematic replacement of filter cartridges is not observed, an additional increase in air filter pressure drop occurs, resulting in a significant decrease in engine power, an increase in fuel consumption and a decrease in the dynamics of vehicle movement.

The paper [65] presents the characteristics of the filling ratio $\eta_v = f(n)$, power $N_e = f(n)$ and torque $M_o = f(n)$ o and specific fuel consumption $g_e = f(n)$ of the six-cylinder naturally aspirated ($V_{ss} = 6.842$ dm^3) 359M Diesel engine with a classic injection system, for three values of the air filter pressure drop $\Delta p_f = 2.3, 6, 12$ kPa [65]. As the engine speed increases in the range of $n = 1200$–2800 rpm, regardless of the value of the air filter pressure drop, the characteristics of filling $\eta_v = f(n)$, power $N_e = f(n)$ and torque $M_o = f(n)$ shift almost in parallel toward lower values, and specific fuel consumption $g_e = f(n)$ toward higher values. The increase in the air filter pressure drop in the range of 2.3–12 kPa, when the engine is operated at engine speed $n = 2800$ rpm and at 100% load, causes: a decrease in fill factor by 25.7%, a decrease in power by 7.16% and an increase in specific fuel consumption by 8.49%.

An increase in the air filter pressure drop by 1 kPa causes an average decrease in the filling factor by 2.65%, a decrease in power by 0.739% and an increase in specific fuel consumption by 0.876%. The author did not conduct any research on changes in exhaust emissions.

The results of investigations into the effect of the air filter pressure drop Δp_f on the characteristics of the external effective power N_e and specific fuel consumption g_e of the Diesel engine of a special vehicle are presented in [66]. A 12-cylinder (V-system) naturally aspirated engine with a displacement of 38.88 dm^3 and rated power of 430 kW (580 hp) at n = 2000 rpm, with a classical injection system and multi-range speed controller (direct fuel injection) was studied. The increase in air filter pressure drop was modeled in the range Δp_f = 3–30 kPa. The effect of air filter pressure drop is not apparent until Δp_f = 6 kPa is reached. A further increase in air filter pressure drop already causes a significant decrease in engine power and an increase in specific fuel consumption, as well as a parallel shift in the external characteristics of power and specific fuel consumption toward lower values of N_e power and fuel consumption g_e with a simultaneous shift toward lower rotational speeds. For air filter pressure drop Δp_f = 26.7 kPa, the decreases in power N_e take the values: 11.75% at 2000 rpm, 20.6% at n = 1400 rpm and 32.7% for 1200 rpm. The author has not carried out any studies on changes in exhaust emissions.

Dziubak and Trawiński [67] presented experimental research on the effect of air filter pressure drop in the range of Δp_f = 3.1–24.7 kPa on the filling factor and smoke opacity of the turbocharged, six-cylinder (V_{ss} = 6 dm^3) Diesel T359E engine with a classical injection system. The tests were carried out using the method of free acceleration of the engine loaded with its own resistance torque and its total moment of inertia, hereinafter referred to as the "dynamic characteristics" method, which makes it possible to determine the course of the instantaneous value of the engine torque without loading it on the dynamometer bench. In the final stage of the engine acceleration process for the air filter with a clean filter cartridge (Δp_f = 3.1 kPa), the filling factor has a value of 1.02. With the increase in the filter pressure drop, the filling factor takes on smaller and smaller values, respectively: 0.90; 0.81; 0.75. Thus, an increase in the air filter pressure drop Δp_f by 1 kPa causes a decrease in the filling ratio by 1.49%, 1.29% and 1.23% on average. The value of the air filter pressure drop Δp_f = 24.7 kPa, i.e., eightfold increase in the air filter pressure drop Δp_f over the value of the initial air filter pressure drop Δp_{f0}, causes a twofold increase in the smoke opacity of the T359E engine (increase in the light absorption coefficient) to k = 0.81 m^{-1}. This does not cause the T359E engine to exceed its maximum value–k_{max} = 3.0 m^{-1}.

The results of the study of the effect of the baffle filter on the characteristics of the UTD-20 internal combustion engine of a special vehicle are presented in [69]. A compression-ignition engine of displacement V_{ss} = 15.8 dm^3 and rated power of 226 kW with a classical fuel injection system was used. The effect of the original air filter with a pressure drop of Δp_f = 13.2 kPa and the modernised filter–Δp_f = 4.9 kPa was studied, as well as the modernised filter working with the engine with an increased (about 7%) fuel dose. During the operation of the engine with an increased fuel dose and with the modernised air filter, a significant increase in power and torque was obtained in comparison with the basic variant of the filter: over 2% for 1600 rpm and over 10% for the 2200–2600 rpm range. However, a slight (2%) increase in specific fuel consumption was recorded. The increase in engine power occurred as a result of increased air mass resulting from reduced air filter pressure drop, an increase in engine fill and an increase in fuel delivery. The authors did not study the effect of air filter pressure drop on the composition of exhaust gases.

Yang et al. [70] studied the effect of two different air filter designs characterised by different pressure drop (A—standard filter, B—upgraded filter with lower pressure drop), on the speed characteristics of a single cylinder compression ignition engine. The effective power, torque, specific fuel consumption, smoke, oil temperature and engine exhaust temperature were determined for these two filters and compared with the engine operating parameters without air filter. The engine working without air filter obtained the highest power and torque, and the lowest fuel consumption. The operation of the engine with filter B and A successively causes a shift of the characteristics $N_e = f(n)$ and $M_o = f(n)$ almost in parallel toward smaller values, and the characteristic $g_e = f(n)$ toward larger values, over the entire engine speed range. With the air filter type B, the torque increases its value and is $M_{omax} = 73.6$ Nm at 1500 rpm, and the fuel consumption decreases to $g_e = 230.8$ g/(kWh). This is 1.6% more torque and 1.5% less fuel consumption than with air filter type A (higher pressure drop air filter). Maximum power increases by 1.1% when a lower air filter pressure drop (type B) is used compared to an air filter type A. Over the entire power range, the exhaust emissions are naturally low without air filter and high with air filter type A. The maximum smoke value (light absorption coefficient) increases with an increasing load and during engine operation without air filter, with air filter type A and with air filter type B is: $k = 1.7, 2.36$ and 2.0 m^{-1}, respectively. After replacing type A air filter by type B filter, the reduction in smoke value by 11% was obtained.

Plotnikov et al. [71] presented the results of numerical simulations of the effect of geometrical characteristics of the intake duct on wave phenomena and pressure drop in the intake system of an eight-cylinder turbocharged diesel engine. Numerical studies of the effective parameters of the engine were carried out for the original intake duct length $L = 1625$ mm and diameter $D = 156$ mm and three other internal diameters: $D_1 = 80$ mm, $D_2 = 250$ mm and $D_3 = 330$ mm. It is shown that the geometric dimensions of the intake pipe have a significant effect on the dynamics and pressure drop of the engine intake system. Decreasing the inner diameter to 80 mm leads to significant pressure fluctuations in the intake system, an increase in pressure drop and a decrease in diesel engine power by 0.5–2.5%. Increasing the inside diameter to 250–330 mm leads to a slight smoothing of the pressure amplitude in the inlet pipe, a decrease in pressure drop and, thus, an increase in fill factor of 0.5% on average (Figure 2a). This leads to an increase in engine power by up to 0.7% and a decrease in effective specific fuel consumption by approximately 0.50–0.75 (Figure 2b).

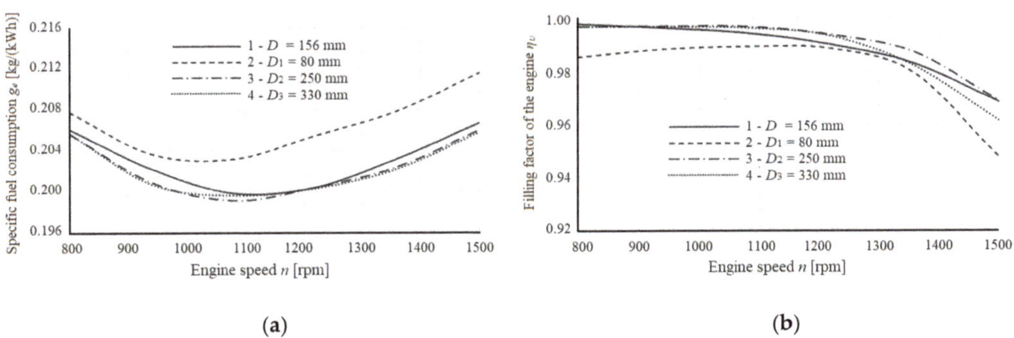

Figure 2. Calculated relationships: (**a**) specific fuel consumption ge, (**b**) filling factor η_v as a function of crankshaft speed n for different intake pipe diameters D. Figures made by the authors based on data from the paper [71].

Abdullah et al. [72] investigated the fuel consumption and exhaust emissions of a carbureted engine as a function of the pressure drop of the intake system. The study was conducted while the engine was running with and without air filter. The hourly fuel consumption with air filter increases from 0.687 dm^3/h to 1.028 dm^3/h, i.e., by 49.6%, when the engine is operated at a constant load in the speed range of 1500 rpm to 2500 rpm. In contrast, when the engine is operated under the same conditions but without an air filter, the hourly fuel consumption takes on a lower value and increases by only 35.2%.

The NO$_x$ content of the exhaust, when the engine is operated without an air filter, is higher than that with an air filter (Figure 3). At speed $n = 1500$ rpm, the NO$_x$ content of the exhaust without air filter is 51.7% higher than when the engine is operated with air filter. For higher speeds of 2000 and 2500 rpm, the NO$_x$ content in the exhaust without air filter is 8.1% and 20.2% higher than that with air filter, respectively. The formation of NO$_x$ is affected by the maximum temperature and pressure of the combustion process. Without air filter, the combustion process is more favourable, which leads to a higher temperature and higher charge pressure, which promotes NO_x formation. The higher pressure drop when the engine is operated with an air filter reduces NO_x formation.

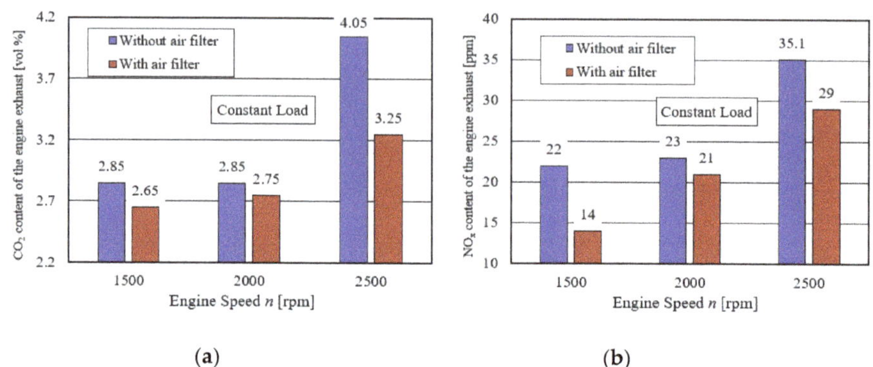

(**a**) (**b**)

Figure 3. Influence of air filter pressure drop on the content in the exhaust gas: (**a**) CO_2, (**b**) NO_x at Constant Load. Figure made by the authors based on data from [72].

Shannak et al. [73] measured the exhaust emissions of a four-cylinder, four-stroke gasoline engine as a function of the pressure drop of the intake system, the value of which was varied using different air inlet pipe diameters: 20, 25, 30, 35, 40 and 63 mm. The tests were conducted at engine speeds ranging from 1000 to 4000 rpm. The results showed that the amounts of hydrocarbons (HC) and carbon monoxide (CO) decrease with an increasing air intake pipe diameter, increasing engine speed and atmospheric pressure and decreasing the altitude at which the engine operates, while CO_2 and oxygen (O_2) remain constant. Increasing the diameter of the inlet pipe from 20 to 63 mm, which is equivalent to a decrease in pressure drop, and increasing the altitude at which the engine operates from 600 to 800 m above sea level, corresponding to a change in atmospheric pressure from 0.94 to 0.91 bar, leads to a reduction in hydrocarbons of about 60% and a reduction in carbon monoxide of about 40% while keeping CO_2 and oxygen constant at about 12.7% and 0.63%, respectively. Increasing the engine speed from 1000 to 4000 rpm leads to a HC reduction of about 45% and a CO reduction of about 25%, while keeping CO_2 and oxygen constant at about 12.7% and 0.63%, respectively.

Thomas et al. [74] studied the effect of air filter pressure drop on emission changes of three truck ZS engines. The results for a Dodge Ram 2500 Truck-6.7 L (2007) with a 6.7 dm^3 inline six-cylinder engine with variable geometry turbocharging, six-speed automatic transmission, a diesel particulate filter (DPF) and a NO_x reduction system (LNT) are shown in Figure 4. An increase in air filter pressure drop from 0.3 kPa to 3.9 kPa and then to 7.6 kPa results in a slight increase in CO, HC and fuel consumption, and a slight decrease in CO_2 and NOx. A significant (from 0.3 to 7.6 kPa) increase in air filter pressure drop causes a slight change in exhaust emissions, which is due to the equipment that this car is equipped with, namely turbocharging to provide adequate airflow, and a diesel particulate filter (DPF) and NO_x mitigation system.

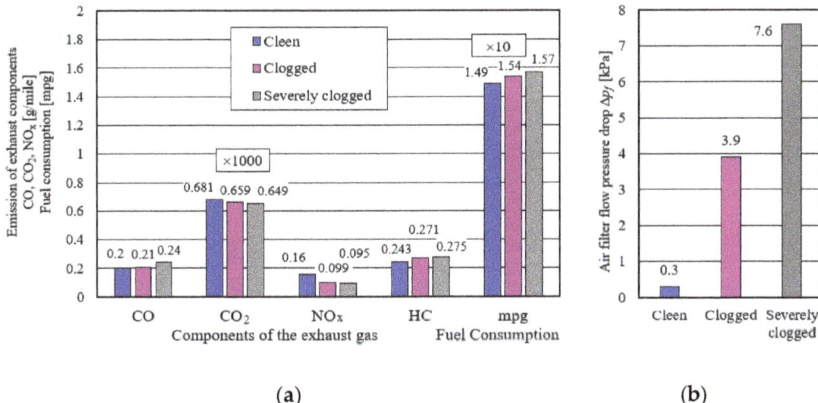

Figure 4. Effect of air filter pressure drop on: (**a**) change in exhaust emissions: CO, CO_2, NO_x and fuel consumption for Dodge Ram 2500 Truck-6.7 L engine, (**b**) air filter pressure drop values. Figure made by the authors based on data from the paper [74].

In order to protect the engine against excessive power reduction and increased exhaust emission caused by the increase in the air filter pressure drop, special sensors of the permissible pressure drop Δp_{fdop} [75,76] are installed in the intake system. Reaching the set value of the flow resistance by the sensor (most often it is at the maximum air flow rate for a given engine) is the signal for air filter servicing—replacement of the filter insert. For trucks and special vehicles, it is assumed that the Δp_{fdop} values are from 6.25 to 7.5 kPa above the flow resistance of a clean air filter [77]. The value of Δp_{fdop} for passenger car engines is assumed to be from 2.5 to 4.0 kPa and, for truck engines, from 4 to7 kPa [78]. For special vehicle engines, the values range from 9 to 12 kPa [79].

The initial efficiency of the air filter (after replacing the filter cartridge with a new one) is low and, depending on the type of filter material, is about 96–98%, but at the end of the service life it increases significantly to over 99.9% [80]. Therefore, it is not advisable to replace the filter cartridge before reaching the established service life. On the other hand, frequent (unjustified) replacement of the filter element may result in premature engine wear. The operation of an air filter is a technical compromise between the increase in air filter pressure drop causing a decrease in engine power output, an increase in fuel consumption and exhaust emissions and the efficiency and accuracy of filtration, factors that determine wear and tear and the durability and reliability of a vehicle engine.

The above analysis shows that, as of today, in the available literature there are no results of investigations of the influence of the pressure drop in the inlet system on the performance of a modern ZS engine used for the propulsion of trucks/truck tractors currently on the roads and constituting the basic means of transport of goods. There is no sufficient data to identify in an unambiguous way the influence of the pressure drop in the inlet system on the composition of the exhaust gases and the emission of toxic compounds. Therefore, it is purposeful to determine experimentally the relation between the parameters of operation of the inlet system of a modern truck engine and the performance of this engine in terms of changes in the emission of individual components of exhaust gases.

3. Experimental Studies on the Influence of Air Filter Pressure Drop on the Performance of the Diesel Engine

3.1. Purpose and Focus of the Study

The aim of the research is an experimental evaluation of the influence of the air filter pressure drop Δp_f on the operation parameters represented by the composition and emission of particular exhaust components, as well as on the opacity of exhaust gases from a modern Diesel engine with an electronic fuel supply control system and a supercharging and air cooling system.

The test object was a six-cylinder, in-line Diesel engine with direct fuel injection—a supply system with electronically controlled injectors, of the Volvo D13C460 EURO V EEV truck with maximum power of 338 kW, being a driving unit of the Volvo FH13 truck tractor with the mileage of 790,500 km. The engine is equipped with four valves per cylinder, which are controlled by hydraulic tappets driven from the central camshaft located on the head. This shaft also drives electronically-controlled pump injectors using a piezo-quartz valve. The basic parameters of the engine are given in Table 1.

Table 1. Basic parameters of the D13C460 EURO V engine [72].

Name of Device	Range
Engine type	D13C460 EURO V
Maximum power at 1400–1900 rpm	460 hp (338 kW)
Maximum engine speed	2100 rpm
Maximum torque at 1000–1400 rpm	2300 Nm
Number of cylinders	6
Cylinder diameter	131 mm
Piston stroke	158 mm
Displacement	12.8 dm^3
Compression ratio	17.8:1

The air supply system consists of an air intake located on the right side of the cabin at its highest height, an external intake duct of nearly rectangular cross-section located on the rear wall of the cabin, a rubber folding element connecting the external duct with the air filter intake duct and a single-stage (baffle) air filter. The filtering element is a cylindrical insert made of pleated filtering paper with active surface A_c = 13.72 m^2. On the outlet pipe from the air filter there is a sensor of acceptable pressure drop set at Δp_{fdop} = 4.8–5.0 kPa. Suitable filling of engine cylinders is assured by turbocharger and charge air cooler operating in an "air-to-air" system. A detailed description of the engine and intake system can be found in [69].

3.2. Test Methodology and Conditions

Tests were carried out on a standard dynamometer bench. The motor was loaded with a water brake type Zöllner PS1-3812/AE with a maximum power of 1250 kW. The torque M_o generated by the engine was measured with a strain gauge transducer connecting the swinging brake housing to the foundation. Hourly fuel consumption G_e was measured using an AVL fuel balance with a 5 s time interval and then averaged over a 60 s interval. The coolant temperature t_{ch} during engine testing was set equal to the operating temperature of t_{ch} = 87–92 °C and was maintained using an external heat exchanger. The opacity of the exhaust gases was determined with the AVL 439 OPACIMETER, which works on the principle of measuring the light absorption coefficient

Flue gas composition was measured using an Atmos FIR analyser operating on a hot 180 °C sample using Fourier Transform Infrared Spectroscopy (FTIR) and with a zirconia cell oxygen analyser built into the measuring chamber. The device has a Certificate confirming compliance with the requirements of the EN 15267-1:2009, EN 15267-2:2009, EN 15267-3:2007 standards. In the used configuration, the device allowed to measure the following components of the exhaust gases: NO, NO_2, N_2O, SO_2, CO, CH_4, CO_2, O_2, H_2O. The measurement results were automatically converted to normal conditions (p_n = 101 325 Pa, T_n = 273.15K). The spectral operating range of the analyser was 485–5500 cm^{-1}.

The volumetric air demand Q_s by the engine was recorded with a thermo-aerometric flow transducer. The air filter pressure drop Δp_f was determined as the pressure difference p_1 before and p_2 after the air filter, with the use of a TESTO 400 differential pressure gauge. The detailed test methodology was described in [69].

During the tests, for each speed of rotation, the following parameters of engine operation and parameters of the air flow in the intake system were measured directly:

- engine torque, M_o [Nm],
- engine rotational speed, n [rpm],
- hourly fuel consumption, G_e [kg/h],
- engine air demand, Q_s [m^3/h],
- exhaust gas temperature, t_s [°C],
- exhaust gas opacity—light absorption coefficient, k [m^{-1}],
- air pressure before p_1 and after air filter p_2, [kPa],
- charge air pressure, p_d, [kPa],
- exhaust gas components: NO, NO_2, N_2O, SO_2, CO, CO_2, O_2, H_2O.

Based on the directly measured values of engine operating parameters, the following were determined:

- effective engine power, N_e [kW],
- specific fuel consumption, g_e [g/(kWh)],
- air filter pressure drop Δp_f [kPa].
- emission of individual exhaust components: NO, NO_2, N_2O, SO_2, CO, CO_2, O_2, H_2O,
- relative change in emission of exhaust gas components.

The emission of particular exhaust components was determined according to requirements contained in [81]. NO_x concentration was determined as the sum of $NO+NO_2$ components. All tests were repeated in duplicate to eliminate coarse errors that could lead to incorrect inference. Control of engine load, engine speed and brake load was performed from the control and measurement cabin, where gauges for measuring engine performance are also located. During the tests the engine operation was continuously controlled by using the diagnostic interface NAVIGATOR TXTs with the software IDC 5 TRUCK. The schematic diagram of the test stand is shown in Figure 5. The applied measurement equipment and its accuracy are presented in Table 2.

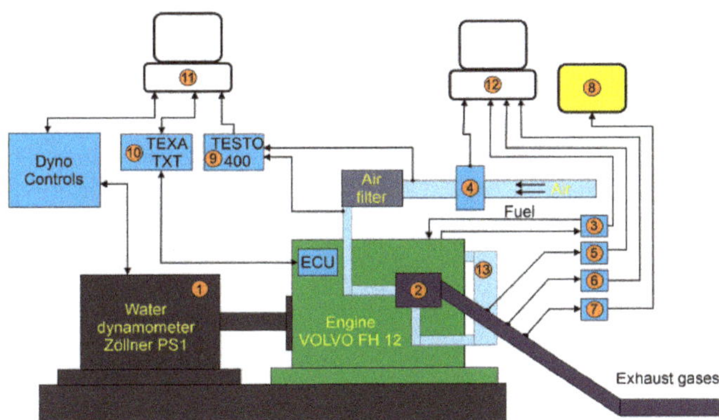

Figure 5. Diagram of the dynamometer stand with Volvo D13C460 EURO V engine: 1—water brake, 2—turbocharger, 3—fuel consumption measurement system, 4—air consumption measurement system, 5—exhaust smoke measurement system, 6—exhaust gas temperature measurement, 7—exhaust gas analyser sampling system, 8—FTIR type exhaust gas analyser, 9—engine intake system pressure measurement system, 10—TEXA TXT diagnostoscope, 11—computer controlling the operation of the dynamometer brake and recording engine operating parameters, 12—computer controlling the operation of the measurement systems, 13—charge air cooler.

Table 2. List of investigation equipment used during investigation.

No.	Name of Device/Measured Quantity	Type	Range	Accuracy
1.	Water dynamometer • torque—M_o • rotated speed—n	Zöllner PS1-3812/AE	M_o = (0–7000) Nm n = (0–3000) rpm Ne = (0–1250) kW	± 1 Nm ± 1 rpm ± 1 kW
2.	Fuel weight-meter (diesel)—G_e	AVL 733S Fuel Balance	(0–200) kg/h	± 0.005 kg/h
3.	Smoke concentration—extinction coefficient of light radiation—k	AVL Opacimeter 4390	(0.001–10.0) m^{-1}	± 0.002 m^{-1}
4.	Exhaust analyser—measuring of toxic elements concentration in exhaust gases • carbon dioxide (CO_2) • carbon monoxide (CO) • nitrogen oxides (NO) • nitrogen diooxide (NO_2) • oxygen (O_2) • hydrocarbons (HC) • water (H_2O)	Atmos FIR emissions monitoring FTIR systems	CO_2 (0.01-23) % CO (1.0–11,000) ppm NO (1.0–6000) ppm NO_2 (1.0–6000) O_2 (0.1–21) % HC (1.0–5000) H_2O (0.25–25) %	± 0.1% measured quantity
5.	Thermocouple—measuring of exhaust temperature—t_s	NiCr—NiAl (K)	(–50–1100) °C	± 1 °C
6.	Mass air consumption—Q_s	SensyMaster FMT430 Thermal Mass Flowmeter	(100–6000) m^3/h	± 1.0 m^3/h
7.	Vacuum in the intake system	TESTO 400	(–100–200) hPa	0.3 Pa + 1% measured quantity

The results of engine operating parameters: power and hourly fuel consumption, obtained during the tests, were reduced to normal conditions in accordance with PN-ISO 15550:2009 [82].

3.3. Analysis of Research Results

During the experimental research, the influence of four (New, A-33, B-66, C-90) technical states of the same air filter on the external characteristics of the Volvo D13C460 EURO V engine, differing in pressure drop, was determined. In each case the same parameters characterising the engine operation were measured.

An increase in the value of the filter pressure drop Δp_f was modelled by means of obscuring a part of the active filtration surface of the cylindrical cartridge. As a result, four technical conditions were obtained differing in the value of the pressure drop of the same air filter. At rotational speed n = 1900 rpm, the pressure drop of the air filter obtained the following values:

- technical condition "New"—filter with clean, brand new, paper air filter cartridge, Δp_f = 0.580 kPa,
- technical condition A-33—air filter with an air filter insert that has had approx. 33% of its active filtration area obscured, Δp_f = 0.604 kPa,
- technical condition B-66—air filter with a filter insert, which is obscured by approx. 66% of the active filtering surface, Δp_f = 0.757 kPa,
- technical condition C-90—air filter with a filter insert, which has approx. 90% of its active filtering surface obscured, Δp_f = 2.024 kPa.

Figure 6 shows the characteristics $\Delta p_f = f(n)$ and engine air demand $Q_s = f(n)$ of the same air filter for four technical states (New, A-33, B-66, C-90) as a function of engine speed n of a Volvo D13C460 EURO V engine. As the engine speed increases in the range of n = 1000–2100 rpm, the pressure drop of the air filter, independently of the pressure drop (percentage obscuration of the active area of the cartridge), increases its value until the engine speed reaches n = 1900 rpm, which is related to the achievement of the maximum air demand of the engine (Figure 6) and the maximum power. Further increase in engine speed causes a decrease in filter pressure drop, which results from a decrease in the air flow rate

Q_s. When the engine was operated with an air filter in A-33 condition (33% obscuration of the active surface of the cartridge), no significant differences were found in the pressure drop values compared to a new filter (New). The recorded differences of the pressure drop are within the limits of the measurement errors and do not significantly affect the other parameters of engine operation presented in the following figures. The analysis of the test results was carried out within the whole operating speed range of the engine, which is the driving unit of the truck tractor.

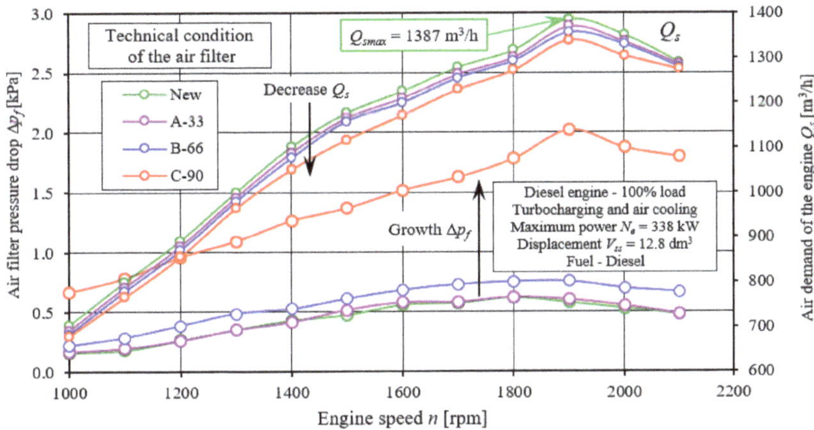

Figure 6. Pressure drop of different air filter states (New, A-33, B-66, C-90) and air demand of Volvo D13C460 EURO V engine as a function of speed n.

Figure 7 shows the charge air pressure pd in the intake manifold of the VOVLO DC13C460 engine as a function of rotational speed n for different technical states of the air filter: New, A-33, B-66, C-90. As the engine speed increases, regardless of the technical state of the filter, the charge air pressure of the engine increases quite rapidly and in the range $n = 1400$–1500 rpm reaches maximum values, after which it systematically decreases. The highest values of the decrease in the charge pressure in the intake manifold, caused by the increase in the air filter pressure drop, were recorded in the range of rotational speeds $n = 1300$–1900 rpm, with the characteristics $p_d = f(n)$ shifted almost in parallel toward smaller values of pd. At rotational speed $n = 1600$ rpm, the decrease in the charge pressure pd, caused by the increase in the air filter pressure drop (A-33, B-66, C-90) over the value of the air filter pressure drop of the new "New" filter, assumes the following values: 0.389%, 1.95%, 4.28%.

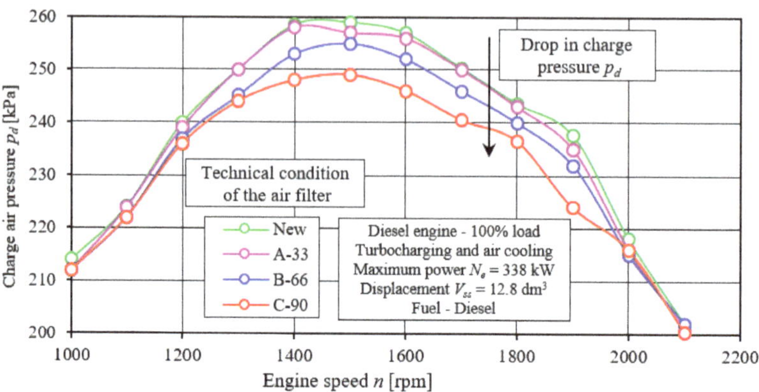

Figure 7. Charge air pressure p_d in the intake manifold of VOVLO DC13C460 engine as a function of engine speed n for different air filter states: New, A-33, B-66, C-90.

Figure 8 shows the filling characteristics $\eta_v = (n)$ as a function of the speed n of the Volvo D13C460 EURO V engine, for four (New, A-33, B-66, C-90), differing in pressure drop, technical states of the air filter. As the engine speed increases, irrespective of the technical condition of the filter (percentage obscuration of the active surface of the cartridge), the engine filling factor increases quite rapidly and, in the range $n = 1400$–1500 rpm, reaches its maximum value ($\eta_v = 2.5$), after which it systematically decreases. When the engine reaches $n = 1900$ rpm, the decrease in the engine fill factor is already steep, which is associated with a sharp drop in boost pressure. As the air filter pressure drop Δp_f increases, the filling characteristics $\eta_v = (n)$, in the speed range $n = 1000$–1900 rpm, are shifted almost parallel toward smaller values. The increase in the air filter pressure drop from a value $\Delta p_f = 0.580$ kPa (technical condition "New") to $\Delta p_f = 2.024$ kPa (C-90) causes a decrease in the maximum value of the filling factor from $\eta_v = 2.5$ do $\eta_v = 2.39$, i.e., by 4.5%. The fill factor η_v was determined from the engine speed, engine displacement, and flushing ratio. Due to the constant, angle of coverage (valve co-opening) independent from the rotational speed, the flushing coefficient was assumed to be constant at the level of 1.0. The valve co-opening angle is the angle of rotation of the crankshaft corresponding to the period during which the intake and exhaust valves of one cylinder are open at the same time.

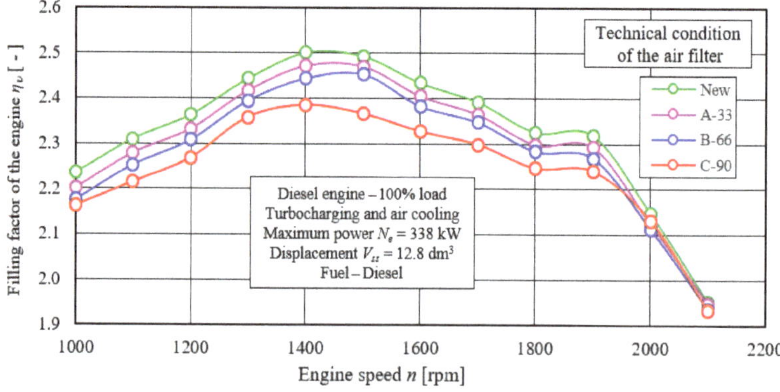

Figure 8. Filling factor η_v of VOVLO DC13C460 motor as a function of speed n for different air filter conditions: New, A-33, B-66, C-90.

The highest smoke opacity was recorded in the speed range $n = 1000$–1100 rpm (Figure 9). However, as the engine speed increased, the smoke level decreased rapidly, irrespective of the technical condition of the air filter, and in the speed range $n = 1100$–1700 rpm it remained stable, after which it increased slightly. However, the increase in air filter pressure drop does not cause significant changes in the degree of smoke opacity in relation to its permissible value, defined in the technical conditions of vehicle operation for this type of vehicles at 1.5 m^{-1} [83].

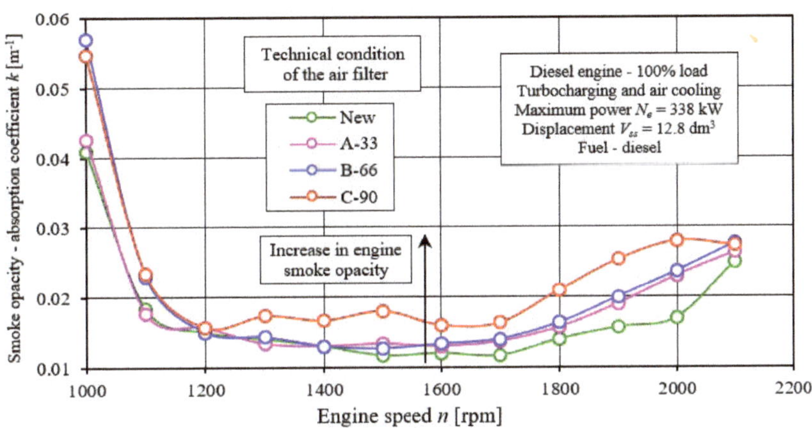

Figure 9. Smoke opacity of VOVLO DC13C460 engine—light absorption coefficient k (absorption) as a function of engine speed n for different air filter states: New, A-33, B-66, C-90.

Figure 10 shows the effect of four (New, A-33, B-66, C-90) air filter conditions, differing in pressure drop, on the characteristics of the effective power $N_e = f(n)$ of the Volvo D13C460 EURO V engine. As the rotational speed increases, the effective engine power Ne, irrespective of the technical condition of the air filter, increases sharply in value until the engine reaches a rotational speed of $n = 1400$ rpm, and then decreases slightly until a rotational speed of $n = 1900$ rpm is reached, after which it loses its value sharply. The use of an air filter Δp_f with an increasing pressure drop, according to the technical states (A-33, B-66, C-90), causes a shift of the characteristics $N_e = f(n)$ in the rotational speed range $n = 1400$–1900 rpm, almost in parallel toward the lower values of the engine power.

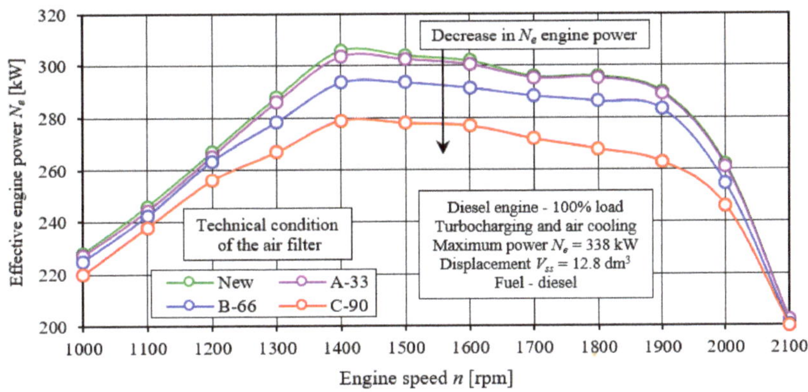

Figure 10. N_e effective power of the VOVLO DC13C460 motor as a function of speed n for different air filter condition k: New, A-33, B-66, C-90.

Figure 11 shows the relative decrease in engine power for each speed caused by successive states of the "k" air filter: A-33, B-66, C-90, in relation to the maximum power of the engine operating with air filter "New". For the engine operating with the air filter "A-33" the power changes are practically imperceptible and oscillate at around 0.5%. Further increase in the resistance of the air filter flow significantly affects the relative decrease in the effective power of N_e engine. In the case of B-66 these changes reach about 4%, while in the case of operation of the engine with the filter in the state C-90, the relative decrease in power exceeds 9%. It can be emphasised that these changes are the greatest in the medium speed range (n = 1300–1900 rpm), in which engines of this type are most often used, which is a very unfavourable phenomenon. The decrease in power due to air filter contamination for a tractor-trailer type vehicle has a very negative effect on its traction characteristics. As the pressure drop increases and the power decreases, the traction properties of the vehicle deteriorate, in particular: the ability to climb a hill in individual gears, the change in the maximum speed value depending on the road inclination angle and the time and distance of acceleration of the vehicle-tractor-trailer combination [84].

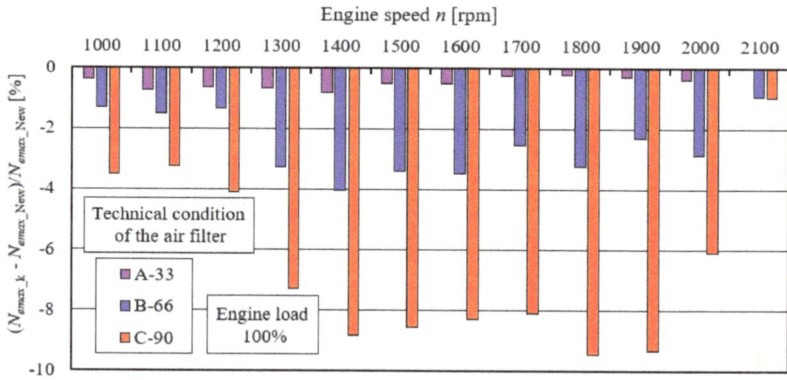

Figure 11. Relative change in engine power for different speeds due to air filter conditions k: A-33, B-66, C-90, relative to power with air filter "New".

Figure 12 shows the proportion of carbon dioxide CO_2 in the exhaust gas of the Volvo D13C460 EURO V engine as a function of the rotational speed n for four different air filter technical states k: New, A-33, B-66, C-90. As the engine speed increases, the CO_2 contribution, regardless of the technical state of the air filter, decreases (almost linearly) its value until the engine reaches the maximum speed n = 2100 rpm. The change of the engine rotational speed in the range of n = 1100–2100 rpm results in a significant decrease in the CO_2 share in the exhaust gases from 10.28% to 6.57%, which is connected to the increase in the mass of the air supplied to the cylinders.

The emission of CO_2 in the engine exhaust gas for particular rotational speeds caused by the technical states of the air filter k: New, A-33, B-66, C-90 is shown in Figure 13, while Figure 14 shows the relative change in the proportion of CO_2 in the engine exhaust for individual speeds caused by the technical states of the air filter k: A-33, B-66, C-90, relative to the emissions obtained when the engine was operated with air filter "New".

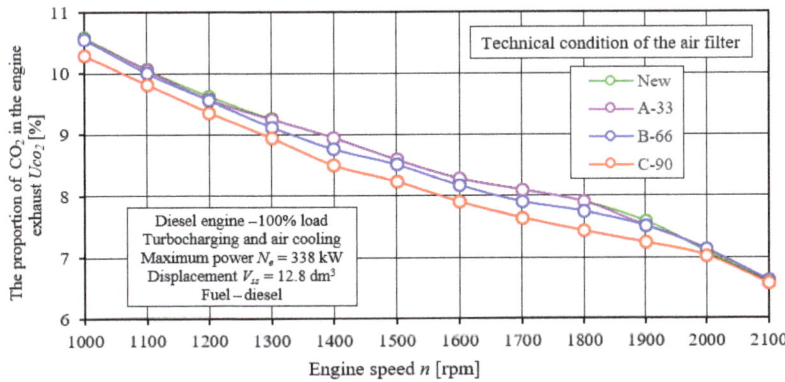

Figure 12. CO_2 contribution to the exhaust of the VOVLO DC13C460 engine as a function of engine speed n for different air filter states k: New, A-33, B-66, C-90.

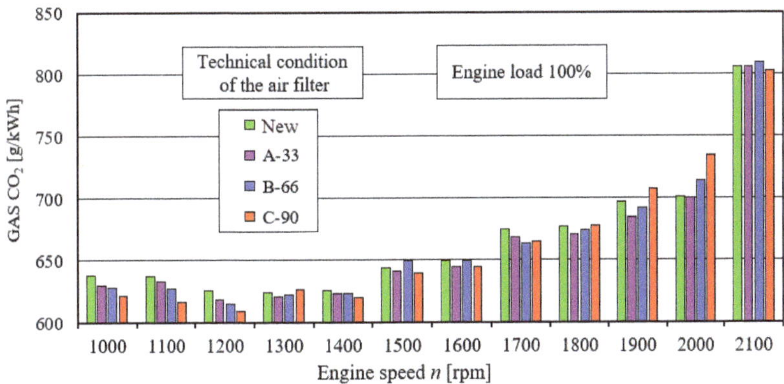

Figure 13. CO_2 emissions in engine exhaust for different engine speeds and successive air filter states k: New, A-33, B-66, C-90.

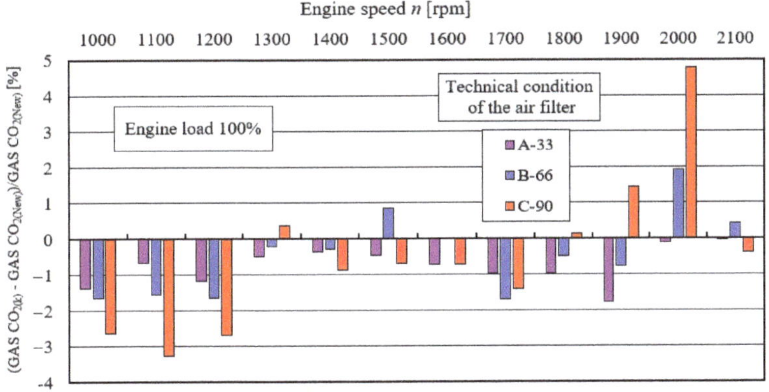

Figure 14. Relative change in the proportion of CO_2 in the engine exhaust gas for different speeds and successive technical states of the air filter k: A-33, B-66, C-90, in relation to the emissions obtained during the operation of the engine with the air filter "New".

Based on the information in Figures 12–14, it can be concluded that air filter pressure drop (filter cartridge fouling) is quite important for CO_2 emissions. In the low-speed range of 1000–1200 rpm, the deterioration of the filter element causes a reduction in CO_2 emissions. This is due to a reduction in the amount of fuel fed to the cylinder by the ECU engine control system, due to a reduction in boost pressure.

As the engine speed increases above 1200 rpm, the effect of filter condition on CO_2 emissions is reduced. This is a result of the engine controller adjusting the fuel delivery to the current conditions in the engine intake system. The relative changes in CO_2 emissions do not exceed 1%. At high engine speeds $n = 2000$–2100 rpm a negative effect of the technical condition of the air filter on the CO_2 emissions can be observed. This phenomenon is a result of increasing damping of the air flow on the filter baffle. At high engine speeds, the engine control system tries to ensure the required charge pressure by changing the dose and angle of the fuel feed in such a way to generate a larger amount of exhaust gases necessary to increase the turbocharger rotational speed and consequently increase the charge air pressure. Such action is connected with the necessity of delivering a larger amount of fuel with the same mass of air to the cylinder, which results in increased CO_2 emission. This phenomenon is unfavourable from the point of view of atmosphere pollution; however, in the range of these rotational speeds, the engine does not run very often.

In the case of operation of the engine with the air filter in the technical condition A-33 (33% obscuration of the active surface of the cartridge), no significant differences were found (about 1.25%) in the change of the share of CO_2 in the exhaust gases. Increasing the pressure drop of the air filter (B-66, C-90) already causes a significant decrease in the proportion of CO_2 in the exhaust, but there is not such a significant decrease in CO_2 emissions (Figure 14).

This can be explained by the fact that the increase in air filter pressure drop causes a decrease in the air flow supplied to the engine Q_s and the boost pressure. Lower charge pressure is treated by the engine ECU as a signal to reduce (correct) the power generated by the engine by delivering less fuel, which is reflected in a decrease in hourly fuel consumption and power. The reduction in CO_2 concentration in the exhaust gases is correlated with the reduction in useful power resulting in small changes in CO_2 emissions. On the basis of the results presented in Figures 13 and 14, it can be concluded that the increase in the pressure drop in the inlet system of a modern ZS truck engine has no significant effect on the CO_2 emission into the atmosphere.

Figure 15 shows the proportion of carbon monoxide CO in the exhaust gases of the Volvo D13C460 EURO V engine as a function of rotational speed n for four air filters differing in pressure drop (New, A-33, B-66, C-90). With increasing engine rotational speed, the concentration of carbon monoxide CO, regardless of the technical condition of the air filter, decreases its value until the engine reaches the maximum rotational speed $n = 2100$ rpm. The greatest changes (decrease) were registered in rotational speeds in the range of $n = 1000$–1200 rpm. The high concentration of carbon monoxide in the range of $n = 1000$–1100 rpm results from the low filling ratio of the engine (Figure 8), and the high dose of fuel fed to the cylinder resulting from the necessity to generate the necessary effective engine power required to overcome the vehicle motion resistance during the start-up and hill climbing.

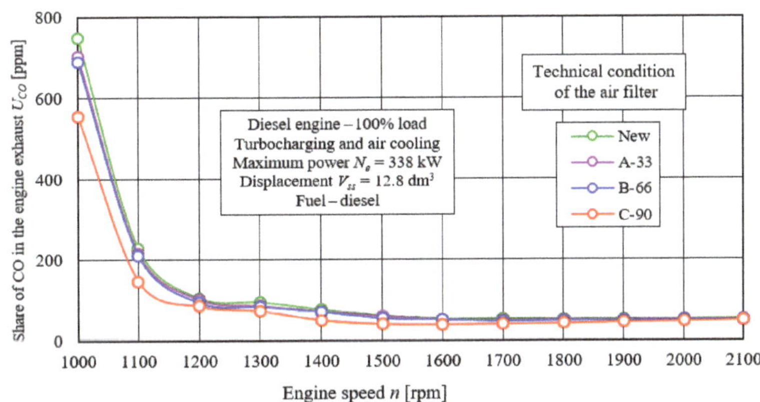

Figure 15. Contribution of CO in the exhaust of the VOVLO DC13C460 engine as a function of engine speed n for different states of the air filter k: New, A-33, B-66, C-90.

Figure 16 shows the specific CO in the engine exhaust for each speed caused by air filter states k: New, A-33, B-66, C-90. The relative change in the proportion of CO in the engine exhaust for individual engine speeds is caused by air filter conditions k: A-33, B.-66, C-90, in relation to the emission obtained during operation of the engine with air filter "New" is shown in Figure 16.

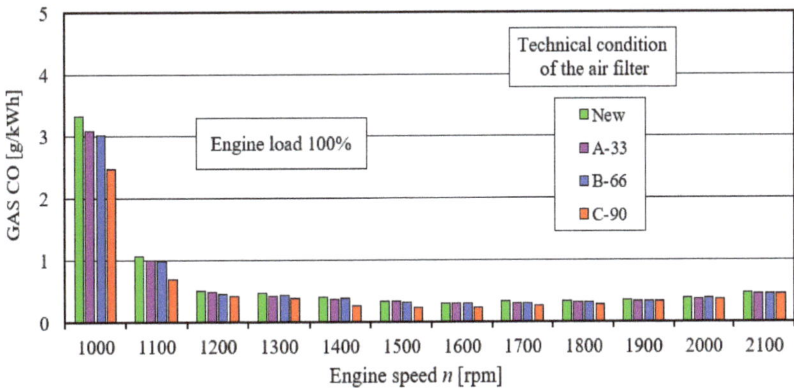

Figure 16. Unit CO emissions in engine exhaust for different speeds due to air filter conditions k: New, A-33, B-66, C-90.

When analysing the influence of the filter condition on CO emission, it should also be remembered that the measured values in the medium and high-speed range are very small—at the level of a dozen or so ppm, i.e., this is the measuring range of the analyser burdened with a large measurement uncertainty. Therefore, the nature of the changes should be interpreted qualitatively rather than strictly quantitatively.

Based on the information provided in Figures 15–17, it can be concluded that the condition of the air filter (pressure drop) is important for carbon monoxide—CO emissions. In the low to medium speed range of 1000–1700 rpm, increasing the pressure drop of the filter element results in a decrease in CO emissions. This is related to a decrease in the dose of fuel fed to the engine cylinders by the engine control system ECU, as a result of reducing the boost pressure and in an effort to reduce the increase in emission of toxic components of the exhaust gases.

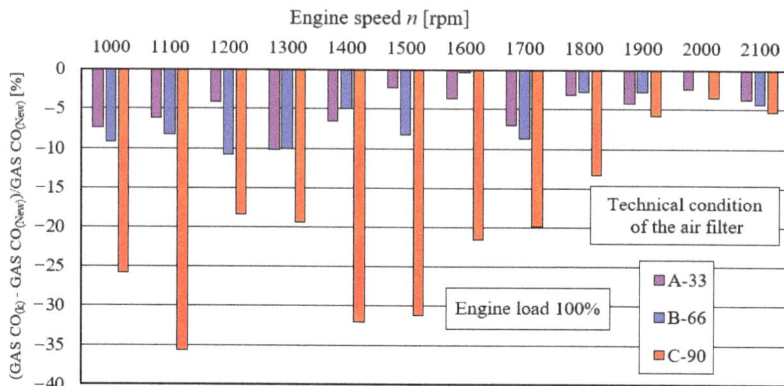

Figure 17. Relative change in the proportion of CO in the engine exhaust gas for different rotational speeds and successive technical states of the air filter k: A-33, B-66, C-90, in relation to the emission obtained during engine operation with the air filter "New".

As the engine speed increases above 1600 rpm, the effect of air filter pressure drop on CO emissions is reduced. This is a result of actions resulting from the adjustment of the fuel dose by the engine controller to the currently prevailing conditions in the engine intake system—boost pressure. This problem is described in more detail during the analysis of the influence of the air filter pressure drop on the CO_2 emission.

In the case of operation of an engine with an air filter in the A-33 state (33% of the active surface of the filter cartridge is covered), the changes in the CO emission amount from 3 to 10%, depending on the rotational speed. Increasing the air filter pressure drop (B-66, C-90) already results in a significant decrease in CO concentration in the exhaust gases; moreover, there is a significant decrease in the CO emission—Figure 17. This phenomenon is similar to the one described for the CO_2 emission.

Figure 18 shows, for four air filters differing in pressure drop (New, A-33, B-66, C-90), NO nitric oxide concentration, and Figure 19 shows GASNO nitric oxide emissions as a function of speed n of the Volvo D13C460 EURO V engine. The relative variation of the NO contribution to the engine exhaust for different speeds due to air filter conditions k: A-33, B-66, C-90, in relation to the emission obtained during engine operation with air filter "New" is shown in Figure 20.

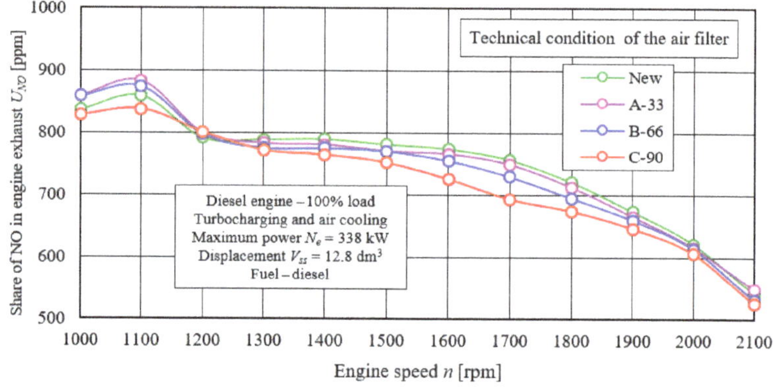

Figure 18. Contribution of NO in the exhaust of the VOVLO DC13C460 engine as a function of engine speed n for different states of the air filter k: New, A-33, B-66, C-90.

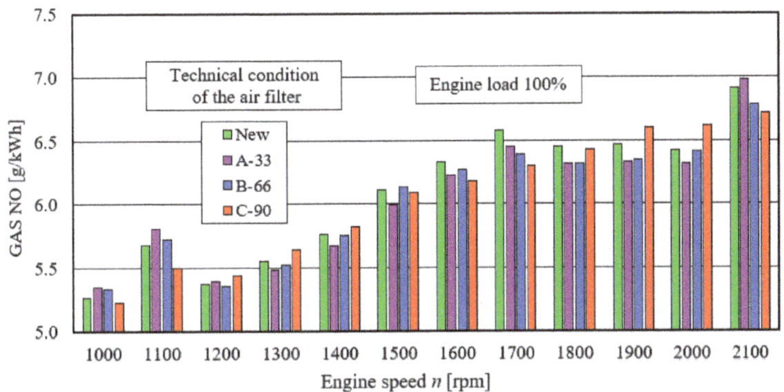

Figure 19. Unit NO emissions in engine exhaust for different engine speeds and successive air filter states *k*: New, A-33, B-66, C-90.

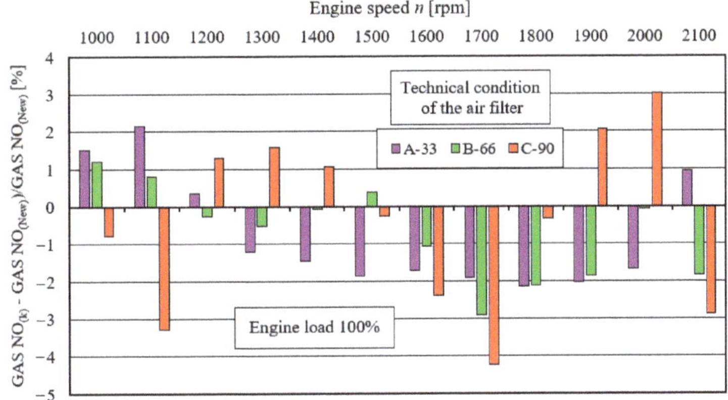

Figure 20. Relative change in the proportion of NO in the engine exhaust gas for different speeds and successive states of the air filter *k*: A-33, B-66, C-90, in relation to the emissions obtained when the engine was operated with air filter "New".

In the low-speed range, the NO concentration remains high at 830–880 ppm regardless of the filter condition. This phenomenon is a result of low boost pressure (Figure 7), low air filling of the engine (Figure 8) and quite a high fuel dose necessary to generate the set torque. Satisfying the above engine operating conditions results in increased exhaust temperature, which promotes the formation of oxides of nitrogen—Figures 18–20. Increasing the engine speed increases the boost pressure and consequently increases the amount of oxygen in the fuel–air mixture, which results in a decreased combustion temperature. The reduction in combustion temperature results in a reduction in the concentration of NO (Figure 18), NO_2 (Figure 21), and NO_x as the sum of NO and NO_2 in the exhaust gas while increasing the generated useful power. The rate of increase in the generated effective power is greater than the decrease in the concentration of NO and NO_2 in the exhaust gas, resulting in a decrease in NO (Figure 19), NO_2 (Figure 22), and NO_x as the sum of NO and NO_2.

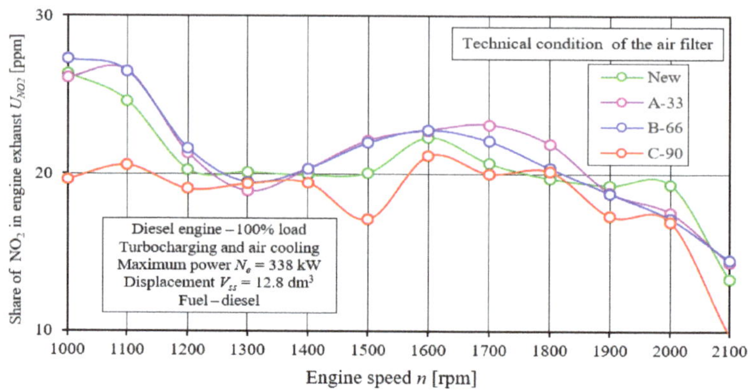

Figure 21. Contribution of NO_2 in the exhaust of VOVLO DC13C460 engine as a function of engine speed n for different states of air filter k: New, A-33, B-66, C-90.

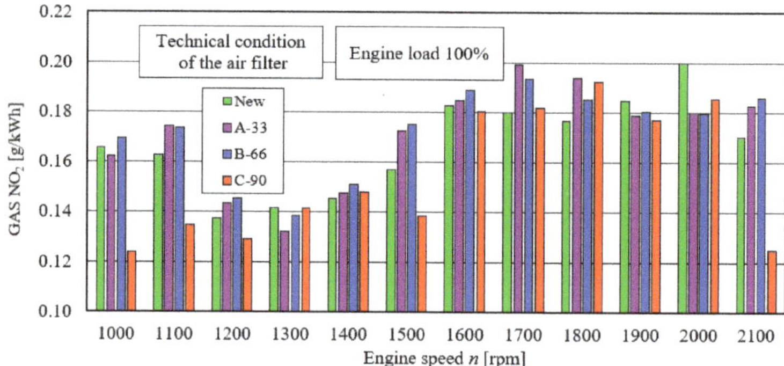

Figure 22. Unit NO_2 emissions in engine exhaust for different engine speeds and successive air filter conditions k: New, A-33, B-66, C-90.

Based on the information in Figures 19 and 22, it can be concluded that the condition of the air filter has no significant effect on NO, NO_2, NOx emissions. The changes in NO and NO_x emissions range from +2 to −3%, depending on the engine operating point (speed) and do not have a clearly identified nature of change. Changes in NO_2 emissions are slightly larger and range from +10 to −25%; however, NO_2 emissions are negligible relative to NO emissions, so its variations do not significantly affect total NO_x emissions (Figure 23).

In the low-speed range of 1000–1100 rpm, the deterioration of the filter element (increase in air filter pressure drop) results in reduced NO_x emissions. This is the result of a reduction in fuel delivery to the cylinder by the ECU engine management system, due to a reduction in boost pressure. In the mid-range, and as engine speed increases above 1200 rpm, the effect of air filter condition (effect of pressure drop) on NO_x emissions decreases. The relative changes in NO_x emissions do not exceed 1%.

When the engine was operated with the air filter being in the condition A-33 and B-66 (33% and 66% of the filter cartridge active area obscured), no significant differences (about 2.5%) were found in the change of NO_x concentration and emission. Further increasing the air filter pressure drop (C-90) only causes a significant decrease in the NO_x concentration in the exhaust gas; however, there is not such a significant decrease in NO_x emissions.

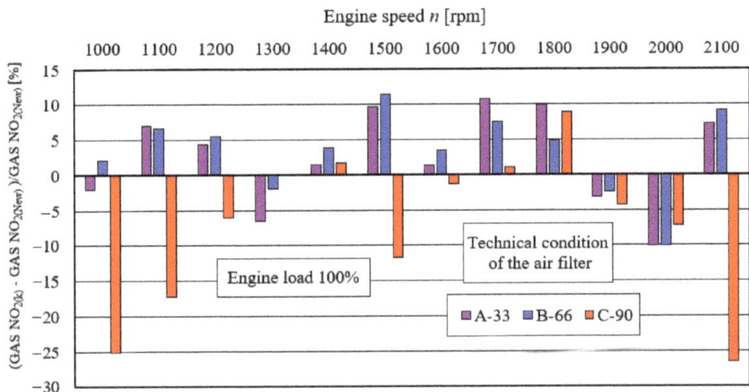

Figure 23. Relative change in the proportion of NO_2 in the engine exhaust for different speeds and successive states of the air filter k: A-33, B-66, C-90, in relation to the emissions obtained when the engine was operated with air filter "New".

Based on the results in Figures 24–26, it can be concluded that the increase in pressure drop in the intake system of a modern truck Diesel engine has no significant effect on NO_x emissions to the atmosphere.

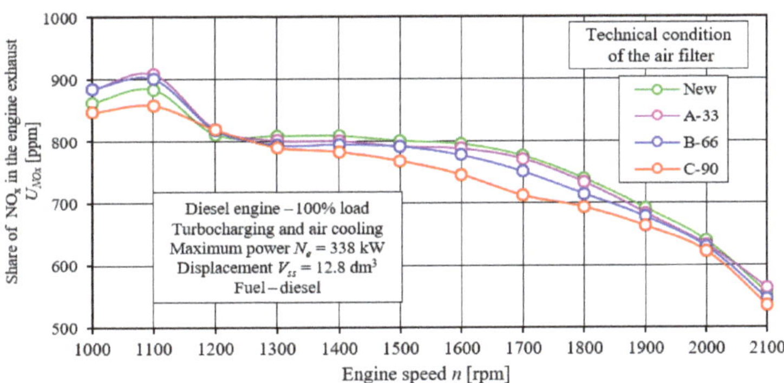

Figure 24. Contribution of NO_x in the exhaust of the VOVLO DC13C460 engine as a function of engine speed n for different air filter states k: New, A-33, B-66, C-90.

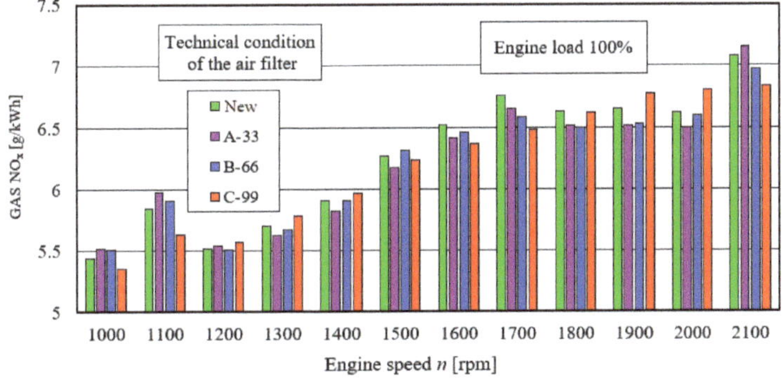

Figure 25. Unit NO_x emissions in engine exhaust for different engine speeds and successive air filter condition k: New, A-33, B-66, C-90.

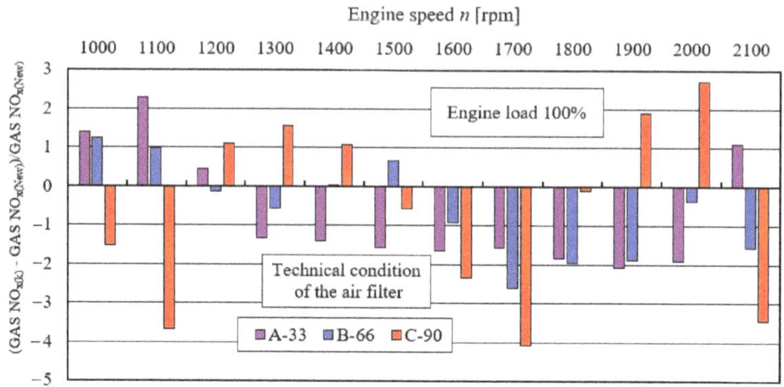

Figure 26. Relative change in the proportion of NO_x in the engine exhaust for different speeds and successive states of the air filter k: A-33, B-66, C-90, in relation to the emissions obtained when the engine was operated with the "New" air filter.

Figure 27 shows the exhaust gas temperature t_s as a function of engine speed n of a Volvo D13C460 EURO V engine for four different air filter technical states (New, A-33, B-66, C-90). As the engine speed increases, the temperature of t_s, exhaust gas, irrespective of the technical condition of the air filter, slowly but systematically (almost linearly) decreases its value until the engine reaches the rotational speed $n = 1800$ rpm, after which it increases slightly and then decreases sharply. Operation of the engine with an air filter with increasing pressure drop Δp_f, according to the technical states (A-33, B-66, C-90), causes a shift of the characteristics $t_s = f(n)$ almost in parallel toward lower values of the engine temperature. Technical states A-33 and B-66 do not cause significant changes in exhaust gas temperature. Operation of the engine with air filter with pressure drop $\Delta p_f = 2.024$ kPa (C-90) causes an already significant, drop of about 20 °C in exhaust temperature t_s in comparison with the operation of the engine with "New" filter. This phenomenon is connected with changes in boost pressure resulting from changes in pressure drop in the inlet system. Reducing the boost pressure causes a decrease in the pressure and temperature of the end of combustion, which results in a decrease in NO_x contributions (Figure 24). In a turbocharged compression ignition engine, the combustion is close to complete and total, reducing the boost pressure causes the ECU to reduce the maximum dose of fuel fed to the cylinder during the working stroke, which consequently contributes to a decrease in the combustion temperature.

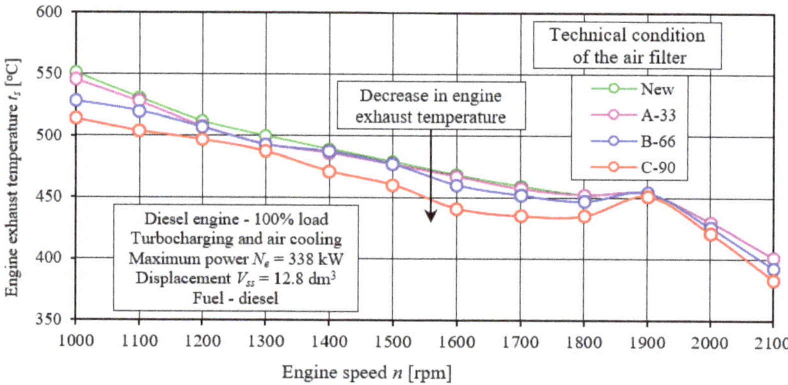

Figure 27. Exhaust gas temperature t_s at the turbine exit of the turbocharging unit of the VOVLO DC13C460 engine as a function of engine speed n for different air filter states: New, A-33, B-66, C-90.

Figure 28 shows the HC hydrocarbon concentration as a function of speed n of the Volvo D13C460 EURO V engine for four air filters differing in pressure drop (New, A-33, B-66, C-90). When analysing the effect of filter condition on HC emissions, we were limited to assessing the effect of the nature of HC concentration changes as a function of air filter condition. This action was dictated by the very low HC concentrations of several ppm which, given the high uncertainty of the analyser measurements in the range up to 25 ppm, could result in incorrect inferences.

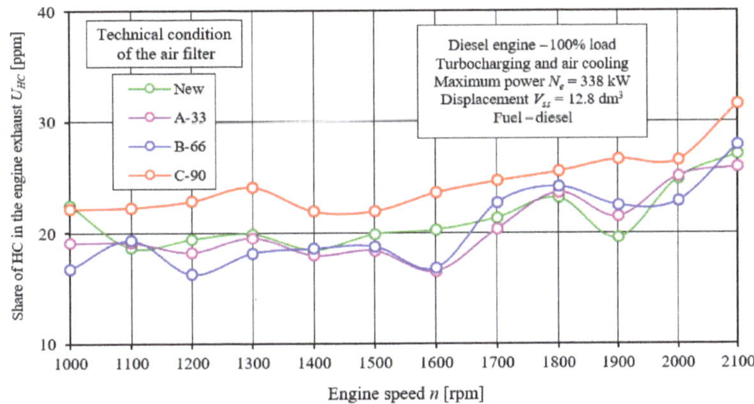

Figure 28. Contribution of HC in the exhaust of VOVLO DC13C460 engine as a function of engine speed *n* for different states of air filter *k*: New, A-33, B-66, C-90.

At the same time, the study shows that the condition of the air filter (pressure drop) is not significant for HC concentration. A significant increase in pressure drop results in an increase in HC concentration by a few ppm. The determined average HC emission for the engine operating with the air filter in the "New" state (Δp_f = 0.58 kPa) was 0.0547 g/kWh, while for the filter in the C-90 state (Δp_f = 2.024 kPa) it was 0.0653 g/kWh, respectively. These values are small enough that they do not significantly affect the environmental performance of the tested engine.

4. Conclusions

1. In the available literature there are no results of research on the influence of the inlet system flow resistance on the emission of toxic components of exhaust gases from a modern Diesel engine used for driving trucks (truck tractors) currently travelling on the roads and constituting the basic means of transporting goods.

2. The conducted research has shown that the increase in flow resistance in the inlet system of the modern ZS truck engine has no significant effect on the NO_x emission into the atmosphere, and does not cause any significant changes in the degree of smoke opacity of exhaust gases in relation to its acceptable value specified in the technical conditions for this type of vehicle.

3. The observed effect of the increase in air filter flow resistance of the modern ZS truck engine is a decrease in air demand by the engine, decrease in the boost pressure and, as a result, the decrease in the filling ratio η_v which, in connection with the fuel dose reduction, causes the decrease in the engine power. An increase in air filter flow resistance by an average of 1 kPa results in a decrease in engine power of over 6%. In the conditions of vehicle use, this is associated with a reduction in the ability to climb a hill in individual gears and an increase in the time and distance of acceleration of the vehicle-tractor-trailer combination. As the flow resistance increases, the emissions of exhaust components change: NO, NO_2, NO_x, CO, HC and CO_2. These values are so small that they do not significantly affect the ecological properties of the tested engine.

4. The results obtained show the effect of the air filter flow resistance of a modern truck engine on its performance, and in particular on the changes in the emission of the individual components of exhaust gases. It is advisable to continue the work; the final effect of which should be a determination of the maximum permissible flow resistance, the exceeding of which should eliminate the vehicle from further operation because of deterioration of the economic, energetic, ecological and traction properties of the vehicle.

5. It is advisable to extend the presented research with the analysis of the influence of increased flow resistance in the air filtration system on the change of traction properties of the vehicle equipped with this type of inlet system.

Author Contributions: Conceptualisation, T.D. and M.K.; methodology, M.K.; software, T.D.; validation, T.D. and M.K.; formal analysis, T.D. and M.K.; investigation, T.D. and M.K.; data curation, T.D. and M.K.; writing—original draft preparation, T.D. and M.K.; writing—review and editing, T.D.; visualisation, T.D. and M.K.; supervision, T.D. and M.K.; project administration, M.K.; funding acquisition, T.D. and M.K. All authors have read and agreed to the published version of the manuscript.

Funding: The article was written as part of the implementation of the university research grant No. UGB 758/2022.

Institutional Review Board Statement: Not applicable.

Informed Consent Statement: Not applicable.

Data Availability Statement: Data is contained within the article.

Conflicts of Interest: The authors declare no conflict of interest.

References

1. Kalghatgi, G. Is it really the end of internal combustion engines and petroleum in transport? *Appl. Energy* **2018**, *225*, 965–974. [CrossRef]

2. Soudagar, M.E.M.; Nik-Ghazali, N.-N.; Abul Kalam, M.; Badruddin, I.A.; Banapurmath, N.R.; Akram, N. The effect of nano-additives in diesel-biodiesel fuel blends: A comprehensive review on stability, engine performance and emission characteristics. *Energy Convers. Manag.* **2018**, *178*, 146–177. [CrossRef]

3. Chłopek, Z.; Lasocki, J.; Strzałkowska, K.; Zakrzewska, D. Impact of pollutant emission from motor vehicles on air quality in a city agglomeration. *Combust. Engines* **2019**, *177*, 7–11. [CrossRef]

4. Kuppili, S.K.; Dheeraj Alshetty, V.; Diya, M.; Shiva Nagendra, S.M.; Ramadurai, G.; Ramesh, A.; De Vito, L. Characteristics of real-world gaseous exhaust emissions from cars in heterogeneous traffic conditions. *Transp. Res. Part D Transp. Environ.* **2021**, *95*, 102855. [CrossRef]

5. Lal, S.; Patil, R.S. Monitoring of atmospheric behaviour of nox from vehicular traffic. *Environ. Monit. Assess.* **2001**, *68*, 37–50. [CrossRef]

6. Li, N.; Xia, T.; Nel, A.E. The role of oxidative stress in ambient particulate matter-induced lung diseases and its implications in the toxicity of engineered nanoparticles. *Free Radic. Biol. Med.* **2008**, *44*, 1689–1699. [CrossRef]

7. Rokosch, U. *Układy Oczyszczania Spalin i Pokładowe Systemy Diagnostyczne Samochodów OBD*; WKŁ: Warszawa, Poland, 2007. (In Polish)

8. Merkisz, J. *Ekologiczne Problemy Silników Spalinowych, T.1. Wydawnictwo Politechniki Poznańskiej*; Poznań University of Technology: Poznan, Poland, 1998. (In Polish)

9. Worek, J.; Badura, X.; Białas, A.; Chwiej, J.; Kawoń, K.; Styszko, K. Pollution from Transport: Detection of Tyre Particles in Environmental Samples. *Energies* **2022**, *15*, 2816. [CrossRef]

10. Wagner, S.; Hüffer, T.; Klöckner, P.; Wehrhahn, M.; Hofmann, T.; Reemtsma, T. Tire wear particles in the aquatic environment—A review on generation, analysis, occurrence, fate and effects. *Water Res.* **2018**, *139*, 83–100. [CrossRef]

11. Panko, J.; Hitchcock, K.; Fuller, G.; Green, D. Evaluation of Tire Wear Contribution to PM2.5 in Urban Environments. *Atmosphere* **2019**, *10*, 99. [CrossRef]

12. Bänsch-Baltruschat, B.; Kocher, B.; Stock, F.; Reifferscheid, G. Tyre and road wear particles (TRWP)—A review of generation, properties, emissions, human health risk, ecotoxicity, and fate in the environment. *Sci. Total Environ.* **2020**, *733*, 137823. [CrossRef]

13. Prenner, S.; Allesch, A.; Staudner, M.; Rexeis, M.; Schwingshackl, M.; Huber-Humer, M. Static modelling of the material flows of micro- and nanoplastic particles caused by the use of vehicle tyres. *Environ. Pollut.* **2021**, *290*, 118102. [CrossRef]

14. Leifheit, E.F.; Kissener, H.L.; Faltin, E.; Ryo, M.; Rillig, M.C. Tire abrasion particles negatively affect plant growth even at low concentrations and alter soil biogeochemical cycling. *Soil Ecol. Lett.* **2021**. [CrossRef]

15. Chasapidis, L.; Grigoratos, T.; Zygogianni, A.; Tsakis, A.; Konstandopoulos, A.G. Study of Brake Wear Particle Emissions of a Minivan on a Chassis Dynamometer. *Emiss. Control Sci. Technol.* **2018**, *4*, 271–278. [CrossRef]
16. Hagino, H.; Oyama, M.; Sasaki, S. Airborne brake wear particle emission due to braking and accelerating. *Wear* **2015**, *334–335*, 44–48. [CrossRef]
17. Farwick zum Hagen, F.H.; Mathissen, M.; Grabiec, T.; Hennicke, T.; Rettig, M.; Grochowicz, J.; Benter, T. On-road vehicle measurements of brake wear particle emissions. *Atmos. Environ.* **2019**, *217*, 116943. [CrossRef]
18. Agarwal, A.K.; Singh, A.P.; Maurya, R.K.; Chandra Shukla, P.; Dhar, A.; Srivastava, D.K. Combustion characteristics of a common rail direct injection engine using different fuel injection strategies. *Int. J. Therm. Sci.* **2018**, *134*, 475–484. [CrossRef]
19. Gupta, T.; Kothari, A.; Srivastava, D.K.; Agarwal, A.K. Measurement of number and size distribution of particles emitted from a mid-sized transportation multipoint port fuel injection gasoline engine. *Fuel* **2010**, *89*, 2230–2233. [CrossRef]
20. Kittelson, D.B. Engines and nanoparticles. *J. Aerosol. Sci.* **1998**, *29*, 575–588. [CrossRef]
21. Kang, S.; Lee, S.; Bae, C. Effects of Multi-Stage Split Injection on Efficiency and Emissions of Light-Duty Diesel Engine. *Energies* **2022**, *15*, 2219. [CrossRef]
22. Fayad, M.A. Effect of fuel injection strategy on combustion performance and NO_x/smoke trade-off under a range of operating conditions for a heavy-duty DI diesel engine. *SN Appl. Sci.* **2019**, *1*, 1088. [CrossRef]
23. Ewphun, P.P.; Otake, M.; Nagasawa, T.; Kosaka, H.; Sato, S. Investigation on effect of offset orifice nozzle under multi pulse ultrahigh pressure injection and PPC combustion conditions. *Int. J. Automot. Eng.* **2020**, *11*, 20204127. [CrossRef]
24. Dond, D.; Gulhane, N. Effect of Injection System Parameters on Performance and Emission Characteristics of a Small Single Cylinder Diesel Engine. *Int. J. Automot. Mech. Eng.* **2021**, *18*, 8790. [CrossRef]
25. Noroozi, M.J.; Seddiq, M. Effects of Post-Injection Characteristics on the Combustion, Emission, and Performance in a Diesel-Syngas Reactivity Controlled Compression Ignition Engine. *Int. J. Automot. Mech. Eng.* **2021**, *18*, 9006–9021. [CrossRef]
26. Hotta, Y.; Inayoshi, M.; Nakakita, K.; Fujiwara, K.; Sakata, I. *Achieving Lower Exhaust Emissions and Better Performance in an HSDI Diesel Engine with Multiple Injection*; SAE Technical Paper 2005-01-0928; SAE International: Warrendale, PA, USA, 2005.
27. Kumar, K.S.; Raj, R.T.K. Effect of Fuel Injection Timing and Elevated Intake Air Temperature on the Combustion and Emission Characteristics of Dual Fuel Operated Diesel Engine. *Procedia Eng.* **2013**, *64*, 1191–1198. [CrossRef]
28. Fayad, M.A.; Fernández-Rodríguez, D.; Herreros, J.M.; Lapuerta, M.; Tsolakis, A. Interactions between aftertreatment systems architecture and combustion of oxygenated fuels for improved low temperature catalysts activity. *Fuel* **2018**, *229*, 189–197. [CrossRef]
29. Lakkireddy, V.R.; Mohammed, H.; Johnson, J.H. *The Effect of a Diesel Oxidation Catalyst and a Catalyzed Particulate Filter on Particle Size Distribution from a Heavy Duty Diesel Engine*; SAE Technical Paper 2006:0148–7191; SAE International: Warrendale, PA, USA, 2006.
30. Chen, H.-Y.; Mulla, S.; Weigert, E.; Camm, K.; Ballinger, T.; Cox, J.; Blakeman, P. Cold Start Concept (CSC™): A Novel Catalyst for Cold Start Emission Control. *SAE Int. J. Fuels Lubr.* **2013**, *6*, 372–381. [CrossRef]
31. Thangaraja, J.; Kannan, C. Effect of exhaust gas recirculation on advanced diesel combustion and alternate fuels—A review. *Appl. Energy* **2016**, *180*, 169–184. [CrossRef]
32. Mossa, M.A.A.; Hairuddin, A.A.; Nuraini, A.A.; Zulkiple, J.; Hasyuzariza, M.T. Effects of Hot Exhaust Gas Recirculation (EGR) on the Emission and Performance of a Single-Cylinder Diesel Engine. *Int. J. Automot. Mech. Eng.* **2019**, *16*, 6660–6674. [CrossRef]
33. García, A.; Monsalve-Serrano, J.; Villalta, D.; Lago Sari, R. Performance of a conventional diesel aftertreatment system used in a medium-duty multi-cylinder dual-mode dual-fuel engine. *Energy Convers. Manag.* **2019**, *184*, 327–337. [CrossRef]
34. Geng, P.; Cao, E.; Tan, Q.; Wei, L. Effects of alternative fuels on the combustion characteristics and emission products from diesel engines: A review. *Renew. Sustain. Energy Rev.* **2017**, *71*, 523–534. [CrossRef]
35. Zhang, B.; Zuo, H.; Huang, Z.; Tan, J.; Zuo, Q. Endpoint forecast of different diesel-biodiesel soot filtration process in diesel particulate filters considering ash deposition. *Fuel* **2020**, *272*, 117678. [CrossRef]
36. Paykani, A.; Chehrmonavari, H.; Tsolakis, A.; Alger, T.; Northrop, W.F.; Reitz, R.D. Synthesis gas as a fuel for internal combustion engines in transportation. *Prog. Energy Combust. Sci.* **2022**, *90*, 100995. [CrossRef]
37. Hoseini, S.S.; Najafi, G.; Ghobadian, B.; Mamat, R.; Sidik, N.A.C.; Azmi, W.H. The effect of combustion management on diesel engine emissions fueled with biodiesel-diesel blends. *Renew. Sustain. Energy Rev.* **2017**, *73*, 307–331. [CrossRef]
38. Knothe, G.; Razon, L.F. Biodiesel fuels. *Prog. Energy Combust. Sci.* **2017**, *58*, 36–59. [CrossRef]
39. Kalghatgi, G.; Levinsky, H.; Colket, M. Future transportation fuels. *Prog. Energy Combust. Sci.* **2018**, *69*, 103–105. [CrossRef]
40. Othman, M.F.; Adam, A.; Najafi, G.; Mamat, R. Green fuel as alternative fuel for diesel engine: A review. *Renew. Sustain. Energy Rev.* **2017**, *80*, 694–709. [CrossRef]
41. How, C.B.; Taib, N.M.; Mansor, M.R.A. Performance and Exhaust Gas Emission of Biodiesel Fuel with Palm Oil Based Additive in Direct Injection Compression Ignition Engine. *Int. J. Automot. Mech. Eng.* **2019**, *16*, 6173–6187. [CrossRef]
42. Daukšys, V.; Račkus, M.; Zamiatina, N. Energy Efficiency Improvement Adding Various Amounts of CNG in the Naturally Aspirated Compression Ignition Engine. *Procedia Eng.* **2017**, *187*, 222–228. [CrossRef]
43. Karczewski, M.; Chojnowski, J.; Szamrej, G. A Review of Low-CO_2 Emission Fuels for a Dual-Fuel RCCI Engine. *Energies* **2021**, *14*, 5067. [CrossRef]
44. Kukharonak, H.; Ivashko, V.; Pukalskas, S.; Rimkus, A.; Matijošius, J. Operation of a Spark-ignition Engine on Mixtures of Petrol and N-butanol. *Procedia Eng.* **2017**, *187*, 588–598. [CrossRef]

45. Freitas, E.S.d.C.; Guarieiro, L.L.N.; da Silva, M.V.I.; Amparo, K.K.d.S.; Machado, B.A.S.; Guerreiro, E.T.d.A.; de Jesus, J.F.C.; Torres, E.A. Emission and Performance Evaluation of a Diesel Engine Using Addition of Ethanol to Diesel/Biodiesel Fuel Blend. *Energies* **2022**, *15*, 2988. [CrossRef]

46. Yang, W.; Zhang, L.; Ma, F.; Xu, D.; Ji, W.; Zhao, Y.; Zhang, J. Simulation about the Effect of the Height-to-Stroke Ratios of Ports on Power and Emissions in an OP2S Engine Using Diesel/Methanol Blends. *Energies* **2022**, *15*, 2942. [CrossRef]

47. Andrych-Zalewska, M.; Chlopek, Z.; Merkisz, J.; Pielecha, J. Research on Exhaust Emissions in Dynamic Operating States of a Combustion Engine in a Real Driving Emissions Test. *Energies* **2021**, *14*, 5684. [CrossRef]

48. Pielecha, J.; Skobiej, K.; Kubiak, P.; Wozniak, M.; Siczek, K. Exhaust Emissions from Plug-in and HEV Vehicles in Type-Approval Tests and Real Driving Cycles. *Energies* **2022**, *15*, 2423. [CrossRef]

49. Youn, I.; Jeon, J. Combustion Performance and Low NO_x Emissions of a Dimethyl Ether Compression-Ignition Engine at High Injection Pressure and High Exhaust Gas Recirculation Rate. *Energies* **2022**, *15*, 1912. [CrossRef]

50. Karczewski, M.; Szczęch, L. Influence of the F-34 unified battlefield fuel with bio components on usable parameters of the IC engine. *Maint. Reliab.* **2016**, *18*, 358–366. [CrossRef]

51. Agarwal, A.K.; Gupta, J.G.; Dhar, A. Potential and challenges for large-scale application of biodiesel in automotive sector. *Prog. Energy Combust. Sci.* **2017**, *61*, 113–149. [CrossRef]

52. Lanjekar, R.D.; Deshmukh, D. A review of the effect of the combustion of biodiesel on NO_x emission, oxidative stability and cold flow properties. *Renew. Sustain. Energy Rev.* **2016**, *54*, 1401–1411. [CrossRef]

53. Kegl, B. Effects of biodiesel on emissions of a bus diesel engine. *Bioresour. Technol.* **2008**, *99*, 863–873. [CrossRef]

54. Khond, V.W.; Kriplani, V.M. Effect of nanofluid additives on performances and emissions of emulsified diesel and biodiesel fueled stationary CI engine: A comprehensive review. *Renew. Sustain. Energy Rev.* **2016**, *59*, 1338–1348. [CrossRef]

55. Shaafi, T.; Sairam, K.; Gopinath, A.; Kumaresan, G.; Velraj, R. Effect of dispersion of various nano additives on the performance and emission characteristics of a CI engine fueled with diesel, biodiesel and blends—A review. *Renew. Sustain. Energy Rev.* **2015**, *49*, 563–573. [CrossRef]

56. Ettefaghi, E.; Ghobadian, B.; Rashidi, A.; Najafi, G.; Khoshtaghaza, M.H.; Rashtchi, M.; Sadeghian, S. A novel bio-nano emulsion fuel based on biodegradable nanoparticles to improve diesel engines performance and reduce exhaust emissions. *Renew. Energy* **2018**, *125*, 64–72. [CrossRef]

57. Saxena, V.; Kumar, N.; Saxena, V.K. A comprehensive review on combustion and stability aspects of metal nanoparticles and its additive effect on diesel and biodiesel fueled C.I. engine. *Renew. Sustain. Energy Rev.* **2017**, *70*, 563–588. [CrossRef]

58. Damanik, N.; Ong, H.C.; Tong, C.W.; Mahlia, T.M.I.; Silitonga, A.S. A review on the engine performance and exhaust emission characteristics of diesel engines fueled with biodiesel blends. *Environ. Sci. Pollut. Res.* **2018**, *25*, 15307–15325. [CrossRef] [PubMed]

59. Kegl, T.; Kovač Kralj, A.; Kegl, B.; Kegl, M. Nanomaterials as fuel additives in diesel engines: A review of current state, opportunities, and challenges. *Prog. Energy Combust. Sci.* **2021**, *83*, 100897. [CrossRef]

60. Jaroń, A.; Borucka, A.; Sobecki, G. Assessment of the possibility of using nanomaterials as fuel additives in combustion engines. *Combust. Engines* **2022**, *189*, 103–112. [CrossRef]

61. Fayad, M.A.; AL-Ogaidi, B.R.; Abood, M.K.; AL-Salihi, H.A. Influence of post-injection strategies and CeO_2 nanoparticles additives in the C30D blends and diesel on engine performance, NO_X emissions, and PM characteristics in diesel engine. *Part. Sci. Technol.* **2021**, 1–14. [CrossRef]

62. Tkaczyk, M.; Sroka, Z.J.; Krakowian, K.; Wlostowski, R. Experimental Study of the Effect of Fuel Catalytic Additive on Specific Fuel Consumption and Exhaust Emissions in Diesel Engine. *Energies* **2021**, *14*, 54. [CrossRef]

63. Adzmi, M.A.; Abdullah, A.; Abdullah, Z.; Mrwan, A.G. Effect of Al_2O_3 and SiO_2 Metal Oxide Nanoparticles Blended with POME on Combustion, Performance and Emissions Characteristics of a Diesel Engine. *Int. J. Automot. Mech. Eng.* **2019**, *16*, 6859–6873. [CrossRef]

64. Wróblewski, P.; Drożdż, W.; Lewicki, W.; Dowejko, J. Total Cost of Ownership and Its Potential Consequences for the Development of the Hydrogen Fuel Cell Powered Vehicle Market in Poland. *Energies* **2021**, *14*, 2131. [CrossRef]

65. Dziubak, T. Analiza wpływu oporu przepływu filtru powietrza na napełnienie tłokowego silnika spalinowego. *Zesz. Nauk. Politech. Szczec.* **1992**, *19*, 33–40.

66. Dziubak, T. Diagnostics of military vehicle engine air filters. *Bull. Mil. Univ. Technol.* **1998**, *10*, 73–89.

67. Dziubak, T.; Trawiński, G. The experimental Research of the air filter flow drag influence on the T359E engine operation parameters. *Bull. Mil. Univ. Technol.* **2001**, *4*, 135–149.

68. Dziubak, T.; Trawiński, G. The experimental assessment of air supply system modification on inlet air filtration efficiency and military vehicle engine effectiveness improvement. *J. KONES Powertrain Transp.* **2010**, *17*, 79–86.

69. Dziubak, T.; Karczewski, M. Experimental Study of the Effect of Air Filter Pressure Drop on Internal Combustion Engine Performance. *Energies* **2022**, *15*, 3285. [CrossRef]

70. Yang, A.-J.; Song, Z.-J.; Yang, J.-W.; Zhao, C.-J. Effect of Air Filter on Performance of S1110 Diesel Engine. *IOP Conf. Ser. Mater. Sci. Eng.* **2019**, *688*, 022050. [CrossRef]

71. Plotnikov, L.V.; Bernasconi, S.; Brodov, Y.M. The Effects of the Intake Pipe Configuration on Gas Exchange, and Technical and Economic Indicators of Diesel Engine with 21/21 Dimension. *Procedia Eng.* **2017**, *206*, 140–145. [CrossRef]

72. Abdullah, N.R.; Shahruddin, N.S.; Mamat, A.M.I.; Kasolang, S.; Zulkifli, A.; Mamat, R. Effects of Air Intake Pressure to the Fuel Economy and Exhaust Emissions on a Small SI Engine. *Procedia Eng.* **2013**, *68*, 278–284. [CrossRef]

73. Shannak, B.; Damseh, R.; Alhusein, M. Influence of air intake pipe on engine exhaust emission. *Forsch. Im Ing.* **2005**, *70*, 128–132. [CrossRef]
74. Thomas, J.; West, B.; Huff, S. *Effect of Air Filter Condition on Diesel Vehicle Fuel Economy*; SAE Technical Paper 2013-01-0311; SAE International: Warrendale, PA, USA, 2013. [CrossRef]
75. Bugli, N.; Dobert, S.; Flora, S. *Investigating Cleaning Procedures for OEM Engine Air Intake Filters*; SAE Technical Paper 2007-01-1431; SAE International: Warrendale, PA, USA, 2007. [CrossRef]
76. Bugli, N. *Automotive Engine Air Cleaners—Performance Trends*; SAE Technical Paper 2001-01-1356; SAE International: Warrendale, PA, USA, 2001. [CrossRef]
77. Norman, K.; Huff, S.; West, B. *Effect of Intake Air Filter Condition on Vehicle Fuel Economy*; U.S. Department of Energy (DOE) Information Bridge: Washington, DC, USA, 2009.
78. Barris, M.A. *Total Filtration^{TM}: The Influence of Filter Selection on Engine Wear, Emissions and Performance*; SAE Technical Paper: 952557; SAE International: Warrendale, PA, USA, 1995.
79. Bugli, N.; Green, G. *Performance and Benefits of Zero Maintenance Air Induction Systems*; SAE Technical Paper 2005-01-1139; SAE International: Warrendale, PA, USA, 2005. [CrossRef]
80. Dziubak, T.; Dziubak, S.D. Experimental Study of Filtration Materials Used in the Car Air Intake. *Materials* **2020**, *13*, 3498. [CrossRef]
81. Regulation No 83 of the Economic Commission for Europe of the United Nations (UN/ECE)—Uniform Provisions Concerning the Approval of Vehicles with Regard to the Emission of Pollutants According to Engine Fuel Requirements, Supplement 1 to the 06 Series of Amendments—Date of Entry into Force: 23 June 2011, Official Journal of the European Union, 15 February 2012. Available online: https://op.europa.eu/en/publication-detail/-/publication/2f8f0ce5-66fb-4a38-ae68-558ae1b04a5f/language-en (accessed on 22 May 2022).
82. *PN-ISO 15550:2009*; Silniki Spalinowe Tłokowe—Określanie i Metoda Pomiaru Mocy Silnika—Wymagania Ogólne. PKN: Warszawa, Poland, 2009. (In Polish)
83. Rozporządzenie Ministra Infrastruktury z Dnia 31 Grudnia 2002 r. w Sprawie Warunków Technicznych Pojazdów Oraz Zakresu ich Niezbędnego Wyposażenia (Dz. U. z 2016 r. poz. 2022, z 2017 r. poz. 2338 oraz z 2018 r. poz. 855). Available online: file:///C:/Users/MDPI/Downloads/D20220122.pdf (accessed on 25 May 2022). (In Polish)
84. Karczewski, M.; Wieczorek, M. Assessment of the Impact of Applying a Non-Factory Dual-Fuel (Diesel/Natural Gas) Installation on the Traction Properties and Emissions of Selected Exhaust Components of a Road Semi-Trailer Truck Unit. *Energies* **2021**, *14*, 8001. [CrossRef]

Article

Influence of the Working Parameters of the Chassis Dynamometer on the Assessment of Tuning of Dual-Fuel Systems

Janusz Chojnowski and Mirosław Karczewski *

Faculty of Mechanical Engineering, Military University of Technology in Warsaw, 00-908 Warsaw, Poland; janusz.chojnowski@wat.edu.pl
* Correspondence: miroslaw.karczewski@wat.edu.pl; Tel.: +48-261-837-754

Abstract: The article presents the justification for the necessity to use chassis dynamometers in the tuning process of dual-fuel trucks. The research system used and the research methodology are presented. The research results present the approach to solving problems related to setting the technical (physical) data of the tested vehicle on the dynamometer, selection of the vehicle engine operation range, the impact of the value of the forced load on the vehicle drive axle, selection of the dyno operation mode for the expected tasks and the impact of the correctness of the selection of the scope of the analysis of data on losses in the drive system. The article shows the above-mentioned influence on the test results on the dynamometer and on the tuning results. The article closes with a conclusion detailing prospects for developing the presented results.

Keywords: chassis dynamometer; semi-truck; dual-fuel; tuning

Citation: Chojnowski, J.; Karczewski, M. Influence of the Working Parameters of the Chassis Dynamometer on the Assessment of Tuning of Dual-Fuel Systems. *Energies* **2022**, *15*, 4869. https://doi.org/10.3390/en15134869

Academic Editor: Massimiliano Renzi

Received: 31 May 2022
Accepted: 28 June 2022
Published: 2 July 2022

1. Introduction

Modern digitally controlled gas fueling systems require appropriate adjustment called calibration, i.e., proper selection of gaseous fuel injection times, depending on the rotational speed and engine load [1]. Most of the commercially offered gas installation controllers are equipped with the auto-calibration function, which allows the appropriate settings to be selected automatically [2]. It is most often performed at engine idling, which does not take into account the variable load, the operation of additional engine-driven systems and the related need to increase or decrease the dose of gaseous fuel, and a number of other factors that have a significant impact on the engine performance [3]. In most popular vehicles, the auto-calibration procedure is sufficient for the proper operation of the gas supply system; however, there are cars in which this is insufficient, and it is required to perform a more accurate calibration of the engine running on gas fuel under load [4]. In such cases, the auto-calibration serves only to prepare the engine for operation and start it in the "gas" mode [2,3,5,6]. Subsequently, a road test and continuing tuning under road conditions is suggested.

Semi-tractor engines using dual-fuel installations (pilot diesel dose starting the combustion process of a compressed gas dose in the engine cylinder) do not always have the possibility of auto-calibration, because they result from the inability (or incomplete ability) to work in the dual-fuel mode at engine idle speed (with no injection a suitably small pilot dose, e.g., in engines using unit injectors) [6]. Turbocharged diesel engines used in trucks are not able to generate the full boost pressure, and thus the maximum torque values, without an appropriate load [7]. In addition, semi-trucks, due to the specificity of their work, generate very high torque values on the wheels, which is necessary to enable the movement of heavy loads while maintaining appropriate dynamics and in changing road conditions (e.g., changing the elevation of the ground or wind direction). It is impossible or very difficult to force the engine to work in extreme conditions during road tests without a load. A reasonable solution is therefore to use a chassis dynamometer, which allows you to enforce appropriate operating conditions of the drive system and maintain them for a specific time [5,8].

A chassis dynamometer is a device that is most often used to measure the power and torque of a vehicle engine [1]. The measurement is performed indirectly by registering the torque on the rollers of the chassis dynamometer [1]. Relating the value of this parameter to the rotational speed of the rollers, and then via the drive system to the drive speed of the engine crankshaft, allows the power of the vehicle to be calculated [1]. However, these are not the only possibilities of a chassis dynamometer [9]. It also allows you to simulate the actual load conditions that occur on the wheels of the vehicle while driving [1,6]. This allows, for example, fuel consumption to be measured, exhaust emissions to be checked while driving, or the behavior of various engine systems to be tested under load [5,8]. It is also possible to measure the flexibility of the car's engine within a certain speed range, to determine the correctness of the speedometer and tachograph indications, and to check the hill-climbing ability. The ability to simulate various load conditions and maintain certain conditions over a longer period of time allows for accurate observation of the gas supply system settings or their changes, and allows for accurate calibration of the gas supply system and detection of faults occurring under strictly defined conditions, e.g., at very high loads or boost pressure [7]. The installer can comfortably observe the indications of measuring instruments using different variants of the dynamometer operation (maintaining a constant load or engine speed), depending on the needs [3,6,10].

Constant engine speed dynamometer mode is very useful in calibrating gas supply systems. Thanks to this, you can adjust the gas injection time and the angle of its injection by moving the controller map so that it reflects the work on the original fuel supply system as much as possible. Under the conditions of the road test, it is impossible to maintain a constant engine speed with increasing load, even when using a hill or a large load for this purpose. That is why the chassis dynamometer is a device ideally suited for the calibration of gas installations for trucks [3,11]

2. Materials and Methods

2.1. Research Setup—Materials

This work deals with the aspects and correctness of the use of a chassis dynamometer in the process of tuning dual-fuel CNG/diesel installations in a 2014 Volvo FH 13 semi-truck and the impact on the tuning results and their correctness. The vehicle's engine was converted to dual-fuel operation by adding a prototype sequential indirect gas injection system. The tests were carried out on a mobile dynamometer device built on the structure of a 6 ft container which is shown in Figure 1 below.

(a)

(b)

Figure 1. A mobile dynamometer device: (**a**) front view, (**b**) view of the interior of the dynamometer (authors' source).

Table 1 presents the technical data of the mobile dynamometer device declared by the manufacturer. Table 2 presents the technical data of the tested tractor unit.

Table 1. Selected declared parameters of the mobile dynamometer device according to the design assumptions and the manufacturer (manufacturer data).

Parameter	Value/Name/Type
Mode of Operation	Load Type
Brakes type	Frenelsa FF16—eddy current 2 × 3300 (Nm)
Number of measuring axle	1
Max/min wheel track	2700/1000 (mm)
Rollers diameter	320 (mm)
Minimum wheel diameter	700 (mm)
Maximum load on the measuring axle	15,000 (kg)
V-max	200 (km/h)
Maximum measurement error	1 (%)
Permissible ambient temperature (operation)	od−10 do +30 (°C)
Permissible ambient temperature (storage)	od −40 do +50 (°C)
Mechanical synchronization of rotational speed of dynamometer rollers	Yes

Table 2. Selected factory technical data of the tested vehicle [12].

Parameter	Value/Name/Type
Vehicle Brand/Model/Year of Production	Volvo/FH13/2014
Engine type	Turbodiesel, D13C460, EU5SCR-M
Engine displacement/Number of cylinders/ Cylinder diameter/Piston stroke/ Compression ratio	12.8 L/6 cylinders/131 mm/158 mm/17.8:1
Max power	460 hp (338 kW) at 1400–1900 rpm
Max engine speed	2100 RPM
Economic speed	1000–1500 RPM
Optimal speed range	1150–1400 RPM
Gearbox type	VTO2214B, Automatic with manual selection
Number of gears	14 gears

Due to the nature of the research, the measured parameters should be divided into the following groups of parameters characterizing:

- The operation of the dual fuel (CNG/diesel) system controller;
- The work of the chassis dynamometer;
- The operation of the semi-tractor engine control system.

The parameters for each group that should be measured, and the necessary measuring equipment are presented below.

In the tested case, the engine was converted to dual-fuel operation by adding a prototype sequential indirect gas injection system. In this system, pilot diesel fuel acts as the ignition source, the system has an ability to "cut" and "exchange" the diesel oil dose to a gas fuel up to 95%, and it uses 2 gas injectors per cylinder [5,13]. Diagram of operation is presented in Figure 2 below.

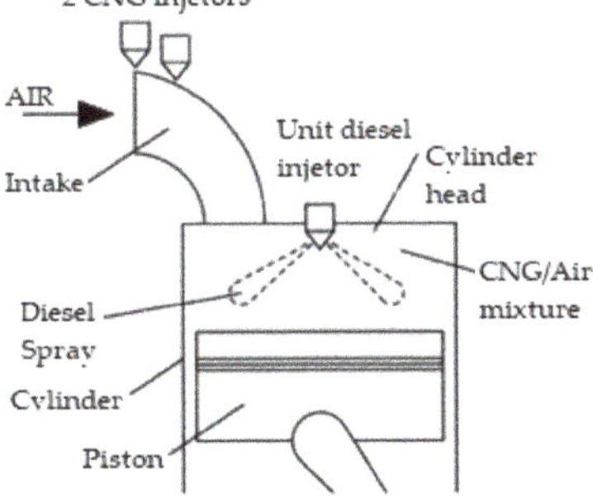

Figure 2. Diagram of operation of a dual-fuel supply system (authors' source based on [14]).

Parameters characterizing the operation of the dual-fuel supply system, recorded with the use of the gas fuel supply system controller are shown in Table 3 below.

Table 3. Selected parameters characterizing the operation of the gas fuel supply system (manufacturer's data).

Parameter	Value/Name/Type
Time [s]	Time from the Beginning of the Mileage Recording, the Basis of the Recorded Mileage
Engine speed [RPM]	Engine's crank speed registered by gas supply controller
Instantaneous CNG consumption [kg/h]	Gas mass fed to the inlet manifold,
Instantaneous diesel oil consumption [kg/h]	Instant consumption of diesel oil—diesel mass fed to the cylinders
Diesel actual [%]	Actual instant share of diesel oil in the dose of fuel delivered to the engine
Diesel averaged [%]	The average periodic share of diesel oil in the fuel dose delivered to the engine
Diesel dose [mg]	Instantaneous dose of diesel fuel delivered to one cylinder for one work cycle
Gas time [ms]	Instantaneous gas injector opening time for one/particular cylinder for one work cycle
Calculated gas time [ms]	Instantaneous gas injector opening time for one/particular cylinder for one work cycle after taking into account the correction parameters
Temp. reducer [°C]	Temperature of the gas pressure reducer
MAP pressure [bar]	Air pressure in the intake manifold
Gas pressure [bar]	Gas pressure on the gas pressure reducer
Pressure on gas rail [bar]	Gas pressure on the injection rail
Lambda	Signal from the oxygen concentration sensor in the exhaust gas—lambda probe
System faults	Monitor of faults generated by the gas supply system

Parameters characterizing the work of the chassis dynamometer measured with the use of the dynamometer controller—software are shown in Table 4 below.

Table 4. Selected parameters characterizing the chassis dynamometer measured with the use of the dynamometer controller (manufacturer's data).

Parameter	Value/Name/Type
Selected Time [s]	Time from the Beginning of the Test
Engine/roller speed ratio	Speed transmission ratio—motor/rollers
Engine speed [RPM]	Calculated engine speed
Engine speed rotational acceleration [RPM/s]	The determined speed of the engine rotational speed change during acceleration
Road speed [km/h]	Calculated linear driving speed of vehicle
Rollers received power [hp]	The power received on each of the roller
Rollers inertial torque [Nm]	The moment of inertia on each of the roller
Rollers brakes torque [Nm]	The braking torque for each roller
Power on wheels [hp]	Calculated power on vehicle wheels
Loss power [hp]	Total power lost resulting from speed changes and load changes
Drivetrain inertial power [hp]	Power of inertia of the drive system
Engine inertial torque [Nm]	Motor moment of inertia
Engine inertial power [hp]	Motor inertia power
Engine power [hp]	Calculated engine power
Engine corr. power [hp]	Corrected power of the engine
Engine torque [Nm]	Engine torque
Engine corr. torque [Nm]	Corrected motor torque
Exhaust gas temperature [°C]	Exhaust gas temperature measured by external thermocouple
Ambient temperature [°C]	Ambient temperature
Ambient pressure [hPa]	Measured by the system of the built-in weather station of the dynamometer
Ambient humidity [%]	Measured by the system of the built-in weather station of the dynamometer

Parameters characterizing the operation of the tractor engine control system. Measurement using a TEXA TXT diagnoscope with IDC5—Truck software:

- Engine rotational speed [RPM];
- Vehicle travel speed [km/h];

- Boost pressure [bar];
- Torque [Nm];
- Hourly basic fuel consumption—diesel fuel [kg/h];
- Engine load [%];
- Engine management system errors.

The elements listed above have been compiled in a research arrangement. Figure 3 below shows a block diagram of the test stand. Figure 4 shows the vehicle undergoing tests.

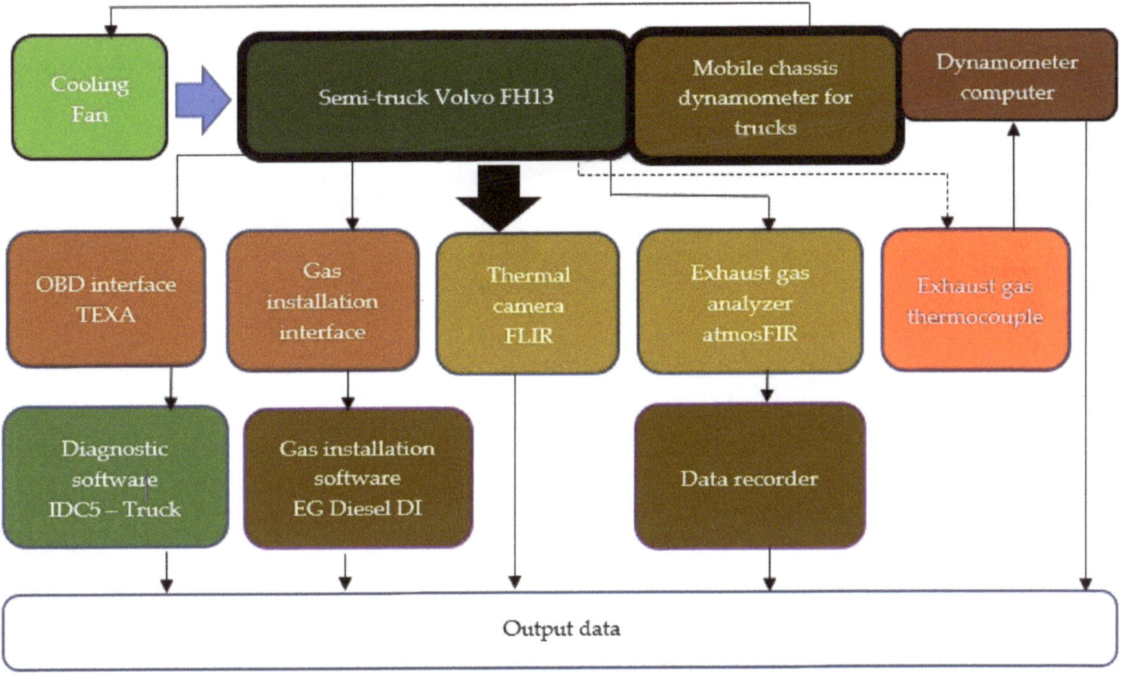

Figure 3. Block diagram of the test stand (authors' source).

(a)

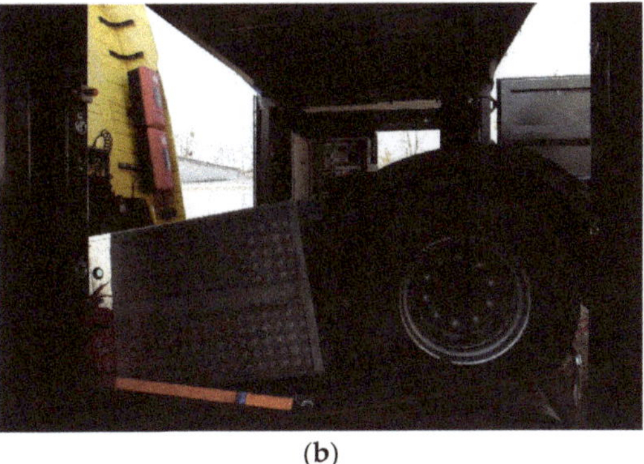

(b)

Figure 4. Volvo FH13 on mobile dynamometer: (**a**) front view, (**b**) view of the vehicle drive axle on dyno rollers (authors' source).

2.2. Methodology

The main goals of the research were to tune the Volvo FH13 semi-truck equipped with D13C engine to work on dual-fuel CNG-diesel mode and also to test the mobile semi-truck dynamometer device. The titular research on the influence of the aspects of the dynamometer operation on the tuning of dual-fuel installations is a derivative of the main research. The research methodology dealing with the above subject matter has been developed empirically and includes an approach to solving problems related to:

- Preliminary data settings for the technical (physical) data of the test vehicle on the dynamometer;
- Selection of the operating range of the vehicle's engine (gear selection);
- The influence of the value of the forced load of the vehicle driving axle;
- Selection of the dyno working mode appropriate to the expected tasks;
- The impact of the correctness of the selection of the scope of the analysis of data on losses in the drive system.

Before the tests, the vehicle was always secured in accordance with the manufacturer's instructions and safety guidelines. The car's operating fluids were heated to normal operating temperature, and the tires at the beginning of each test had a temperature of 50 °C. In order to ensure repeatability of the tests, a number of parameters such as temperature, pressure and speed in the vehicle were monitored in real time.

3. Results

3.1. Settings for the Technical (Physical) Data of the Test Vehicle on the Dynamometer

The main task before starting the test and tuning on the dynamometer is the need to configure the vehicle in the dynamometer software in a way that best suits the actual technical conditions of the vehicle. In the software of the dynamometer on which the tests were carried out, in order to validate the tests, when starting a "new project", the vehicle data such as losses in the drive system, engine and drivetrain system inertia should be correctly configured. At this point, it should be noted that the tool for estimating the inertia of the drive system built into the dynamometer software, calculating this value from the geometric, physical data used on the drive wheels on the vehicle (shown in Table 5), differed from the actual value calculated empirically. When determining the mass moment of inertia of its wheel in relation to its axis of rotation, the method presented in [15] was used. The tested wheel is suspended (Figure 5) on 3 vertical steel ropes $l = 6.664$ m long, evenly spaced along the circumference with a radius of $r = 0.1405$ m (the radius value results from the diameter of the hole in the wheel rim).

Table 5. Data table for calculating the inertia of the vehicle propulsion system tested on the dynamometer (authors' source).

Tire size *	315/70/R22,5
Used wheels count *	4
Calculated wheel inertia **	16.72 kg·m^2
Used wheel inertia **	66.87 kg·m^2
Inertia transferred to roller **	0.66 kg·m^2

* Data entered manually. ** Data calculated by the dynamometer software on the basis of *.

Figure 5. Wheel mounted on the stand for determining the mass moment of inertia.

The vibrating system constructed in this way is set in motion by the rotation of the wheel in the vertical plane by a certain angle alpha (its value cannot exceed 15°). Then, the vibration period T is measured, which is determined by measuring the execution time of 10 complete deflection cycles. The following relationship is used to calculate the mass moment of inertia IY:

$$I_Y = \frac{T^2 \cdot r^2 \cdot m \cdot g}{4\pi^2 \cdot l}, \ \text{kg} \cdot \text{m}^2 \tag{1}$$

After substituting the value of the measured period $T = 15.6$ s, the mass moment of inertia is:

$$I_Y = 19,315 \ \text{kg} \cdot \text{m}^2 \tag{2}$$

This deviates by 15.5%, which may ultimately bend the measurement. The engine inertia is configured on a similar principle, although during research it was unable to verify the correctness of the engine inertia estimation. The range of losses in the drivetrain, entered manually and suggested by the dynamometer software (4–8%), also differs from the value calculated on the basis formulas for the overall efficiency of the drivetrain, according to which the losses are approximately 8–10% depending on the gear ratio used. While during tuning, the final result is not the value of the comparator of performance on diesel only supply and in dual-fuel mode, but the value that approaches the measurement as close to reality as possible, it is worth bearing in mind that the basic settings suggested by the software can really affect the measurement results.

3.2. Selection of the Operating Range of the Vehicle's Engine (Gear Selection)

When tuning the installation, determine the gear ratio that covers the widest range of operating/usable speeds of the engine while maintaining the highest possible speed of the vehicle's drive wheels should be determined (the aim is to reduce the torque on the wheels, the high values torque quickly degrade the tires and increase the measurement error, more on this subject in Section 3.3 of this work [1]). Alternatively, the engine rotational speeds that are of greatest interest to us in the context of the vehicle's operational properties resulting from the specificity of the vehicle's operation or the installation used in it should be determined. For this purpose, it is worth analyzing the factory diagrams of torque and power waveforms, as well as the range of optimized (economic) shaft revolutions. A graph of this type is presented in Figure 6.

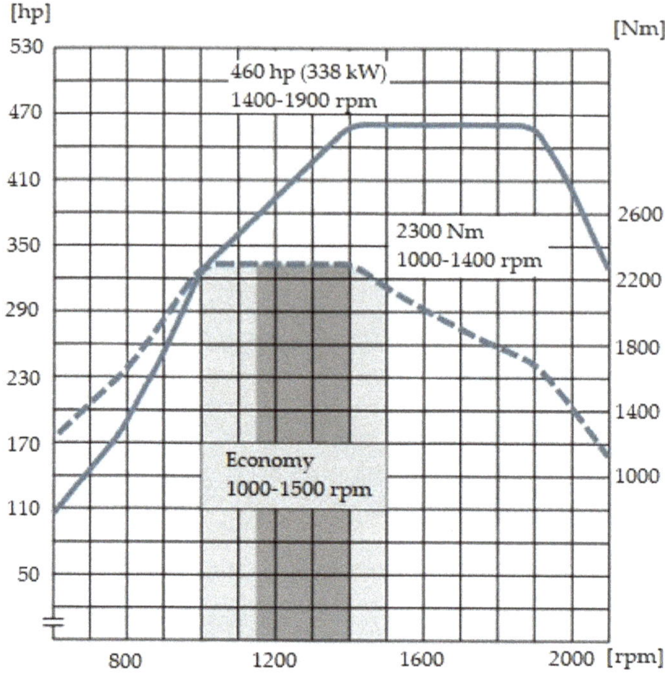

Figure 6. External characteristics of the tested engine [12].

Bearing in mind the manufacturer's chart, it was compared with the results obtained from covering the RPM range for the gears in the test vehicle, which was presented in Figure 7.

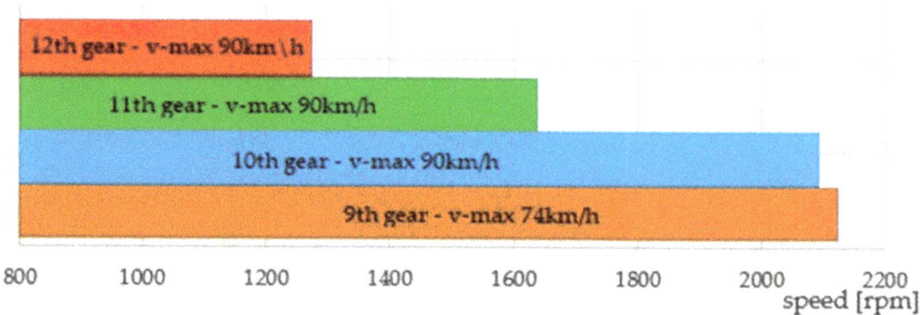

Figure 7. Graph of the coverage of the RPM engine operating range for selected gears (authors' source).

The above data show that the 10th gear is the most appropriate gear for the study of external characteristics and this gear was used to carry out further research presented in this article. In the case of tuning the installation at the engine speed declared by the manufacturer as the most optimal and economical, it should be the 11th gear. The speed range from idle to 800 rpm is irrelevant in the context of these tests, because the gas installation is not able to switch to the dual-fuel mode for technical reasons—the unit injector is not able to generate a correspondingly low pilot dose in this range [11,16].

3.3. The Influence of the Slip of Tires on Rollers

During the research, the influence of slipping of the driving wheels on the rollers of the dynamometer was noticed. This creates a multidimensional problem that can be broken down into several components:

- Dependence of the slip of the driving wheels on the roller due to the load of the driving axle on the rollers;
- Measurement error resulting from the above and from the influence of the engine/roller ratio settings in the dynamometer settings;
- Impact of the tires (heating up due to work on rollers, change in their dynamic radius).

Table 6 summarizes the data from the research on the sliping of the wheels on the rollers for different values of the engine load and different values of the thrust of the driving axle of the tested vehicle.

Table 6. Table of the results of tests of slip of tires on rollers.

Const. Revs. *	1000 RPM (~Max Torque)				1300 RPM (~Eco Revs)				1500 RPM (~Max Power)				1700 RPM (~High Revs)				Axle Load ****
Engine load [%] **	25	50	75	100	25	50	75	100	25	50	75	100	25	50	75	100	
RPM—TEXA **	1001	1022	1039	1066	1299	1323	1349	1380	1489	1519	1545	1574	1679	1710	1737	1768	20,000 N
Dyno (Nm) ***	192	585	935	1310	180	540	886	1275	122	473	799	1110	78	400	700	970	
RPM—TEXA **	992	1009	1028	1045	1288	1309	1331	1353	1483	1503	1524	1549	1673	1697	1717	1741	30,000 N
Dyno (Nm) ***	192	560	917	1293	190	549	870	1250	140	467	790	1117	63	370	688	972	
RPM—TEXA **	993	1005	1019	1041	1289	1306	1324	1345	1481	1500	1520	1542	1670	1692	1714	1733	40,000 N
Dyno (Nm) ***	195	565	920	1280	208	545	880	1250	163	480	790	1125	74	389	700	965	
RPM—TEXA **	996	1015	1036	1061	1292	1320	1344	1375	1490	1514	1541	1570	1676	1707	1733	1760	20,000 N (warm tires) *****
Dyno (Nm) ***	192	583	918	1315	214	559	908	1279	160	480	810	1130	80	405	710	970	

* The test was carried out using the constant revs mode of the dyno. ** Reading made using OBD TEXA TXT diagnoscope with IDC5—Truck software. The correctness of measurements using TEXA was validated on an engine dynamometer. *** Braking force value on a single brake, reading made using dyno software. **** The load on the driving axle of the vehicle was generated by the load adding system with the use of an electric winch, connected with a strain sensor, and the system of mounting to the driving axle of the tested truck—the axle pressure system is shown in Figure 8. The axle load was read using the dyno software. ***** Test started at 70 °C.

(a)

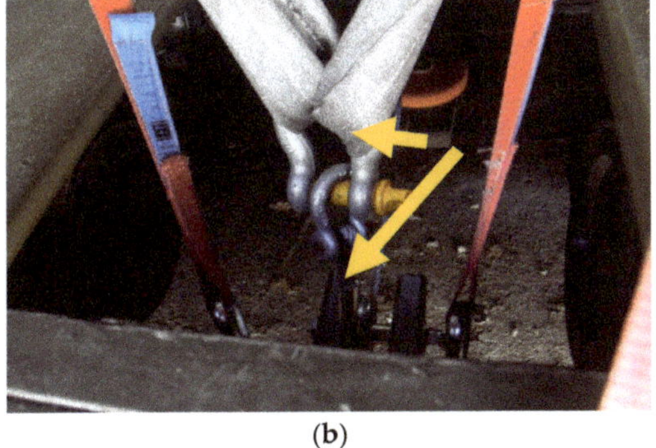

(b)

Figure 8. Vehicle axle load mechanism: (**a**) general view, drive axle slings, (**b**) view of rope linked with an electric winch (authors' source).

Figure 9 summarizes the above data from Table 6 in the values of sliping wheels on rollers for different engine speeds in the context of the engine load and driving axle load (pressure).

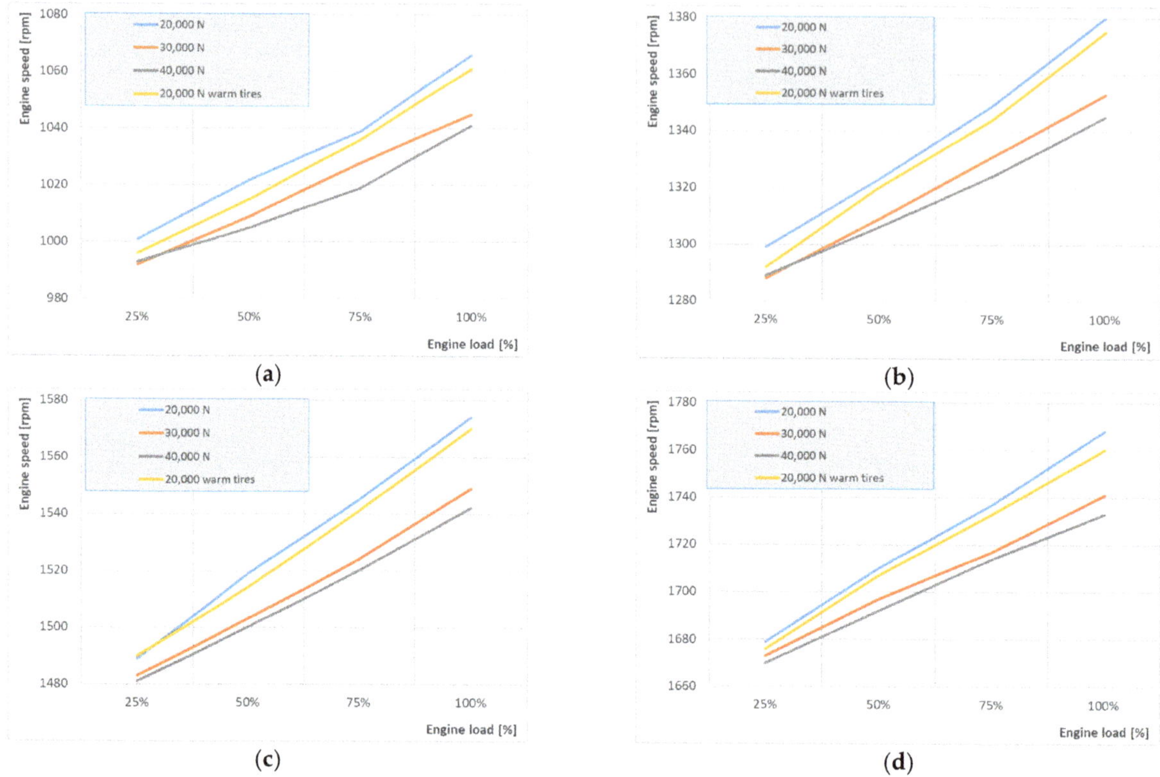

Figure 9. Graphs of differences in engine speed for rollers where n is constant, resulting from wheel slip on rollers for different engine speeds in the context of engine load and drive axle load: (**a**) 1000 rpm, (**b**) 1300 rpm, (**c**) 1500 rpm, (**d**) 1700 rpm (authors' source).

The above charts show that the increase in slip correlates with the increase in engine load (which results from increasing the torque on the wheels of the vehicle [10]). As the engine load increases, the measurement error resulting from the slipping of the wheels on the roller increases, which is also shown in Figure 10. Due to the fact that there is a dynamic difference in the geometric value of the tire diameter [17], all the parameters that affect the tire slip also have an impact on it (axle load, wheel speed, tire temperature and so on). The change in the tire diameter also leads to a negative error at low engine loads [15]. According to [17], measurements on the tire imply that for a high load and low inflation pressure, the deformation in the shoulder area will contribute more to rolling resistance. In an opposite manner, for a low load and high inflation pressure, the deformation in the crown area will contribute more. Meanwhile, the deformation in the shoulder area always contributes more to the rolling resistance for the worn tire. This will in turn accelerate the wear process on the shoulder area. It should also be noted that the change in the dynamic diameter of the tire resulting from the centrifugal force acting on the turning wheel in the context of the thrust of the drive axle generated on the dynamometer is marginal and oscillates around the value of 0.5–1 mm at 90 km/h [18]. The influence of temperature can be observed by analyzing the blue and yellow lines for a load of 20,000 N. The influence of the axle load is more important than the temperature. Considering the necessity of maintaining low tire temperatures (optimally less than 80 °C and no more than 120 °C [19]) while tuning the gas system, in order to prevent their increased wear, it seems reasonable to increase the axle load to 30,000 N while monitoring the temperature. Tire temperature values at the end of each test from all measurement setups are summarized in Figure 10. The graph can be shifted in the vertical axis by modifying the engine/roller ratio in the dynamometer settings. In the context of the data presented above, it can be concluded that it seems reasonable to modify the engine/roller ratio depending on the range of engine loads and thus work in different ranges for the fuel exchange maps in the gas ECU [3]. Setting a constant gear ratio will generate measurement errors and may lead to discrepancies in the readings of rpm values in the dyno and the gas controller.

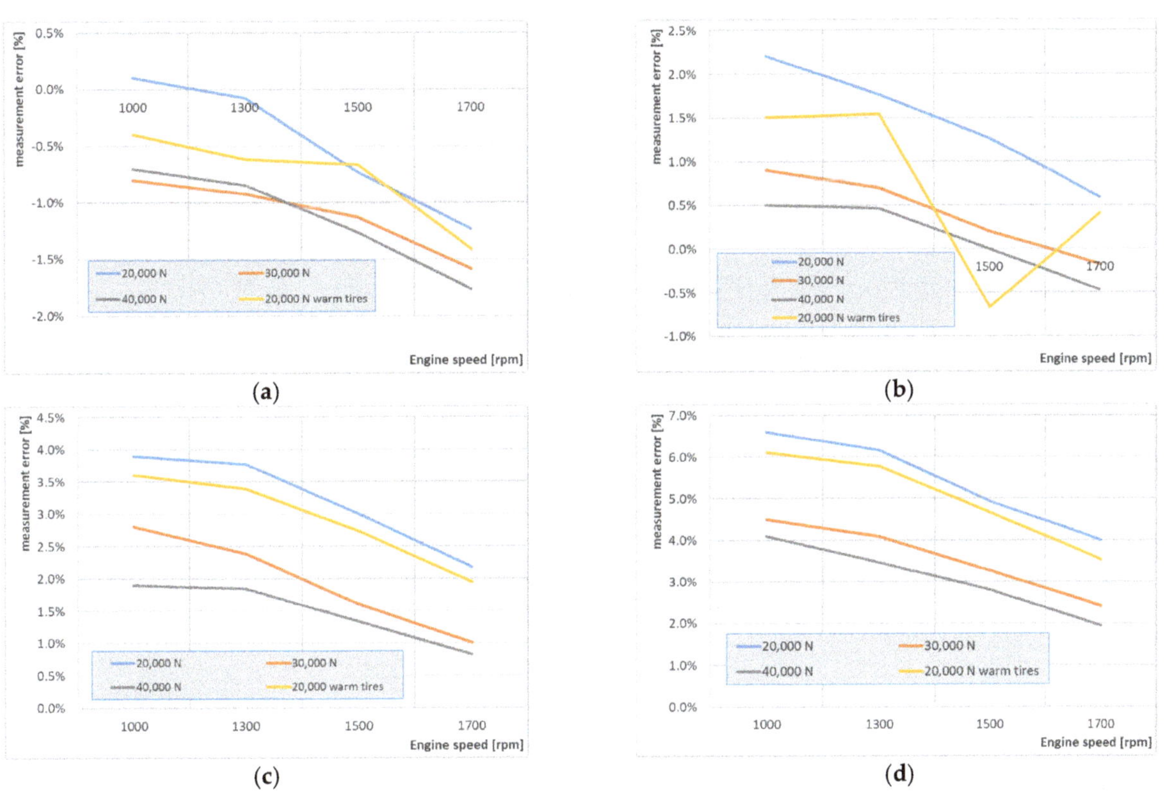

Figure 10. Graphs of errors resulting from the tire slipping on the roll for different rpm depend on engine load: (**a**) 25%, (**b**) 50%, (**c**) 75%, (**d**) 100% (authors' source).

In Figure 11, in the upper right corner of each photo on parameter described as "maks", the highest temperature recorded by the thermal camera during test is shown. It can be noticed that it was always the temperature of the inner wheel near the inside of the tire tread. Figure 11 above shows that it is worth keeping low tire temperature with higher axle pressure during tuning in order to extend the service life of the vehicle tires. Axle load 30,000 N in the case of the tested vehicle seems to be the most appropriate. A possible solution to this problem is adding a ventilator for extra tire cooling.

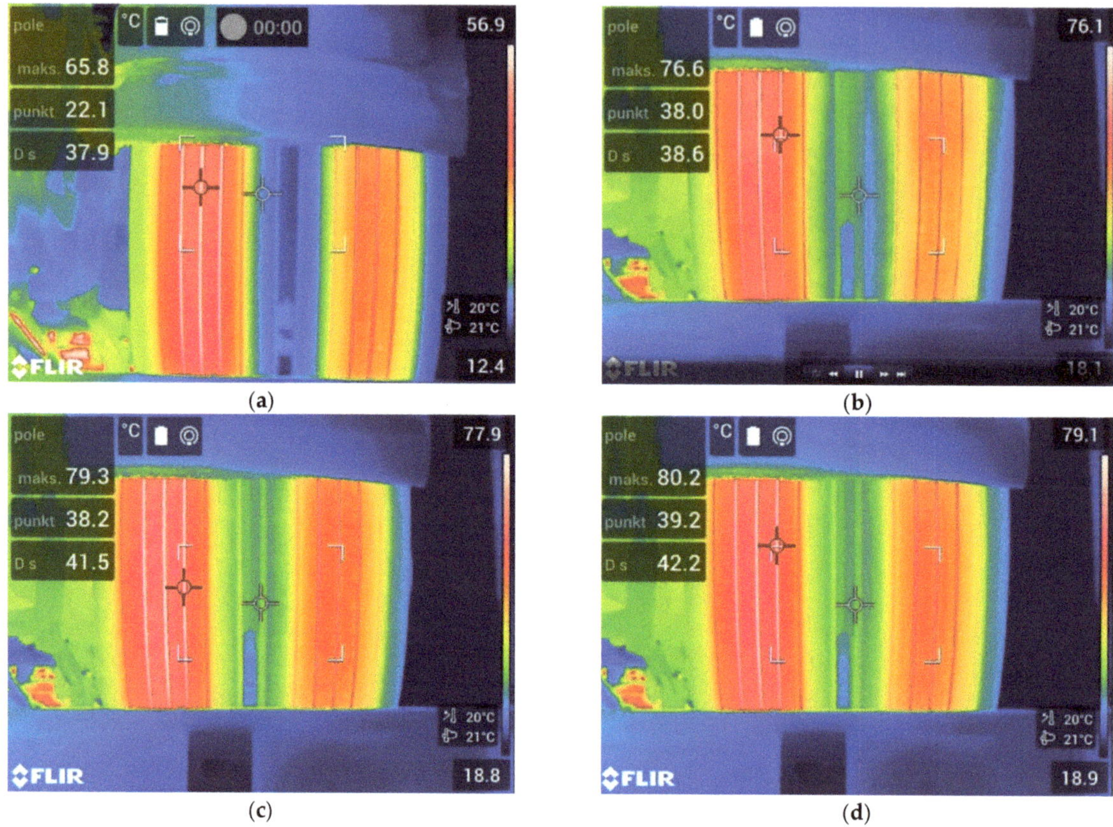

Figure 11. Tire temperature at the end of each test for axle load: (**a**) 20,000 N (**b**) 30,000 N (**c**) 40,000 N (**d**) 20,000 (warm tires) (authors' source).

3.4. Selection of the Optimal Operating Mode of the Chassis Dynamometer

The tuning of dual-fuel installation with the use of a chassis dynamometer in places with partial engine load can be performed in fuel exchange map spots, using the function of maintaining constant revolutions of the dynamometer [20,21]. However, in the case of tuning the external characteristics of the engine (maximum dynamic torque values), it is necessary to use the dyno ramp mode, which allows the vehicle to be braked in the entire range of rotational speeds with an appropriately defined acceleration. The selection of the appropriate acceleration of the vehicle engine on the dynamometer device is of decisive importance in the creation of the boost pressure, and thus the possibility of achieving the maximum value of the torque [7]. In the case of semi- truck, it seems important to generate a sufficiently high braking load on the rollers on the dynamometer, because the semi-truck has a diesel engine with a turbocharger that has a high inertia, which requires a high load to achieve full spool [7,21].

During this test, the highest value of dn/dt as rpm/ sec at which it was possible to correctly determine the external characteristics of the engine (creating highest boost pressure) was experimentally determined. Results are shown in Figure 12 below.

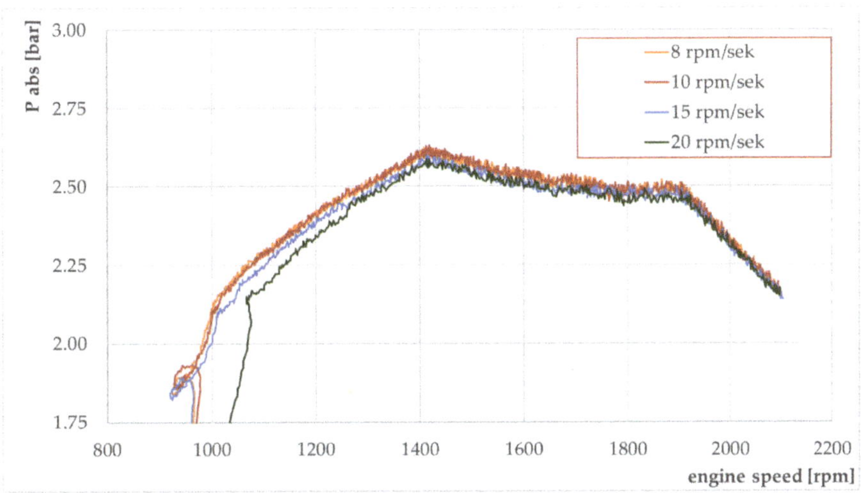

Figure 12. Influence of the coasting-down speed for the test of the external characteristics (maximum torque values) on the boost pressure (authors' source).

The chart above shows that 10 rpm/s meets our boost pressure requirement. Braking at lower acceleration values unnecessarily lengthens the test, which in turn stresses the tires thermally and may lead to more intensive wear. The effects of overheated tires in the form of tearing off parts of the tire tread are shown in Figure 13.

(a)

(b)

Figure 13. Damaged tire due to overheating: (**a**) view of the tire, (**b**) view of the broken part of the tread (authors' source).

3.5. The Impact of the Correctness of the Selection of the Scope of the Analysis of Data on Losses in the Drive System and Repeatability of Tests

As in the case of the tested vehicle, in which an automated gearbox was used, also in other semi-truck, there may be a lack of repeatability in the calculation of the model of losses in the drive system of the vehicle implemented through the free coast of the vehicle drive on the rollers of the dynamometer [1,14]. During the tuning process, in the case of the chart of the maximum performance or the one set for the optimal revolutions, the incorrectly calculated loss model may generate errors in reading the power and torque in dual fuel operation [11]. After the measurement is completed, the automatic gearbox may generate fluctuations in the recorded resistances of the vehicle drive. A similar situation does not take place when the brake is pressed in a car with a manual gearbox. At this point, it is worth considering the repeatability of the selection of the measuring range for the calculation of losses. Whether the operation of the gearbox is repeatable should be noted, as well as whether the scope of the analysis of data for the loss model is appropriately wide and whether it does not include places evidently bending the measurement results. Figure 14 shows a screen from the dyno software, and calculation of the loss model in the drive system.

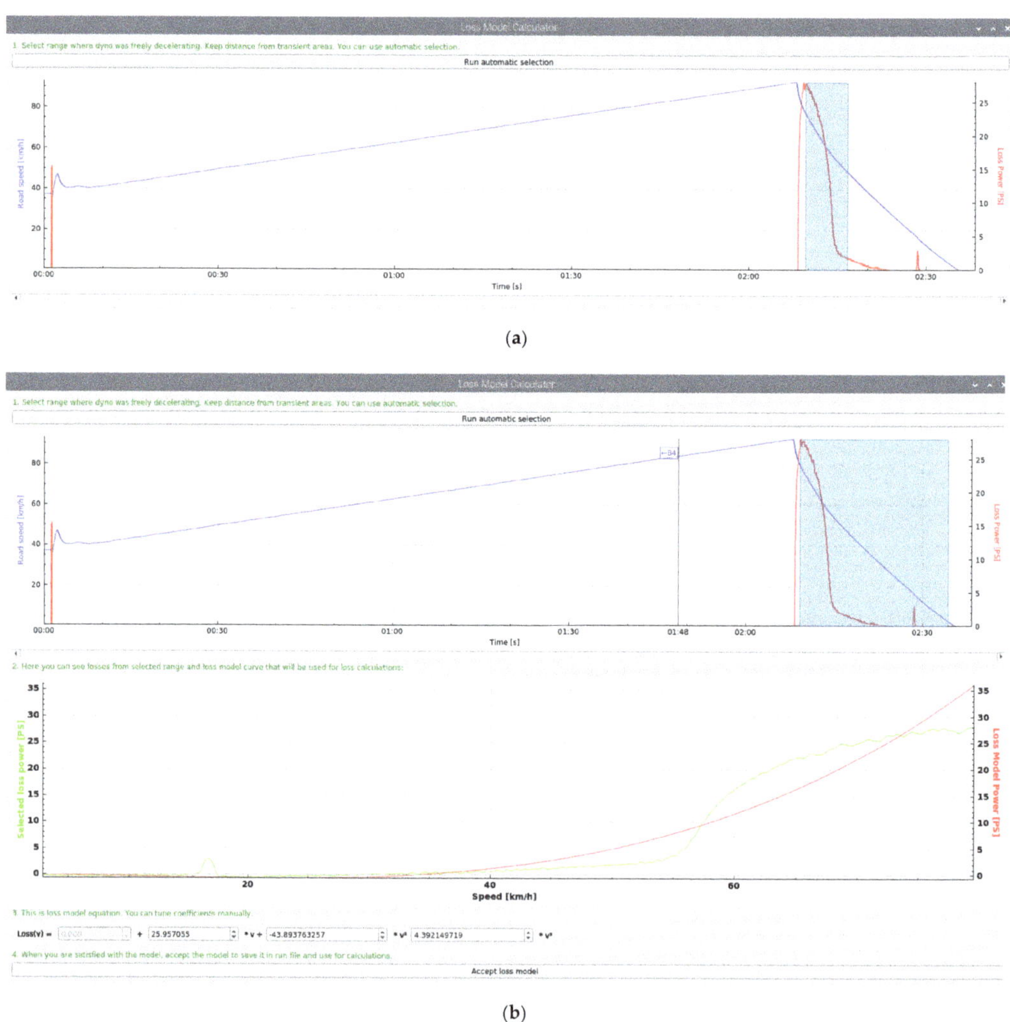

(a)

(b)

Figure 14. Screen from the dyno software, calculation of the loss model in the drive system: (**a**) nominal settings, (**b**) manually selected range of the loss model, and comparison of the repeatability of the automatic vehicle gearbox operation (authors' source).

It is worth remembering about the repeatability of the tests in the case of compiling data before and after conversion to dual fuel power. The key in this aspect is the possible reproduction of the test conditions before and after this conversion. Figure 15 shows the result of the repeatability of the tests for the graph of the boost pressure related to the operating conditions of the fully warmed up and not fully warmed up engine. These types of errors affect a number of elements of the installation (e.g., the temperature of the gas pressure reducer, the possibility of creating the boost pressure).

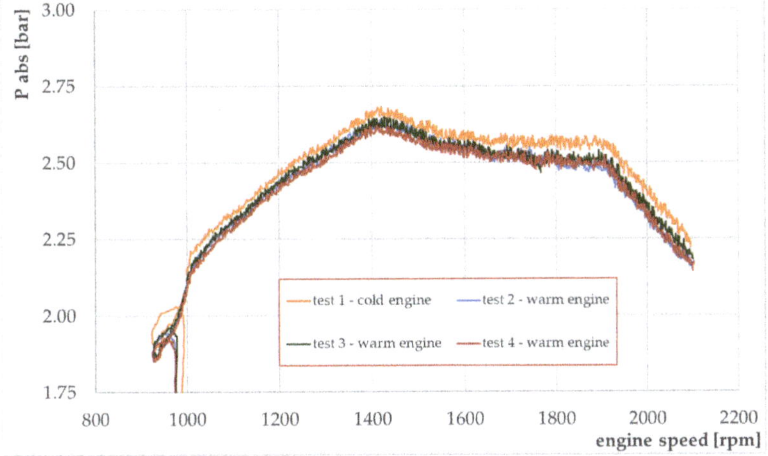

Figure 15. Graph of the repeatability of the boost pressure build-up due to the measurement conditions (engine warm-up) (authors' source).

4. Discussion

All the operating parameters of the chassis dynamometer mentioned in this document affect the assessment and tuning effects of dual-fuel systems in semi-trucks. As part of this study, the aim was to present in general terms the key parameters of the chassis dynamometer that affect the final effect of measurements and tuning for a dual-fuel vehicle. As a result of negligence resulting from the lack of awareness of the above-mentioned critical elements of work with the chassis dynamometer during the tuning of the installation and comparing the effects of this tuning with the single-fuel engine supply, it is possible to present the results of power measurements of the operating parameters for the external characteristics of the engine in combination with a correctly measured and adjusted installation. There is such a statement in Figure 16.

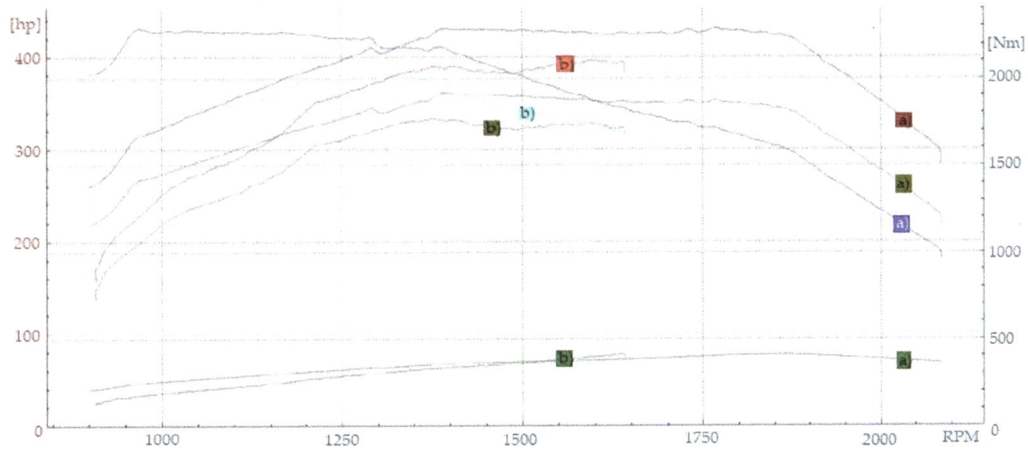

Figure 16. *Cont.*

	Loss power	Wheel power	Engine power	Engine torque	Length	Speed source	Engine/roller ratio	Corr. Value	Amb. Press.	Amb. Hum.	Peak. Eng. speed
a)	77 hp 1874 rpm	361 hp 1389 rpm	432 hp 1770 rpm	2290 Nm 970 rpm	160s	calculated	1.36	0.6%	1014 hPa	23%	2085 rpm
b)	77 hp 1636 rpm	333 hp 1376 rpm	397 hp 1593 rpm	2034 Nm 1218 rpm	113s	calculated	1.08	-1.5%	1024 hPa	16%	1641 rpm

Figure 16. A graph showing the results of external characteristics for the research where: (**a**) attempts were made to reduce the impact of errors, and (**b**) the errors mentioned in this work are combined (authors' source).

The above chart shows the wrongly selected measurement course and the influence of other factors mentioned in this paper, which resulted in the torque and power values lower by 8.2% and 11.2%.

The scale of problems resulting in inaccurate tuning of the dual-fuel system is presented in the map of diesel fuel replacement with gaseous fuel. The higher its value in a given field and the higher the total value of the exchange (while maintaining torque and power waveforms similar to single-fuel operation), the better. Figure 17 below shows the comparison of the results of tuning the map of conversion of diesel fuel to gas fuel. In the example below you can clearly see much worse fuel exchange values, and the overall exchange value decreased by 9.2%.

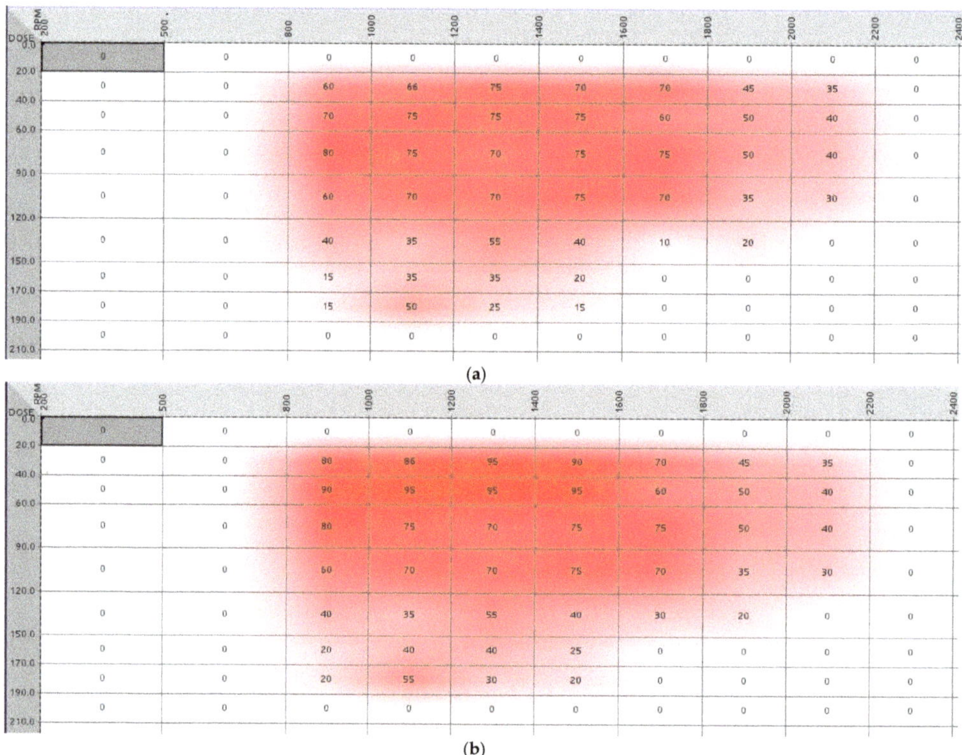

Figure 17. Comparison of the results of tuning the map of conversion of diesel fuel to gas fuel in terms of: (**a**) attempts were made to reduce the impact of errors, and (**b**) where the errors listed in the paper are combined (authors' source).

5. Conclusions

This document presents the influence of the operating parameters of the chassis dynamometer on the evaluation of the tuning of dual-fuel systems. The resulting map of the exchange and the tuning itself are of course influenced by many other elements, such as the condition of the car (e.g., state of filters [22]) or the correct calibration of the dynamometer itself. It is also worth mentioning that the dynamometer presented in the test cannot be used for standardized tests and the effect of the work related to it should always be a comparative effect in which we have an input and final test, then compare both. However, it should be remembered that some of the errors can be effectively eliminated, and this article was also supposed to present this. It is also worth mentioning that the dynamometer presented in the test cannot be used for standardized tests. The effect of work on it should always be a comparative effect in which the received first input and final output are compared with each other. In the context of tuning, this can already give correspondingly good results. Nevertheless, in order to be closer to reality, it seems important to eliminate or minimize the errors described in this work. In the opinion of the authors, further tests in the research system presented in this paper should focus on a more in-depth analysis of the problems presented in this paper. It also seems justified to test in the context of exhaust emissions on the test system presented in this paper, compared to work under standardized engine dynamometer conditions [23–25]. The authors of this work are considering taking up such a topic in the future.

Author Contributions: Conceptualization, M.K. and J.C.; methodology, M.K. and J.C.; validation, M.K. and J.C.; formal analysis, M.K. and J.C.; investigation, M.K. and J.C.; data curation M.K., writing—original draft preparation, J.C.; writing—review and editing, M.K. and J.C.; visualization, M.K.; supervision, M.K.; project administration, M.K.; funding acquisition, M.K. All authors have read and agreed to the published version of the manuscript.

Funding: The publication was created as a result of work under: Name of the competition: Ścieżka dla Mazowsza, Agreement No. MAZOWSZE/0123/19-00, Subject: "Innovative ecological CNG installation for diesel engines limiting the emission of harmful exhaust components together with a mobile diagnostic platform."

Institutional Review Board Statement: Not applicable.

Informed Consent Statement: Not applicable.

Data Availability Statement: Data are contained within the article.

Conflicts of Interest: The authors declare no conflict of interest.

References

1. Hancock, B. *Dyno Testing and Tuning (Performance How-To)*; S-A Design: Santa Fe, NM, USA, 31 July 2020.
2. Yogeeswaran, R.; Subramaniom, S.; Senthil Krishnan, S. *Establishment of Chassis Dynamometers for Commercial Vehicles*; SAE International: Warrendale, PA, USA, 2019.
3. Gazeo.pl. Available online: https://gazeo.pl/biznes/warsztat/Kalibracja-instalacji-LPG-z-wykorzystaniem-hamowni,artykul,6292.html (accessed on 26 May 2022).
4. Dziubak, T.; Bąkała, L.; Karczewski, M.; Tomaszewski, M. Numerical research on vortex tube separator for special vehicle engine inlet air filter. *Sep. Purif. Technol.* **2020**, *237*, 116463. [CrossRef]
5. 4GCS Tuning. Available online: http://gcstuning.pl/strojenie-lpg/ (accessed on 29 May 2022).
6. 544 tuning. Available online: https://44tuning.pl/tuning-lpg-wieksza-moc-na-gazie-niz-benzynie-wysoko-oktanowej (accessed on 28 May 2022).
7. Heywood, J.B. *Internal Combustion Engine Fundamentals*, 2nd ed.; McGraw-Hill Education: New York, NY, USA, 2018.
8. Dąbrowski, T.; Kurczyński, D.; Łagowsk, P.; Warianek, M. The scientific analysis of the influence of petrol and LPG gas powered engines dynamic indicators on active road safety. In Proceedings of the 2018 XI International Science-Technical Conference Automotive Safety, Casta, Slovakia, 18–20 April 2018.
9. Öberg, P.; Nyberg, P.; Nielsen, L. A New Chassis Dynamometer Laboratory for Vehicle Research. *SAE Int. J. Passeng. Cars–Electron. Electr. Syst.* **2013**, *6*, 152–161. [CrossRef]
10. Galindo, E.; Blanco, D.; Brace, C.; Chappell, E.; Burke, R. *Chassis Dynamometer Testing: Addressing the Challenges of New GlobalLegislation (WLTP and RDE)*; SAE International: Warrendale, PA, USA, 2017; ISBN 978-0-7680-8278-4.

11. Karim, G.A. *Dual-Fuel Diesel Engines*; CRC Press: New York, NY, USA, 2015.
12. Volvo Trucks Website. Available online: https://stpi.it.volvo.com/STPIFiles/Volvo/FactSheet/D13C460,%20EU5SCR-M_Eng_02_1144722.pdf (accessed on 30 May 2022).
13. Chojnowski, J.; Karczewski, M.; Szamrej, G. The phenomenon of knocking combustion and the impact on the fuel exchange and the output parameters of the diesel engine operating in the dual-fuel mode (Diesel-CNG). In *NAŠE MORE 2021*; University of Dubrovnik: Dubrovnik, Croatia, 2021; pp. 55–62.
14. SAE J2951 Drive Quality Evaluation for Chassis Dynamometer Testing J2951_201401 2014. Available online: https://www.sae.org/standards/content/j2951_201401/ (accessed on 16 March 2022).
15. Zając, M. *Układy Przeniesienia Napędu Samochodów Ciężarowych*; Wydawnictwa Komunikacji i Łączności: Warsaw, Poland, 2014; pp. 277–279.
16. Li, W.; Liu, Z.; Wang, Z. Experimental and theoretical analysis of the combustion process at low loads of a diesel natural gas dual-fuel engine. *Energy* **2016**, *94*, 728–741. [CrossRef]
17. Xiong, X.; Tuononen, A. Rolling deformation of truck tires: Measurement and analysis using a tire sensing approach. *J. Terramechanics.* **2015**, *61*, 33–42, ISSN 0022-4898. [CrossRef]
18. Anghelache, G.; Moisescu, R. The measurement of dynamic radii for passengercar tyre. *IOP Conf. Ser. Mater. Sci. Eng.* **2017**, *252*, 012014. [CrossRef]
19. Technical Data—Pirelli Truck Tyres. Available online: https://www.pirelli.com/asset/index.php?idelement=20899 (accessed on 29 May 2022).
20. Kuiken, K. Gas and dual-fuel engines for ship propulsion, power plants and cogeneration. In *Book I: Engine Systems and Environment*; PJ Onnen.: Antwerp, The Netherlands, 2016.
21. Luft, S. *Dwupaliwowy Silnik o Zapłonie Samoczynnym Zasilany Mieszaniną Gazów Propan-Butan (LPG) I Olejem Napędo-Wym*; Monografie; Politechnika Radmska, Wydawnictwo: Radom, Poland, 2002; pp. 22–24.
22. Dziubak, T. The effects of dust extraction on multi-cyclone and non-woven fabric panel filter performance in the air filters used in special vehicles. *Eksploat. I Niezawodn. Maint. Reliab.* **2016**, *18*, 348–357. [CrossRef]
23. Pielecha, J. *Badania Emisji Zanieczyszczeń Silników Spalinowych*; WPP: Poznań, Poland, 2017; pp. 155–162.
24. Bielaczyc, P.; Klimkiewicz, D.; Woodburn, J.; Szczotka, A. Exhaust Emission Testing Methods–Bosmal's Legislative and Development Emission Testing Laboratories. *Combust. Engines* **2019**, *178*, 88–98. [CrossRef]
25. Karczewski, M.; Wieczorek, M. Assessment of the Impact of Applying a Non-Factory Dual-Fuel (Diesel/Natural Gas) Installation on the Traction Properties and Emissions of Selected Exhaust Components of a Road Semi-Trailer Truck Unit. *Energies* **2021**, *14*, 8001. [CrossRef]

 *energies*

Article

Determining the Unit Values of the Allocation of Greenhouse Gas Emissions for the Production of Biofuels in the Life Cycle

Mariusz Niekurzak

Faculty of Management, AGH University of Science and Technology, 30-059 Krakow, Poland; mniekurz@zarz.agh.edu.pl

Abstract: Thanks to the allocation methods, i.e., the division of the total GHG emissions between each of the products generated in the production of biofuels, it is possible to reduce the emissions of these gases by up to 35% in relation to the production and combustion of fuels derived from crude oil. As part of this study, the biodiesel production process was analyzed in terms of greenhouse gas (GHG) emissions. On the basis of the obtained results, the key factors influencing the emissions level of the biodiesel production process were identified. In order to assess the sensitivity of the results of the adopted allocation method, this study included calculations of GHG emissions with an allocation method based on mass, energy, and financial shares. The article reviews recent advances that have the potential to enable a sustainable energy transition, a green economy, and carbon neutrality in the biofuels sector. The paper shows that the technology used for the production of biodiesel is of great importance for sustainable development. The possibility of using renewable raw materials for the production of fuels leads to a reduction in the consumption of fossil fuels and lower emission of pollutants. It showed that during the combustion of biodiesel, the percentages of released gas components, with the exception of nitrogen oxides, which increased by 13%, were significantly lower: CO_2—78%, CO—43%, SO_2—100%, PM10—32%, and volatile hydrocarbons—63%. Moreover, it was found that biodiesel undergoes five times faster biodegradation in the environment than diesel oil.

Keywords: biofuel; greenhouse gas emissions; reduction in GHG emissions

Citation: Niekurzak, M. Determining the Unit Values of the Allocation of Greenhouse Gas Emissions for the Production of Biofuels in the Life Cycle. *Energies* **2021**, *14*, 8394. https://doi.org/10.3390/en14248394

Academic Editors: Tadeusz Dziubak and Diego Luna

Received: 17 November 2021
Accepted: 8 December 2021
Published: 13 December 2021

Publisher's Note: MDPI stays neutral with regard to jurisdictional claims in published maps and institutional affiliations.

1. Introduction

The progressive exploitation of non-renewable resources, such as coal, oil, or gas, leads to the excessive use of these raw materials and the exhaustion of stocks. The beginning of the industrial era based on energy-intensive systems has increased the demand for energy. For several years, an increase in interest in the production of fuels from organic sources has been observed in the world [1]. This is a result of the overlapping of several factors: high oil prices, individual countries' striving for energy sovereignty, counteracting global warming, and the limited resources of non-renewable resources. In order to meet the challenges faced by the energy sector and meet environmental protection requirements, the development of renewable energy sources is essential [2–4]. Biofuels are all fuels that are produced from biomass. Biomass is considered to be all biodegradable animal and plant matter, as well as their metabolic products. Biofuels can be in the form of: gaseous, solid, or liquid. The representative of the first group is biogas obtained in the process of anaerobic fermentation [5–8]. Liquid biofuels are mainly: bioethanol (ethyl alcohol produced from plants in fermentation and distillation processes) and biodiesel (chemically processed vegetable oil). Solid biofuels are processed and unprocessed biomass, as well as a biodegradable fraction of municipal waste. All the mentioned biofuels are used in heating and power engineering [9–11].

Many researchers have reported that different blends of biodiesel and diesel can be effective in reducing CO, HC, and PM emissions such as cooking oil waste biodiesel [12], jatropha oil biodiesel [13], caranja biodiesel [14], biodiesel from rapeseed oil [15], soybean oil biodiesel [16], and palm oil biodiesel [17]. Many researchers have also studied the effect of biofuels on the performance of internal combustion and diesel engines [18–27]. Few studies have been carried out on the calculation of GHG emissions with the use of the allocation method based on mass, energy, and financial contributions of biodiesel. Therefore, this article fills this research gap, and the applied methods have a potential application value for further analysis of the physicochemical properties of, for example, PM particles emitted from diesel engines in the future.

Poland, like other EU Member States, is obliged to implement the provisions of EU directives, including Directive 2009/28/EC [1] promoting renewable energy sources (RED Directive). It is a comprehensive document with a lot of attention to the assessment of biofuels and bioliquids and the need to demonstrate that they meet the sustainability criteria. Confirmation of this fact is to be obtained by the supplier of an appropriate certificate under the selected certification system. One of the elements of the audit is the assessment of the determination of the value of greenhouse gas emissions over the life cycle. The correctness of the determination of this value is therefore extremely important, and it is influenced by many factors, including the method of allocating GHG emissions, as well as the calculation tools used.

The new directive on renewable energy, introduced in 2021, provides for a reduction in greenhouse gas emissions by 40% by 2030, compared to the result from 1990. In addition, the standard provides for a 32% energy share renewable in final energy consumption [28]. One of the most important changes envisaged by the RED II directive is the fact that not only the rules relating to the biofuel production chain will be implemented in the EU-wide sustainable development, but also for biomass fuels that are used in the electricity sector, as well as in the heating and cooling sector. In the context of biofuels themselves, the new regulations put more emphasis on reducing greenhouse gas emissions. In this regard, it sets new criteria to reduce gas emissions by 65–70% for installations that will start operating after 1 January 2021, and 80% for installations that will start operating after 1 January 2026 [29]. First generation fuels, produced on the basis of agricultural raw materials, will be additionally burdened with indirect land-user risk indicators. The assumptions of the latest regulations prioritize the development of advanced biofuels because they assume an increase in the share of these fuels from 0.5%, which took place in 2020, to the forecasted 3.5% in 2030. As part of advanced biofuels, the production of biodiesel from used cooking oils will continue to play a significant role [30,31]. In connection with the above activities, greater supervision and monitoring of entities dealing with their greenhouse gas emissions for the production of biofuels in the life cycle is expected.

2. Materials and Methods

The issues of the allocation method and other factors influencing the GHG emissions result for biodiesel are presented in this paper. In addition, all the components of GHG emissions generated during the cultivation of the raw material used to produce the final biocomponent were determined and their legitimacy was determined. This article analyzes rapeseed–agricultural raw materials most often used in Poland for the production of vegetable oil, from which methyl esters of fatty acids are produced at a later stage. The calculations made in this study were based on real data obtained from various entities. The obtained data was averaged and served as input data for the calculations of greenhouse gas emissions.

Due to the fact that for the purposes of meeting the requirements of Directive 2009/28/EC [1], the GHG emissions for the cultivation stage is given in g CO_2eq per 1 MJ of the obtained biofuel, the obtained emissions for one ton of agricultural raw material should be recalculated taking into account all successive conversion processes. In the case of oilseed rape, these are the most common processes leading to the production of fatty acid methyl esters (FAME). For this purpose, calculations are made with the use of conversion factors for a given treatment process and emission allocation factors are applied taking into account the type of obtained products: main and by-products. As a result, the final result may be influenced by the selection of conversion factors used in the calculations for oil pressing and FAME production, as well as allocation factors depending on the mass share of individual process products and their calorific value.

2.1. Overall Mass Balance for the Entire Process

In order to analyze GHG emissions on the basis of various allocation methods, the overall mass balance of the entire production process was calculated according to individual stages (Figure 1):

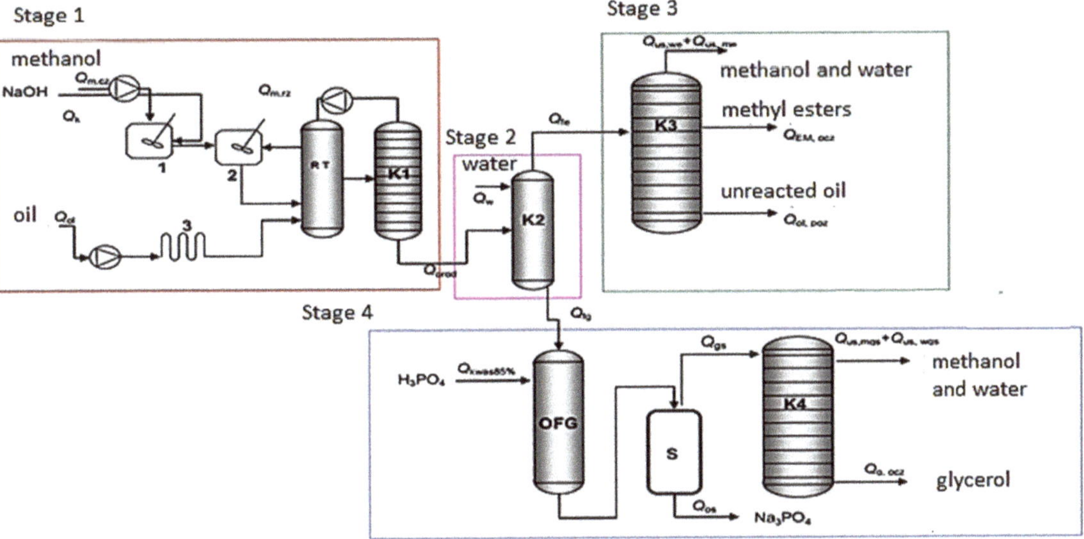

Figure 1. Diagram of biodiesel production by transesterification with the use of a basic catalyst in a flow system. Source: own study.

Figure 1 shows a technological scheme for the production of biodiesel from rapeseed oil by transesterification with the use of a basic catalyst (NaOH), including 4 stages:

(1) Transesterification and recovery of methanol;
(2) Separation of methyl esters and glycerin fractions;
(3) Purification of methyl esters;
(4) Purification of the glycerin fraction.

The production of biodiesel as part of the research consisted in directing the stream of crude rapeseed oil, after increasing the pressure and heating (temperature 60 °C and pressure 4 bar) for transesterification. Fresh methanol and catalyst (NaOH) are then routed to the mixer, to which also the methanol recovered from transesterification is recirculated. The resulting sodium methoxide is successively directed to the column where the transesterification is carried out, after which the mixture is sent to the distillation column. Methanol is recovered at the temperature of 150 °C, and then, after cooling down to the temperature of 60 °C, it is returned to the process. The remaining components, after cooling in a heat exchanger, are directed to the separation, where the ester phase is separated from the glycerin phase and impurities. Then, the esters are routed to purification. Initially, in the countercurrent reactor they are rinsed with water at 25 °C to remove soaps. From the countercurrent column, they are directed to a centrifuge, where they separate from impurities, and then pass them to vacuum distillation to dry them. In this way, products with a purity of 99.8% are obtained. Subsequently, the glycerin phase is directed to the tank where H_3PO_4 is introduced in order to neutralize the basic catalyst. After centrifugation in the centrifuge, the Na_3PO_4 sediment formed is treated as waste. Then, after the distillation of crude glycerin, technical glycerin is obtained. The advantages of the process include: complexity of the system, high efficiency, and high quality of esters.

Process description:

Stage 1

- Combination of sodium base (Q_k) with methanol ($Q_{m,\,cz}$) in the mixer (1);
- Feeding the obtained mixture and recirculated methanol ($Q_{m,\,rz}$) to the mixer (2);
- Supplying oil (Q_{ol}) and methanol with the catalyst to the RT reactor for transesterification;
- Feeding the transesterified mixture to the distillation column (K1) to recover the methanol;
- Recirculation of the recovered condensed methanol ($Q_{m,\,rz}$) to the mixer (2).

Stage 2

- Feeding the transesterification products: esters, glycerin, unreacted oil, catalyst, and water as a washing substance to the washing column (K2);
- Separation of the ester phase (Q_{fe}) from the glycerin phase (Q_{fg}) in the K2 column.

Stage 3

- Directing impure esters (Q_{fe}) to column K3 in order to remove from them methanol ($Q_{us,\,me}$), water ($Q_{us,\,we}$) and unreacted oil ($Q_{ol,\,poz}$);
- Collection of purified methyl esters ($Q_{EM,\,eyes}$) in the tank.

Stage 4

- Directing the contaminated glycerin phase (Q_{fg}) to a neutralization reactor (OFG) to remove the catalyst, methanol, and water;
- Feeding phosphoric acid to the reactor;
- Directing the products resulting from the neutralization to the separator (S) in order to remove the sediment (Q_{osad});
- Crude glycerin (Q_{gs}) is directed to the distillation column (K4) to remove water ($Q_{us,\,wg}$) and methanol ($Q_{us,\,mg}$);
- Purified glycerin ($Q_{g,\,ocz}$) is formed in the K4 column.

2.2. GHG Emissions Allocation

In the production of biofuels, in addition to the main product, there are also by-products and waste. In line with the methodology set out in the RED directive, the GHG emissions generated during production are allocated to the main product and by-products. Emissions are not allocated to waste if it is used for other purposes (e.g., energy). Then the emissions amount for the generation step is assumed to be zero. The way in which the resulting GHG emissions are "split" between the produced biofuel and by-products will have an impact on the final result of the biofuel's ability to reduce greenhouse gas emissions. Emissions allocation should be carried out at the production stage, which produces the biofuel, bioliquid, or by-product suitable for storage and sale. The allocation of GHG emissions can be carried out at individual stages of the production of the final product and by-products, after which these products are still processed in subsequent stages. If the subsequent stages of production (products and by-products) are related to the previous ones (material or energy factors), the allocation should be made at the moment when these stages become separate processes, not related in any way to the previous ones.

The total GHG emissions and allocation to the main product and by-product were calculated on the basis of the following formulas [32,33]:

$$C_t = C_f + C_m + C_e \tag{1}$$

where

C_t—total emissions related to all inputs, CO_2eq,

C_f—emissions contributed with the raw material, CO_2eq,

C_m—emissions brought in with other materials, CO_2eq,
C_e—emissions related to energy consumption, CO_2eq.

The allocation of GHG emissions to biofuels/bioliquids and to the by-product was calculated from the following formulas:

$$C_1 = C_t \cdot Q_1 \cdot \frac{LHV_1}{Q_1 \cdot LHV_1 + Q_2 \cdot LHV_2}$$

$$C_2 = C_t \cdot Q_2 \cdot \frac{LHV_1}{Q_1 \cdot LHV_1 + Q_2 \cdot LHV_2} \tag{2}$$

where

C_t—total emissions related to all inputs, CO_2eq,
C_1—GHG emissions allocation to biofuel/bioliquid, CO_2eq,
C_2—allocation of GHG emissions to the by-product, CO_2eq,
$Q_{1/2}$—the quantity of the product 1/2, expressed in mass units,
$LHV_{1/2}$—calorific value of product 1/2, expressed as a unit of energy per unit mass.
As part of the research, the allocation was carried out:

- On the basis of physical quantities (mass, energy content).

This method is based on assigning GHG emissions to each of the resulting products and by-products in direct proportion to their obtaining (based on the mass or energy balance) [34]. If the allocation method is adopted based on the mass balance, the mass of the main products, and by-products was initially calculated. Then, based on their percentages of the total mass of production (sum of the masses of the main product and the by-product), they were assigned an emissions percentage.

- On the basis of economic figures.

Allocation based on economic quantities gives the least stable and less comparable results. The allocation can be made based on the market prices of raw materials and finished products, production costs, storage, transportation of the final product and by-products. Analyzes carried out in different regions of the world may differ from each other, because the prices of raw materials and by-products, as well as production costs, can vary significantly depending on the economic policy of a country and on the location of the region.

- Based on an extensive system.

The allocation made by the extended system method is used especially by scientists from the USA. According to the concept of this method, the system boundaries are extended to include additional alternative products. The activities not related to the life cycle of a given product are also included in the calculations. First, you need to define the amount of biofuel produced and the by-products and products that are on the market that can be replaced by biofuel by-products. Next, the ratio to which the products in question can be replaced by by-products of the biofuel production process is calculated and the environmental impact of the products to be replaced is determined. It may turn out that you replace existing products on the market with products byproducts of the biofuel production process will reduce the negative environmental impact of the biofuel life cycle.

2.3. GHG Emissions Calculation Method

According to the RED Directive, greenhouse gas emissions from the production and use of transport fuels, biofuels and bioliquids are calculated from the formula [32,33]:

$$E = e_{ec} + e_l + e_p + e_{td} + e_u - e_{sca} - e_{ccs} - e_{ccr} - e_{ee} \tag{3}$$

where

E—total emissions caused by the use of fuel,
e_{ec}—emissions caused by the extraction or cultivation of raw materials,

e_l—annual emissions caused by changes in the amount of the carbon element in connection with the change in land use,

e_p—emissions caused by technological processes,

e_{td}—emissions from transport and distribution,

e_u—emissions caused by the fuel used,

e_{sca}—emissions saving value due to carbon accumulation in the soil thanks to better farming,

e_{ccs}—reduction in emissions due to carbon capture and storage in deep geological structures,

e_{ccr}—emissions reduction due to carbon capture and replacement,

e_{ee}—emissions reduction due to increased electricity production from cogeneration.

3. Results and Discussion

3.1. Overall Mass Balance for the Entire Process

Technological calculations for the four stages of biodiesel production.

Stage 1. The alcohol transesterification and recovery process include the calculation of methanol and catalyst charge and calculations related to methanol recovery and recirculation (Table 1).

Table 1. Technological assumptions for methanol transesterification and recovery.

Parameter	Symbol	Unit	Assumed Value
oil flow rate	Q_{ol}	kg/h	1050
methanol concentration	$\eta_{m,e}$	% weight of raw material	11
catalyst concentration	η_k	% weight of raw material	1.0
alcohol density	ρ_m	g/cm^3	0.797
oil density	ρ_{ol}	g/cm^3	0.899
content of triacylglycerols	η_{Ac}	%	~100
transesterification temperature	T_e	°C	60
transesterification pressure	P_e	kPa (atm)	400 (4.07)
yield of transesterification	η_e	%	95
alcohol distillation temperature	T_{dest}	°C	150
alcohol distillation pressure	P_{dest}	kPa (atm)	30 (3.06)
alcohol recovery efficiency	η_{dest}	%	94

Source: own study based on [32,33].

- Supply of raw materials

The calculations concern the required amounts of methanol and catalyst to carry out the transesterification [33,35]:

Amount of methanol needed for transesterification $Q_{m,t}$ (kg/h):

$$Q_{m,t} = Q_{ol} \cdot \frac{\eta_{m,e}}{100} = 1050 \cdot \frac{11}{100} = 115.5 \left[\frac{kg}{h}\right] \tag{4}$$

The amount of methanol supplied to the reactor Q_m (kg/h), with its twofold excess:

$$Q_m = 2 \cdot Q_{m,t} = 2 \cdot 115.5 = 231 \left[\frac{kg}{h}\right] \tag{5}$$

Required amount of catalyst Q_k (kg/h):

$$Q_k = Q_{ol} \cdot \frac{\eta_k}{100} = 1050 \cdot \frac{1.0}{100} = 10.5 \left[\frac{kg}{h}\right] \tag{6}$$

- Recovery of methanol

The calculations concern the possibility of methanol recovery and its reuse in the transesterification process and the amount of pure methanol supplied to the process.

Amount of methanol theoretically recoverable $Q_{m,teor}$ (kg/h):

$$Q_{m,teor} = Q_m - \left(Q_{m,t} \cdot \frac{\eta_e}{100}\right) = 231 - \left(115.5 \cdot \frac{95}{100}\right) = 121.275 \left[\frac{kg}{h}\right] \qquad (7)$$

Actual amount of methanol recovered $Q_{m,rz}$ (kg/h):

$$Q_{m,rz} = Q_{m,teor} \cdot \frac{\eta_{dest}}{100} = 121.275 \cdot \frac{94}{100} = 113.9985 \left[\frac{kg}{h}\right] \qquad (8)$$

Amount of pure methanol to be fed to the reactor, taking into account its recirculation, $Q_{m,cz}$ (kg/h):

$$Q_{m,cz} = Q_m - Q_{m,rz} = 231 - 113.9985 = 117.0015 \left[\frac{kg}{h}\right] \qquad (9)$$

Amount of methanol remaining after distillation in the stream of transesterification products $Q_{m,poz}$ (kg/h):

$$Q_{m,poz} = Q_{m,teor} - Q_{m,rz} = 121.275 - 113.9985 = 7.2762 \left[\frac{kg}{h}\right] \qquad (10)$$

- Transesterification products

Based on the transesterification equation, when reacting with 100 kg of oil, 100.45 kg of biodiesel and 10.55 kg of glycerol can be obtained (assuming the molar weight of the oil is 871.67 g/mol, and the methyl esters are 875.6 g/mol). The material balance of raw materials, products and by-products after the alcohol transesterification and recovery stage is presented in Table 2.

Amount of Q_{ME} methyl esters (kg/h):

$$Q_{ME} = \frac{100.45 \cdot Q_{ol}}{100} \cdot \frac{\eta_e}{100} = \frac{100.45 \cdot 1050}{100} \cdot \frac{95}{100} = 1001.9888 \left[\frac{kg}{h}\right] \qquad (11)$$

Amount of unreacted rapeseed oil $Q_{ol,poz}$ (kg/h):

$$Q_{ol,poz} = Q_{ol} \cdot \left(1 - \frac{\eta_e}{100}\right) = 1050 \cdot \left(1 - \frac{95}{100}\right) = 52.5 \left[\frac{kg}{h}\right] \qquad (12)$$

Amount of glycerol $Q_{glicerol}$ (kg/h):

$$Q_{glicerol} = \frac{10.4 \cdot Q_{ol}}{100} \cdot \frac{\eta_e}{100} = \frac{10.4 \cdot 1050}{100} \cdot \frac{95}{100} = 103.74 \left[\frac{kg}{h}\right] \qquad (13)$$

Stage 2. Separation of methyl esters and glycerin fraction. At the stage of separation of methyl esters and glycerin fractions, the water charge needed for ester washing, methanol and catalyst loads, and water drained from the esters and glycerin phase should be calculated on the basis of the shares of individual components. The technological assumptions for the separation of methyl esters and the glycerol fraction are presented in Table 3.

Table 2. Material balance of raw materials, products, and by-products after the stage of alcohol transesterification and recovery.

Raw Materials			Products		
Type	**Symbol**	**Load (kg/h)**	**Type**	**Symbol**	**Load (kg/h)**
canola oil	Q_{ol}	1050	methyl esters	Q_{ME}	1001.9888
			unreacted oil glycerol	$Q_{ol,poz.}$	52.5
			unreacted oil glycerol	$Q_{glicerol}$	103.7400
catalyst (NaOH)	Q_k	10.5	catalyst (NaOH)	Q_k	10.5
fresh methanol	$Q_{m,cz}$	117.0015	unreacted methanol	$Q_{m,poz.}$	7.2765
Sum				Q_{prod}	1176.0150

Source: own study based on [32,33].

Table 3. Technological assumptions for the separation of methyl esters and glycerin fraction.

Parameter	Symbol	Unit	Value
amount of water for rinsing the methyl esters	η_w	% wag. Q_{prod}	1.0
water fraction (ester fraction/glycerin fraction)	$\eta_{w,e}/\eta_{w,g}$	%	10/90
methanol fraction (ester fraction/glycerol fraction)	$\eta_{m,e}/\eta_{m,g}$	%	60/40
catalyst fraction (NaOH) (ester fraction/glycerol fraction)	$\eta_{k,e}/\eta_{k,g}$	%	0/100
unreacted oil fraction (ester fraction/glycerin fraction)	$\eta_{ol,e}/\eta_{ol,g}$	%	100/0
temperature (separator inlet/outlet)	T_s	°C	50/60
pressure (separator inlet/outlet)	p_s	kPa (atm.)	110/120 1.12/1.22

Source: own study based on [32,33].

Amount of water needed for rinsing methyl esters Q_w (kg/h):

$$Q_w = Q_{prod} \cdot \frac{\eta_w}{100} = 1176.015 \cdot \frac{1.0}{100} = 11.7602 \left[\frac{kg}{h}\right] \tag{14}$$

Amount of water discharged with the ester fraction $Q_{w,e}$ (kg/h):

$$Q_{w,e} = Q_w \cdot \frac{\eta_{w,e}}{100} = 11.7602 \cdot \frac{10}{100} = 1.176 \left[\frac{kg}{h}\right] \tag{15}$$

Amount of methanol discharged with the ester fraction $Q_{m,e}$ (kg/h):

$$Q_{m,e} = Q_{m,poz} \cdot \frac{\eta_{m,e}}{100} = 7.2765 \cdot \frac{60}{100} = 4.3657 \left[\frac{kg}{h}\right] \tag{16}$$

Amount of catalyst (NaOH) discharged with the ester fraction $Q_{k,e}$ (kg/h):

$$Q_{k,e} = Q_k \cdot \frac{\eta_{k,e}}{100} = 10.5 \cdot 0 = 0 \left[\frac{kg}{h}\right] \tag{17}$$

Amount of unreacted oil discharged with the ester fraction $Q_{ol,e}$ (kg/h):

$$Q_{ol,e} = Q_{ol,poz} \frac{\eta_{ol,e}}{100} = 52.5 \cdot \frac{100}{100} = 52.5 \left[\frac{kg}{h}\right] \tag{18}$$

Amounts of water, methanol, catalyst, and unreacted oil discharged with the glycerin fraction, analogous to the ester fraction:

$$Q_{w,g} = Q_w \cdot \frac{\eta_{w,g}}{100} = 11.7602 \cdot \frac{90}{100} = 10.5842 \left[\frac{kg}{h}\right]$$

$$Q_{m,g} = Q_{m,\,poz} \cdot \frac{\eta_{m,g}}{100} = 7.2765 \cdot \frac{40}{100} = 2.9106 \left[\frac{kg}{h}\right]$$

$$Q_{k,g} = Q_{k} \cdot \frac{\eta_{k,g}}{100} = 10.5 \cdot \frac{100}{100} = 10.5 \left[\frac{kg}{h}\right]$$

$$Q_{ol,g} = Q_{ol,poz} \cdot \frac{\eta_{ol,g}}{100} = 52.5 \cdot 0 = 0 \left[\frac{kg}{h}\right] \tag{19}$$

The ester and glycerol fractions discharged from the separator are presented Table 4.

Table 4. Ester and glycerol fractions discharged from the separator.

Parameter	Unit	Value	
		Ester Phase	Glycerin Phase
load of methyl esters/glycerol	kg/h	Q_{ME} 1001.9888	$Q_{glycerol}$ 103.74
water load	kg/h	$Q_{w,e}$ 1.176	$Q_{w,g}$ 10.5842
methanol charge	kg/h	$Q_{m,e}$ 4.3657	$Q_{m,g}$ 2.9106
catalyst load (NaOH)	kg/h	$Q_{k,e}$ 0	$Q_{k,g}$ 10.5
unreacted oil load	kg/h	$Q_{ol,e}$ 52.5	$Q_{ol,g}$ 0
Charge of the ester phase/glycerol phase	kg/h	Q_{fe} 1060.0305	Q_{fg} 127.7348

Source: own study based on [32,33].

Stage 3. Purification of methyl esters. The purification of methyl esters consists in removing water, methanol, and unreacted oils from them based on the degree of removal of individual components from the main product. The technological assumptions for the purification of methyl esters are presented in Table 5.

Table 5. Technological assumptions for the purification of methyl esters.

Parameter	Symbol	Unit	Value
degree of water removal	$\eta_{us\ w,e}$	%	99.6
methanol removal rate	$\eta_{us\ m,e}$	%	99.6
degree of removal of unreacted oil	$\eta_{us\ ol,e}$	%	100
temperature in the distillation column	$T_{dest,c}$	°C	193.7
pressure in the distillation column	$P_{dest,e}$	kPa (atm.)	10 (0.102)

Source: own study based on [32,33].

Removed amount of water $Q_{us\ w,e}$ (kg/h):

$$Q_{us\ w,e} = Q_{w,e} \cdot \frac{\eta_{us\ w,e}}{100} = 1.176 \cdot \frac{99.6}{100} = 1.1713 \left[\frac{kg}{h}\right] \tag{20}$$

Removed amount of methanol $Q_{us\ m,e}$ =(kg/h):

$$Q_{us\ m,e} = Q_{m,e} \cdot \frac{\eta_{us\ w,e}}{100} = 4.3657 \cdot \frac{99.6}{100} = 4.3482 \left[\frac{kg}{h}\right] \tag{21}$$

The amount of unreacted oil removed $Q_{us\ ol,e}$ (kg/h):

$$Q_{us\ ol,e} = Q_{ol,e} \cdot \frac{\eta_{us\ ol,e}}{100} = 52.5 \cdot \frac{100}{100} = 52.5 \left[\frac{kg}{h}\right] \tag{22}$$

Amount of purified methyl esters $Q_{EM,ocz}$ (kg/h):

$$Q_{EM,ocz} = Q_{fe} - Q_{us\ w,e} - Q_{us\ m,e} - Q_{us\ ol,e} = 1060.0305 - 1.1713 - 4.3482 - 52.5 = 1002.0011 \left[\frac{kg}{h}\right] \tag{23}$$

The material balance of raw materials, products and by-products after the purification stage of methyl esters is presented in Table 6.

Table 6. Material balance of raw materials, products, and by-products after the methyl ester purification step.

Raw Materials		Products		Side Products	
Type	**Load (kg/h)**	**Type**	**Load (kg/h)**	**Type**	**Load (kg/h)**
ester fraction	Q_{fe} 1060.0305	purified methyl esters	$Q_{EM,ocz}$ 1002.0011	water removed	$Q_{us\,w,e}$ 1.1713
				methanol removed	$Q_{us\,m,e}$ 4.3482
				unreacted oil removed	$Q_{us\,ol,e}$ 52.5
Sum					58.0195

Source: own study based on [32,33].

Stage 4. Purification of the glycerin fraction. The purification of the glycerol fraction consists in removing the catalyst.

- Catalyst removal (NaOH)

To neutralize 1 kg of NaOH, use 0.81667 kg of pure phosphoric acid. The reaction produces 1.3667 kg of sodium triphosphate and 0.450 kg of water.

Amount of pure phosphoric acid to neutralize the sodium hydroxide $Q_{kwas,100}$ (kg/h):

$$Q_{kwas,100} = 0.81667 \cdot Q_{k,g} = 0.81667 \cdot 10.5 = 8.575 \left[\frac{kg}{h}\right] \tag{24}$$

The amount of phosphoric acid at 85% concentration to neutralize the sodium hydroxide $Q_{kwas,85}$ (kg/h):

$$Q_{kwas,85} = \frac{100}{85} \cdot Q_{kwas,100} = \frac{100}{80} \cdot 8.575 = 10.7188 \left[\frac{kg}{h}\right] \tag{25}$$

Amount of water introduced with 85% phosphoric acid $Q_{w\,kwas,85}$ (kg/h):

$$Q_{w\,kwas,85} = \left(\frac{100-85}{85}\right) \cdot Q_{kwas,\,85} = \left(\frac{100-85}{85}\right) \cdot 10.7188 = 1.8916 \left[\frac{kg}{h}\right] \tag{26}$$

Amount of tri-sodium phosphate formed Q_{osad} (kg/h):

$$Q_{osad} = 1.3667 \cdot Q_{k,g} = 1.3667 \cdot 10.5 = 14.35 \left[\frac{kg}{h}\right] \tag{27}$$

Amount of water formed by the neutralization reaction of sodium hydroxide $Q_{w,z}$ (kg/h):

$$Q_{w,z} = 0.450 \cdot Q_{k,g} = 0.450 \cdot 10.5 = 4.725 \left[\frac{kg}{h}\right] \tag{28}$$

Amount of glycerin fraction after catalyst removal (crude glycerin) $Q_{fg,n}$ (kg/h):

$$Q_{fg,n} = Q_{fg} - Q_{k,g} = 127.7348 - 10.5 = 117.2348 \left[\frac{kg}{h}\right] \tag{29}$$

The material balance of raw materials, products and by-products after the catalyst removal stage from the glycerin fraction is presented in Table 7.

Table 7. Material balance of raw materials, products, and by-products after the catalyst removal step from the glycerin fraction.

Raw Materials		Products		Side Products	
Type	**Load (kg/h)**	**Type**	**Load (kg/h)**	**Type**	**Load (kg/h)**
glycerin fraction	Q_{fg} 127.7348	glycerin fraction after catalyst removal	$Q_{fg,n}$ 117.2348	tri-sodium phosphate (precipitate)	Q_{osad} 14.35
pure phosphoric acid	$Q_{kwas,100}$ 8.575	water formed by the neutralization reaction	$Q_{w,z}$ 4.725		
water introduced from acid. phosphorus 85%	$Q_{kwas,85}$ 1.8916	water introduced from phosphoric acid 85%	$Q_{w\,kwas,85}$ 1.8916		
Sum		Q_{gs} = 123.8514			

Source: own study based on [32,33].

- Purification of crude glycerin

The purification of crude glycerin is based on the removal of water and methanol based on the degree of removal of these components. Technological assumptions for the purification of raw glycerin is presented in Table 8.

Table 8. Technological assumptions for the purification of raw glycerin.

Parameter	Symbol	Unit	Value
degree of water removal	$\eta_{us\,w,gs}$	%	33.7
methanol removal rate	$\eta_{us\,m,gs}$	%	100

Source: own study based on [32,33].

Removed amount of water $Q_{us\,w,gs}$ (kg/h):

$$Q_{us\,w,gs} = Q_{gs} - Q_{fg,n} \cdot \frac{\eta_{us\,w,gs}}{100} = 123.8514 - 117.2348 \cdot \frac{33.7}{100} = 84.3433 \left[\frac{kg}{h}\right] \quad (30)$$

Removed amount of methanol $Q_{us\,m,gs}$ (kg/h):

$$Q_{us\,m,gs} = Q_{m,g} \cdot \frac{\eta_{us\,m,gs}}{100} = 2.9106 \cdot \frac{100}{100} = 2.9106 \left[\frac{kg}{h}\right] \quad (31)$$

Amount of purified glycerin $Q_{g,ocz}$ (kg/h):

$$Q_{g,ocz} = Q_{gs} - Q_{us\,w,gs} - Q_{us\,m,gs} = 123.8514 - 84.3433 - 2.9106 = 36.5975 \left[\frac{kg}{h}\right] \quad (32)$$

Material balance of raw materials, products, and by-products after the crude glycerin purification step is presented in Table 9.

Table 9. Material balance of raw materials, products, and by-products after the crude glycerin purification step.

Raw Materials		Products		Side Products	
Type	**Load (kg/h)**	**Type**	**Load (kg/h)**	**Type**	**Load (kg/h)**
raw glycerin	Q_{gs} 123.8514	purified glycerin	$Q_{g,ocz}$ 36.5975	water removed	$Q_{us\,w,gs}$ 84.3433
				methanol removed	$Q_{us\,m,gs}$ 2.9106
Suma					84.1049

Source: own study based on [32,33].

Overall mass balance. On the basis of mass balances prepared for each stage of biodiesel production, a general balance was prepared for an hour of the entire technological process. Material balance of raw materials, products, and by-products after taking into account all four stages of the technological process is presented in Table 10.

Table 10. Material balance of raw materials, products, and by-products after taking into account all four stages of the technological process.

Raw Materials		Products		Side Products	
Type	**Load (kg/h)**	**Type**	**Load (kg/h)**	**Type**	**Load (kg/h)**
canola oil	1050	methyl esters purified	1002.0011	unreacted oil	52.5
catalyst	10.5			purified glycerin	36.5975
methanol	117.0015			methanol	4.3482
water	11.7602			water	1.1713
phosphoric acid	8.575			tri-sodium phosphate	14.35
Sum	1197.837		1002.0011		108.867

Source: own study based on [32,33].

Table 11 presents the annual mass balance of the transesterification process. It was assumed that the process installation works 8000 h a year, and to facilitate the calculations, a ton was taken as the basic unit. The annual material balance of raw materials, products, and by-products of the transesterification process is presented in Table 11.

Table 11. The annual material balance of raw materials, products, and by-products of the transesterification process.

Raw Materials		Products		Side Products	
Type	**Load (t/Number of Hours a Year (8000 h)**	**Type**	**Load (t/Number of Hours a Year (8000 h)**	**Type**	**Load (t/Number of Hours a Year (8000 h)**
canola oil	8400	methyl esters purified	8016	unreacted oil	420
catalyst	84			purified glycerin	292.8
methanol	936			methanol	346.4
water	94.08			water	9.6
phosphoric acid	688			tri-sodium phosphate	115.2
Sum	10,202.08		8016		1184

Source: own study based on [32,33].

3.2. GHG Emissions Allocation

To assess the impact of emissions allocation in biodiesel production, it was assumed that the emissions would be split between the main product—biodiesel (purified methyl esters) and the by-product—glycerin. Three different ways of allocating emissions were carried out on the basis of mass balance, financial, and calorific values. Allocation based on the mass balance is presented in Table 12.

Table 12. Allocation based on the mass balance.

Product	Annual Production (t)	Total Weight of Products	% of Assigned Emissions
biodiesel	8016	8308.8	96.5
glycerin	292.8		3.5

Source: own study based on [32,33].

- Allocation based on the mass balance of the installation

The allocation on the basis of a mass balance showed that 96.5% of the emissions are attributed to biodiesel and the remaining 3.5% to glycerin. Allocation of emissions taking into account the market value of the resulting products is presented in Table 13.

Table 13. Allocation of emissions taking into account the market value of the resulting products.

Product	Annual Production (t)	Value (EUR/ton)	Product Value (EUR)	The Total Value of EUR	% of Assigned Emissions
biodiesel	8016	800	6,273,391.31	6507146.67	98.50%
glycerin	292.8	322.23	94,346.67		1.50%

Source: own study based on [32,33].

- Financial allocation

The financial allocation of the issue takes into account the market values of the products. The prices of biodiesel and glycerin were taken from internet sources. In this case, the attributed emissions for biodiesel is 98.6%, while for glycerin only 1.5% (this is due to the high price of biodiesel in relation to glycerin and higher annual biofuel production). Allocation of emissions based on the calorific value of products is presented in Table 14.

Table 14. Allocation of emissions based on the calorific value of products.

Product	Annual Production (t)	Calorific Value (GJ/t)	Energy Contained in Product (GJ)	Total Energy (GJ)	% of Assigned Emissions
biodiesel	8016.0	37.5	300,600	306,543.84	98
glycerin	292.8	20.3	5943.84		2

Source: own study based on [32,33].

- Allocation based on the calorific value of the products

Taking into account the calorific values of the products, the emissions assigned to biodiesel is 98% and to glycerin 2%. This is due to almost twice the calorific value of biofuel as glycerin. As can be seen from the above calculations, the allocation method is important in estimating GHG emissions for the main product (i.e., *biodiesel*). The attributed GHG emissions to the biofuel ranges from 96.5% (allocation based on a mass balance) to 98.5% (financial allocation).

3.3. GHG Emissions in the Life Cycle of Biodiesel with Different Allocation Factors for the Transesterification Stage

The final result of GHG emissions is also influenced by the values of emitted pollutants obtained in the entire process (cultivation, storage, and transport). An analysis was performed to assess the impact of the adopted method of GHG emissions allocation at the production stage on the final value. It was based on the GHG emissions values (converted into GJ of energy contained in the biofuel) presented in the Biograce calculator. According to Biograce, the emissions allocation factor for biodiesel is 95.7% and is close to the calculated results. The calculations assume that the land-use change emissions and the brownfield rehabilitation bonus are zero and are not taken into account in the analysis. GHG emissions in the biofuel life cycle based on mass share is presented in Table 15.

For the allocation factor of 96.50%, the GHG emissions result is 51.92 g CO_2 eq/MJ for the transesterification process. GHG emissions in the biofuel life cycle based on financial allocation is presented in Table 16.

For the allocation factor of 98.50%, the GHG emissions result is 52.27 g CO_2 eq/MJ for the transesterification process. GHG emissions in the life cycle of a biofuel based on the energy content is presented in Table 17.

Table 15. GHG emissions in the biofuel life cycle based on mass share.

Stage	Issue without Taking into Account the Allocation (g CO_2 eq/MJ)	Allocation Factor	Issue after Taking into Account the Allocation (g CO_2 eq/MJ)	Share of Emissions GHG
		stage$_{ec}$		
Cultivation	48.35	58.60%	28.33	54.57%
Storage	0.72	58.60%	0.42	0.81%
		stage$_p$		
Oil extraction	6.5	58.60%	3.81	7.34%
Refining	1.06	95.70%	1.01	1.95%
Transesterification	17.51	96.50%	16.90	32.54%
		stage$_{td}$		
Rapeseed transport	0.3	58.60%	0.18	0.34%
Rapeseed oil transport	0	95.70%	0.00	0.00%
Transport of biodiesel to the warehouse	0.47	100.00%	0.47	0.91%
Transport to petrol stations	0.8	100.00%	0.80	1.54%
Sum	75.71		51.92	100.00%

Source: own study based on [32,33].

Table 16. GHG emissions in the biofuel life cycle based on financial allocation.

Stage	Issue without Taking into Account the Allocation (g CO_2 eq/MJ)	Allocation Factor	Issue after Taking into Account the Allocation (g CO_2 eq/MJ)	Share of Emissions GHG
		stage$_{ec}$		
Cultivation	48.35	58.60%	28.33	54.20%
Storage	0.72	58.60%	0.42	0.81%
		stage$_p$		
Oil extraction	6.50	58.60%	3.81	7.29%
Refining	1.06	95.70%	1.01	1.94%
Transesterification	17.51	98.50%	17.25	33.00%
		stage$_{td}$		
Rapeseed transport	0.30	58.60%	0.18	0.34%
Rapeseed oil transport	0.00	95.70%	0.00	0.00%
Transport of biodiesel to the warehouse	0.47	100.00%	0.47	0.90%
Transport to petrol stations	0.80	100.00%	0.80	1.53%
Sum	75.71		52.27	100.00%

Source: own study based on [32,33].

For an allocation factor of 98%, the GHG emissions result is 52.18 g CO_2 eq/MJ for the transesterification process. The GHG emissions result for the esterification process ranges between 52.27 and 51.92 g CO_2 eq/MJ. In the case of financial allocation, 52.27 g CO_2 eq/MJ was obtained, which is the highest of all GHG emission results. The allocation based on mass shares resulted in a lower emissions result—51.92 g CO_2 eq/MJ. Fluctuations in the final result, depending on the method adopted, amount to a maximum of 0.35 g CO_2 eq/MJ. The lowest GHG emissions result was observed when using the allocation based on mass share and it is lower by 0.26 g CO_2 eq/MJ than the emissions result obtained according to the allocation based on energy content (52.18 g CO_2 eq/MJ). The result based on the financial allocation is 0.09 g CO_2 eq/MJ higher than the allocation based on energy content. The presented calculations show the influence of the adopted allocation method on the final result of the greenhouse gas emissions reduction capacity. The analysis of the obtained GHG emission results shows that for one stage of the biofuel production process (in this case transesterification) the use of different allocation methods does not significantly affect the total GHG emissions result (the maximum difference is 0.35 g CO_2 eq/MJ). It should be remembered that the analysis was carried out only for one production stage, which is the transesterification of rapeseed oil.

Table 17. GHG emissions in the life cycle of a biofuel based on the energy content.

Stage	Issue without Taking into Account the Allocation (g CO_2 eq/MJ)	Allocation Factor	Issue after Taking into Account the Allocation (g CO_2 eq/MJ)	Share of Emissions GHG
		stage$_p$		
Cultivation	48.35	58.60%	28.33	54.29%
Storage	0.72	58.60%	0.42	0.81%
		stage$_p$		
Oil extraction	6.5	58.60%	3.81	7.30%
Refining	1.06	95.70%	1.01	1.94%
Transesterification	17.51	98%	17.16	32.88%
		stage$_{td}$		
Rapeseed transport	0.3	58.60%	0.18	0.34%
Rapeseed oil transport	0	95.70%	0.00	0.00%
Transport of biodiesel to the warehouse	0.47	100.00%	0.47	0.90%
Transport to petrol stations	0.8	100.00%	0.80	1.53%
Sum	75.71		52.18	100.00%

Source: own study based on [32,33].

4. Conclusions

Based on the research, the following conclusions were drawn:

1. The use of biofuels has a better environmental impact than the use of petroleum products, as their combustion emits an average of 35% less greenhouse gases compared to the combustion of diesel fuel.
2. By allocating pollutants, total GHG emissions can be reduced over the life cycle of the main product (biodiesel) by about 31% as emissions are split between it and the by-product (glycerin).
3. The least favorable method of allocating GHG emissions is financial allocation, because its result depends on the prices of raw materials used for production and the prices of final products and by-products, which may differ in individual countries of the world. The high price of biodiesel in relation to the price of glycerin makes the total GHG emissions for the main product the highest.
4. The allocation of pollutants on the basis of mass contributions is the most advantageous method of allocating emissions GHG, as its percentage attribution is calculated on the basis of the quantities actually produced of the main product and the by-product during the year. The total amount of greenhouse gas emissions attributed to the main product is the smallest.
5. Carrying out the allocation of GHG emissions for one stage of the biofuel life cycle-transesterification does not significantly affect the total value of greenhouse gases produced, because this cycle not only consists of the production process, but also the cultivation and storage of raw materials, transport of raw materials to the plant, and transport final products to recipients.

Funding: This research received no external funding.

Conflicts of Interest: The author declares no conflict of interest.

Abbreviations

GHG	greenhouse gases
FAME	higher fatty acid methyl esters
RME	rapeseed oil methyl esters
WKT	free fatty acids
ppm	parts per million
$\eta_{m,e}$	methanol concentration
η_k	catalyst concentration
η_e	transesterification efficiency
η_{des}	alcohol recovery efficiency
Q_{ol}	rapeseed oil flow rate
$Q_{m,t}$	amount of methanol needed for transesterification
Q_m	the amount of methanol fed to the reactor, with its double excess
Q_k	required amount of catalyst
$Q_{m,teor}$	the amount of methanol theoretically possible to recover
$Q_{m,rz}$	actual amount of recovered methanol
$Q_{m,cz}$	the amount of pure methanol to be fed to the reactor, taking into account its recirculation
$Q_{m,poz}$	the amount of methanol remaining in the stream of transesterification products after distillation
Q_{ME}	amount of methyl esters
$Q_{ol,poz}$	quantity of unreacted rapeseed oil
$Q_{glicerol}$	amount of glycerol
Q_{prod}	the amount of esters, glycerin, unreacted oil and catalyst going to the separation of methyl esters and glycerin fraction
Q_w	the amount of water needed to rinse the methyl esters
η_w	the amount of rinsing water methyl esters
$Q_{w,e}$	the amount of water discharged with the ester fraction
$\eta_{w,e}$	water share in the ester fraction
$Q_{m,e}$	amount of methanol discharged with the ester fraction
$\eta_{m,e}$	share of methanol in the ester fraction
$Q_{k,e}$	amount of catalyst discharged with the ester fraction
$\eta_{k,e}$	catalyst share in the ester fraction
$Q_{ol,e}$	the amount of unreacted oil discharged with the ester fraction
$\eta_{ol,e}$	share of unreacted oil in the ester fraction
$Q_{w,g}$	the amount of water discharged with the glycerin fraction
$\eta_{w,g}$	water share in the glycerin fraction
$Q_{m,g}$	amount of methanol discharged with the glycerin fraction
$\eta_{m,g}$	share of methanol in the glycerin fraction
$Q_{k,g}$	the amount of catalyst discharged with the glycerin fraction
$\eta_{k,g}$	catalyst share in the glycerin fraction
$Q_{ol,g}$	the amount of unreacted oil discharged with the glycerin fraction
$\eta_{ol,g}$	share of unreacted oil in the glycerin fraction
Q_{fe}	charge of the ester fraction discharged from the separator
Q_{fg}	charge of glycerin fraction discharged from the separator
$Q_{us\,w,e}$	the amount of water removed from the methyl esters
$\eta_{us\,w,e}$	degree of water removal from esters methyl
$Q_{us\,m,e}$	removed amount of methanol from methyl esters
$\eta_{us\,m,e}$	the degree of methanol removal from methyl esters
$Q_{us\,ol,e}$	the amount of unreacted oil removed from methyl esters
$\eta_{us\,ol,e}$	the degree of removal of unreacted oil from methyl esters
$Q_{ME,ocz}$	amount of purified methyl esters
$Q_{kwas,100}$	the amount of pure phosphoric acid to neutralize the catalyst
$Q_{kwas,85}$	85% phosphoric acid to neutralize the catalyst
$Q_{w\,kwas,85}$	amount of water discharged with 85% phosphoric acid
Q_{osad}	amount of tri-sodium phosphate precipitate
$Q_{w,z}$	the amount of water formed in the catalyst neutralization reaction

$Q_{fg,n}$	the amount of glycerin fraction after catalyst removal
Q_{gs}	the amount of crude glycerin after the catalyst removal step
$Q_{us\,w,gs}$	the amount of water removed from the glycerin fraction
$\eta_{us\,w,gs}$	the degree of water removal from the glycerin fraction
$Q_{us\,m,gs}$	the amount of methanol removed from the glycerin fraction
$\eta_{us\,m,gs}$	the degree of methanol removal from the glycerin fraction
$Q_{g,ocz}$	the amount of purified glycerin

References

1. Dyrektywa Parlamentu Europejskiego i Rady 2009/28/WE z dnia 23.04.2009 r. w Sprawie Promowania Stosowania Energii ze Źródeł Odnawialnych Zmieniająca i w Następstwie Uchylająca Dyrektywy 2001/77/WE oraz 2003/30/WE. Dziennik Urzędowy Unii Europejskiej nr L 140/16 z r. 9 June 2009. Available online: https://eur-lex.europa.eu/legal-content/PL/TXT/PDF/?uri=CELEX:02009L0028-20151005&from=GA (accessed on 8 December 2021).
2. Niekurzak, M. The Potential of Using Renewable Energy Sources in Poland Taking into Account the Economic and Ecological Conditions. *Energies* **2021**, *14*, 7525. [CrossRef]
3. Adams, P.W.R.; McManus, M. Characterisation and variability of greenhouse gas emissions from biomethane production via anaerobic digestion of maize. *J. Clean. Prod.* **2019**, *218*, 529–542. [CrossRef]
4. Olczak, P.; Matuszewska, D.; Kryzia, D. "Mój Prąd" as an example of the photovoltaic one off grant program in Poland. *Energy Policy J.* **2020**, *23*, 123–138. [CrossRef]
5. Niekurzak, M.; Kubińska-Jabcoń, E. Analysis of the return on investment in solar collectors on the example of a household: The case of Poland. *Front. Energy Res.* **2021**, *9*, 660140. [CrossRef]
6. Ledakowicz, S.; Krzystek, L. Wykorzystanie fermentacji metanowej w utylizacji odpadów przemysłu rolno-spożywczego. *Biotechnologia* **2005**, *3*, 165–183.
7. Ardolino, F.; Parrillo, F.; Arena, U. Biowaste-to-biomethane or biowaste-to-energy? An LCA study on anaerobic digestion of organic waste. *J. Clean. Prod.* **2018**, *174*, 462–476. [CrossRef]
8. Kalina, J.; Skorek, J.; Cebula, J.; Latocha, L. Pozyskiwanie i energetyczne wykorzystanie biogazu rolniczego. *Gospodarka Paliwami i Energią* **2003**, *12*, 14–19.
9. Ge, J.; Choi, N. Soot Particle Distribution and Regulated and Unregulated Emissions of a Diesel Engine Fueled with Palm Oil Biodiesel Blends. *Energies* **2020**, *13*, 5736. [CrossRef]
10. Ge, J.C.; Kim, H.Y.; Yoon, S.K.; Choi, N.J. Reducing volatile organic compound emissions from diesel engines using canola oil biodiesel fuel and blends. *Fuel* **2018**, *218*, 266–274. [CrossRef]
11. Ge, J.C.; Kim, H.Y.; Yoon, S.K.; Choi, N.J. Optimization of palm oil biodiesel blends and engine operating parameters to improve performance and PM morphology in a common rail direct injection diesel engine. *Fuel* **2020**, *260*, 116326. [CrossRef]
12. Enweremadu, C.; Rutto, H. Combustion, emission and engine performance characteristics of used cooking oil biodiesel—A review. *Renew. Sustain. Energy Rev.* **2010**, *14*, 2863–2873. [CrossRef]
13. Chauhan, B.S.; Kumar, N.; Cho, H.M. A study on the performance and emission of a diesel engine fueled with Jatropha biodiesel oil and its blends. *Energy* **2012**, *37*, 616–622. [CrossRef]
14. Agarwal, A.; Rajamanoharan, K. Experimental investigations of performance and emissions of Karanja oil and its blends in a single cylinder agricultural diesel engine. *Appl. Energy* **2009**, *86*, 106–112. [CrossRef]
15. Roy, M.M.; Wang, W.; Bujold, J. Biodiesel production and comparison of emissions of a DI diesel engine fueled by biodiesel–diesel and canola oil–diesel blends at high idling operations. *Appl. Energy* **2013**, *106*, 198–208. [CrossRef]
16. Canakci, M.; Van Gerpen, J.H. Comparison of engine performance and emissions for petroleum diesel fuel, yellow grease biodiesel, and soybean oil biodiesel. *Trans. ASAE* **2003**, *46*, 937. [CrossRef]
17. Ng, H.K.; Gan, S. Combustion performance and exhaust emissions from the non-pressurised combustion of palm oil biodiesel blends. *Appl. Therm. Eng.* **2010**, *30*, 2476–2484. [CrossRef]
18. Wróblewski, P.; Iskra, A. *Problems of Reducing Friction Losses of a Piston-Ring-Cylinder Configuration in a Combustion Piston Engine with an Increased Isochoric Pressure Gain*; SAE Technical Paper; SAE: Avelendale, PA, USA, 2020. [CrossRef]
19. Stępień, Z. Geometry of shape of profiles of the sliding surface of ring seals in the aspect of friction losses and oil film parameters. *Combust. Engines* **2016**, *167*, 24–38. [CrossRef]
20. Wróblewski, P. Effect of asymmetric elliptical shapes of the sealing ring sliding surface on the main parameters of the oil film. VII International Congress on Combustion Engines. *Combust. Engines* **2017**, *168*, 84–93. [CrossRef]
21. Wróblewski, P. The effect of the distribution of variable characteristics determining the asymmetry of the sealing rings sliding surfaces on the values of friction loss coefficients and other selected parameters of oil film, VII International Congress on Combustion Engines. *Combust. Engines* **2017**, *171*, 107–116. [CrossRef]
22. Wróblewski, P. An Innovative Approach to Data Analysis in The Field of Energy Consumption and Energy Conversion Efficiency in Vehicle Drive Systems—The Impact of Operational and Utility Factors. In Proceedings of the 37th International Business Information Management Association (IBIMA), Cordoba, Spain, 1–2 April 2021; ISBN 978-0-9998551-6-4.

23. Wróblewski, P. The Impact of the COVID-19 Pandemic on The Development of Electromobility—Analysis of Changes in Purchasing Preferences. In Proceedings of the 37th International Business Information Management Association (IBIMA), Cordoba, Spain, 1–2 April 2021; ISBN 978-0-9998551-6-4.
24. Wróblewski, P.; Kupiec, J.; Drożdż, W.; Lewicki, W.; Jaworski, J. The Economic Aspect of Using Different Plug-in Hybrid Driving Techniques in Urban Conditions. *Energies* **2021**, *14*, 3543. [CrossRef]
25. Wróblewski, P.; Drożdż, W.; Lewicki, W.; Miązek, P. Methodology for Assessing the Impact of Aperiodic Phenomena on the Energy Balance of Propulsion Engines in Vehicle Electromobility Systems for Given Areas. *Energies* **2021**, *14*, 2314. [CrossRef]
26. Wróblewski, P.; Rogólski, R. Experimental Analysis of the Influence of the Application of TiN, TiAlN, CrN and DLC1 Coatings on the Friction Losses in an Aviation Internal Combustion Engine Intended for the Propulsion of Ultralight Aircraft. *Materials* **2021**, *14*, 6839. [CrossRef]
27. Wróblewski, P. Analysis of Torque Waveforms in Two-Cylinder Engines for Ultralight Aircraft Propulsion Operating on 0W-8 and 0W-16 Oils at High Thermal Loads Using the Diamond-Like Carbon Composite Coating. *SAE Int. J. Engines* **2021**, *15*, 2022. [CrossRef]
28. Wróblewski, P.; Koszalka, G. An Experimental Study on Frictional Losses of Coated Piston Rings with Symmetric and Asymmetric Geometry. *SAE Int. J. Engines* **2021**, *15*, 2022. [CrossRef]
29. Nawaz, S.; Ahmad, M.; Asif, S.; Klemeš, J.J.; Mubashir, M.; Munir, M.; Zafar, M.; Bokhari, A.; Mukhtar, A.; Saqib, S.; et al. Phyllosilicate derived catalysts for efficient conversion of lignocellulosic derived biomass to biodiesel: A review. *Bioresour. Technol.* **2021**, *343*, 126068. [CrossRef] [PubMed]
30. Curkowski, A.; Mroczkowski, P.; Oniszk-Popławska, A.; Wiśniewski, G. *Biogaz Rolniczy—Produkcja i Wykorzystanie*; Mazowiecka Agencja Energetyczna Sp. z o.o.: Warszawa, Poland, 2009.
31. Pulka, J. Potencjał biogazu rolniczego na tle innych rodzajów OZE. *Tech. Rol. Ogrod. i Leśna* **2019**, *2*, 15–17.
32. Czyrnek-Delêtre, M.; Rocca, S.; Agostini, A.; Giuntoli, J.; Murphy, J.D. Life cycle assessment of seaweed biomethane, generated from seaweed sourced from integrated multi-trophic aquaculture in temperate oceanic climates. *Appl. Energy* **2017**, *196*, 34–50. [CrossRef]
33. Głaszka, A. *Biogazownie Rolnicze*; Multico Oficyna Wydawnicza: Warszawa, Poland, 2010.
34. Horschig, T.; Adams, P.W.R.; Röder, M.; Thornley, P.; Thrän, D. Reasonable potential for GHG savings by anaerobic biomethane in Germany and UK derived from economic and ecological analyses. *Appl. Energy* **2016**, *184*, 840–852. [CrossRef]
35. Klimiuk, E.; Pawłowska, M.; Pokój, T. Biopaliwa. In *Technologie Dla Zrównoważonego Rozwoju*; Wydawnictwo Naukowe PWN: Warszawa, Poland, 2012.

Article

Experimental Evaluation of the Effect of Replacing Diesel Fuel by CNG on the Emission of Harmful Exhaust Gas Components and Emission Changes in a Dual-Fuel Engine

Mirosław Karczewski * and Grzegorz Szamrej *

Faculty of Mechanical Engineering, Military University of Technology, 2 Gen, Sylwestra Kaliskiego St., 00-908 Warsaw, Poland
* Correspondence: miroslaw.karczewski@wat.edu.pl (M.K.), grzegorz.szamrej@wat.edu.pl (G.S.); Tel.: +48-261837754 (G.S.)

Abstract: The constant development of civilization increases environmental pollution as a result of industrial activity and transport. Consequently, human activity in this area is restricted by regulations governing the permissible emission of harmful substance components into the environment. These include substances emitted by combustion engines, the use of which remains high in many industries. Consequently, research is being conducted to reduce the emissions of harmful exhaust components from existing and newly manufactured internal combustion engines. This research presents a used semi-truck engine, in which an innovative Compressed Natural Gas (CNG) supply system was applied. Using this fuel supply installation allows a mass exchange of the base diesel fuel to natural gas of up to 90%. The study evaluated the effect of the diesel/CNG exchange ratio for different engine operating conditions (engine load, speed) on the concentration of toxic components, such as CO, NO, NO_2, NO_X, as a sum of NO, NO_2, CH_4, C_2H_4, C_2H_6, C_3H_8, NH_3, and CH_2O. The use of a dual-fuel system had a positive effect on the emissions of some harmful exhaust components, even in an engine from a vehicle that had been running for many years on diesel and at high mileage, but, simultaneously, the emissions of some harmful exhaust gas components increased.

Keywords: diesel-CNG; CNG; dual-fuel; alternative fuel; exhaust gas composition; exhaust emissions; CO; NO; NO_2; NO_X CH_4; C_2H_4; C_2H_6; C_3H_8; NH_3; CH_2O

Citation: Karczewski, M.; Szamrej, G. Experimental Evaluation of the Effect of Replacing Diesel Fuel by CNG on the Emission of Harmful Exhaust Gas Components and Emission Changes in a Dual-Fuel Engine. *Energies* 2023, *16*, 475. https://doi.org/10.3390/en16010475

Academic Editor: Constantine D. Rakopoulos

Received: 1 December 2022
Revised: 16 December 2022
Accepted: 26 December 2022
Published: 1 January 2023

1. Introduction

This study is a continuation of an article [1] published by the authors in the *Energies* journal. The second article in the series aims to extend and complete the knowledge on emissions of harmful components of exhaust gases and their changes by dual-fuel engines fuelled with natural gas and diesel fuel. The previous article described in detail the principle of operation of dual-fuel engines running on liquid and gaseous fuels and the test methodology.

Theoretically, dual-fuel engines that succeed in creating a homogeneous mixture of low-reactive fuel and air allow both fuels to be burned with an efficiency comparable to that of an HCCI (Homogeneous Charge Compression Ignition) engine. In the case of dual-fuel combustion, this type of engine operation is considered as a RCCI engine (Reactivity Controlled Compression Ignition), described in detail in other works by the authors [1,2]. The basics of the operation of this type of engine will be cited here as an introduction, a more detailed description can be found in our paper [3] based on dozens of other bibliographic references.

In a dual-fuel Compression Ignition (CI) engine, a motor is fed indirectly by a high-octane-number and low-reactive fuel which, in the presented research, was CNG. Then, after a compression process the high-cetane-number fuel is injected into the combustion chamber which, as a result of self-ignition, leads to the ignition of the liquid fuel and then the gaseous fuel as well. Such a combination is the most popular version of the "Dual-Fuel" CI power supply. There are some good reasons in favor of using such a simply solution, including the high percentage of replacement value and easy modification of the engine [4].

Purpose of the Study and Literature Review

The purpose of the study was to present the emissions and their changes from the tests carried out in order to enable them to be compared with other research works of this type.

There are only a small number of studies in a similar vein in the current literature—most research work is already based on the concept of homogeneous mixture combustion, which allows significantly more efficient use of fuel, and it was not the authors' aim to compare emissions with this type of engine. Such engines are well described in [5], while publications on engine emission tests, in which the combustion chamber of a used engine dismantled from a Euro-3-compliant semi-truck was not interfered with and only the engine feed system was optimized, were not found by the authors. It should be emphasized that this type of modification is mainly carried out by Polish companies that are market leaders with one of the largest production potentials in this field [6–9]. Modifications of this type are currently very strongly developed in the authors' country; however, there are a small number of publications on the subject, so the authors undertook the task of investigating an engine with this type of installation.

In one study [10], emissions were investigated for load characteristic test runs, which cannot be compared with the results published in this article. Study [11] tested engines from newer generation trucks in Steady-State SET Cycle tests, which is not equivalent to the studies published within this article. Although when comparing the general trends in emission changes the directions of changes are in line with the results of our study, the lack of fundamental possibility to compare the values for different types of tests has been proven many times, which was reminiscently examined and presented by Prof. Merkisz in a recent publication [12].

In another study [13], emissions were investigated in a small diesel engine (historically the smallest factory-produced diesel engine designed for cars), which cannot be compared with an engine from a semi-truck. Similar research was also carried out on an older engine [14], also manufactured in Poland. This research is also not equivalent to the research carried out for this article. An even smaller engine was dealt with by researchers in [15], in which emissions were examined in a cursory manner at several operating points with which no meaningful comparison can be made. A similarly sized engine was investigated by [16]. The commonness of research on engines of this type is due to the ease of performing this type of research and the cost effectiveness of conducting this type of research, hence the large number of university publications on small power engines. The largest number of publications on the subject of dual-fuel engines have been presented by Professors Rolf Reids [17] and Garcia [18], and all of their studies as mentioned when citing publication [5] are on higher generation engines, which also makes it impossible to make a clear reference. However, it should also be noted that the paper presents emission results and describes the mechanism of formation and changes of emissions for the following compounds that have not yet been described in the available literature: C_2H_4-ethene, C_2H_6-ethane, C_3H_8-propane, NH_3–ammonia, and CH_2O-formaldehyde. From our point of view, these are unique studies that have not been described in any of the available literature.

It Is also necessary to bear in mind the changes that occur in the engine systems during the test, in which the flow rates may change, due to, for example, a reduction in the capacity of the fuel or air filters [19], which has a significant impact on the engine performance and may affect the differences between individual test runs, and which, due to the difficulty in estimating changes in these values, are not corrected within the calculations for this article [20].

2. Materials and Methods

The degree of substitution/replacement of the base fuel with gaseous fuel can be considered as a percentage degree or energy degree. In the study under consideration, the replacement ratio for the dual-fuel engine was determined as a level ratio of high-octane fuel consumed to the whole fuels used by the engine. This parameter was determined based on Equation (1) [4]:

$$\frac{DIESEL}{CNG} = \frac{Ge_{diesel} - Ge_{diesel\_CNG}}{Ge_{diesel}} * 100\% = 1 - \frac{Ge_{diesel\_CNG}}{Ge_{diesel}} * 100\% \qquad (1)$$

where Ge_{diesel} is diesel fuel hourly engine consumption in kg at the time when engine is running only on diesel, and Ge_{diesel_CNG} is diesel fuel hourly engine consumption in the time when engine is running on both fuels, i.e., CNG and diesel fuel.

This is an effective way of illustrating how much liquid fuel is physically substituted by gaseous fuel, although it is also generally accepted to use the value of replacement ratio calculated from the energy ratio of substituted fuels. This method is used by other researchers [14,21–23], it is also defined by a contemporary standard [24], or book publications [4,25]. It is advisable, however, to determine the value to which we refer in the following part of the study as presented in Formula 1. In this way, we avoid the problem of determining the exact calorific value of the gaseous fuel used. The fractional composition of the gas on the Polish market, from which the fuel was obtained, can be very large, which was already described in detail by the authors in their first joint publication [26]. The assumption of certain rigid defined values could lead to significant cognitive errors, which the authors tried to avoid.

As part of the study, when comparing the emission results of the individual harmful components of exhaust gases, they were also compared to the emission standards currently in force, as defined by regulations in force in the European Union [27,28] and those that are being prepared for introduction as part of further regulations [27–29]. The current standard and the scenarios for the standards to be introduced in the future are presented in Table 1.

Table 1. A tabular overview of the emission limits of the individual exhaust gas components for the different Euro standards, with revisited emission limits according to the alternative scenarios for Euro 7 [27–29].

Euro Scenarios	NOx [mg/kWh]	SPN10 [-]	CO [mg/kWh]	CH$_4$ [mg/kWh]	N$_2$O [mg/kWh]	NH$_3$ [mg/kWh]	THC	NMHC [mg/kWh]
Euro 5 [27]	180	-	500	-	-	-	230-NOx	230-NOx
Euro 6 [28]	80	6×10^{11}	500	-	-	-	170-NOx	170-NOx
Euro 7 A [29]	30	10^{11}	300	10	10	5	55	50
Euro 7 B [29]	10	6×10^{10}	100	5	5	2	30	25
Euro V [27]	2000	-	4000	1100	-	-	1650	550
Euro VI [28]	460	6×10^{11}	4000	500	-	~40	660	160
Euro VII A [29]	120	4×10^{11}	1500	100	50	20	150	50
Euro VII B [29]	40	10^{11}	400	50	25	10	75	25

The emission values specified in the emission standards for the Euro standards cannot be directly related to the results that were achieved in this study, but they can be used for an indicative comparison of the results achieved with the applicable standards. This is because a different type of test is carried out to meet these standards (in this case, the WHTC—the World Harmonized Transient Cycle test [30]), but there are areas of work included in the tests that were also carried out for the present study [31].

The goal of our research was to determine in the experimental way the effect of replacement ratio of diesel fuel by CNG, defined by Equation (1), on the emission of individual exhaust components, CO, NO, NO_2, NO_X CH_4, C_2H_4, C_2H_6, C_3H_8, NH_3, and CH_2O, in fixed operating states in an innovative, non-factory CNG/diesel dual-fuel supply system in a used semi-truck unit engine

All tests were carried out with the semi-truck's VOLVO FH13 D13C460 EU5EEV (338 kW) engine. All-important information about the engine, test bench, and testing parameters are described in our previous article [1].

"The tested engine, according to the approval documents, met the requirements of the EURO V standard when it left the factory," said authors in [32]. The tests and results of useful engine measurements were brought down to normal conditions [33].

Methodology for Determining the Effect of the CNG/ON Substitution Factor on Engine Emission

Diesel fuel EN-590 from one production batch was used for the tests. Lack of reference fuel according to document [34] allows the use of fuel similar to the reference fuel. Therefore, determining the actual values of the engine operating characteristics (operation on commercial fuel from a random petrol station) can be considered an advantage that makes the research and results more realistic.

The tests were carried out on a properly and optimally tuned external natural gas supply system. The power and torque characteristics of the engine in dual-fuel mode largely coincided with the factory characteristics.

Flue gas temperature changes are shown in the test results as shown (Figures 1 and 2), or the emission of individual gaseous components, i.e., CO (Figures 3 and 4), NO and NO_2 (Figures 5–8), NO_X (Figures 9 and 10), CH_4 (Figures 11 and 12), C_2H_6 (Figures 13 and 14), NMOC (Non-Methane Organic compounds on Figures 15 and 16), C_2H_4 (Figures 17 and 18), NH_3 (Figures 19 and 20), and CH_2O (Figures 21 and 22). Every even chart number shows percentage changes of the substance or group of substances measured. In order to simplify the analysis and determine the changes taking place, the results are also presented by comparing the relative results obtained for diesel oil/CNG and diesel fuel only (right side of the figures). The following tests were carried out:

- For two values of load torque in a constant engine speed of 1100 rpm for approx. 520 Nm (61 kW), which was shown on charts as a curve "A", and 1150 Nm (139 kW), which was shown on charts as a curve "B";
- For two values of load torque in a constant engine speed of 1300 rpm for approx. 480 Nm (67 kW), which was shown on charts as a curve "C", and 980 Nm (134 kW), which was showed on charts as a curve "D";
- For two values of load torque in a constant engine speed of 1500 rpm, approx. 500 Nm (79 kW), which was shown on charts as a curve "E", and 1040 Nm (163 kW), which was shown on charts as a curve "F".

These are the conditions in which the engine most frequently operates, as identified during road tests of the semi-truck from which the engine was derived. The operating points correspond to the most operationally relevant operating areas of the engine modified to run on dual fuel. The working points identified here were also consulted with and accepted by the manufacturer involved in the production and further distribution of the dual-fuel controller developed by our research team.

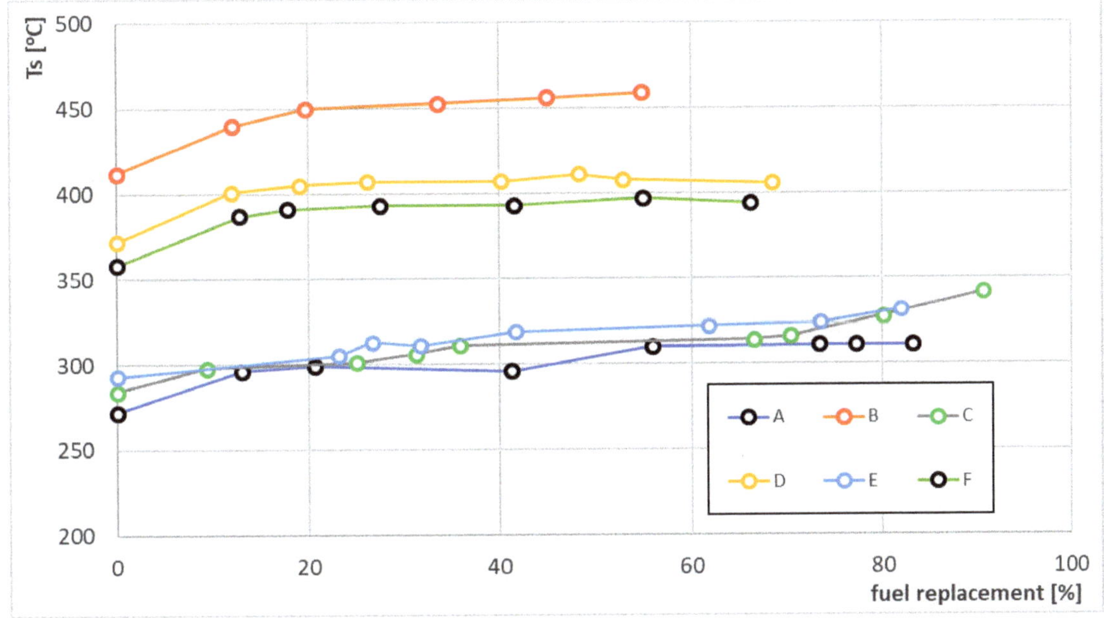

Figure 1. Curves from A to F showing the exhaust gas temperatures for individual engine operating points, where the rotation speed, power, and torque were constant, as a replacement function of diesel fuel replaced by CNG.

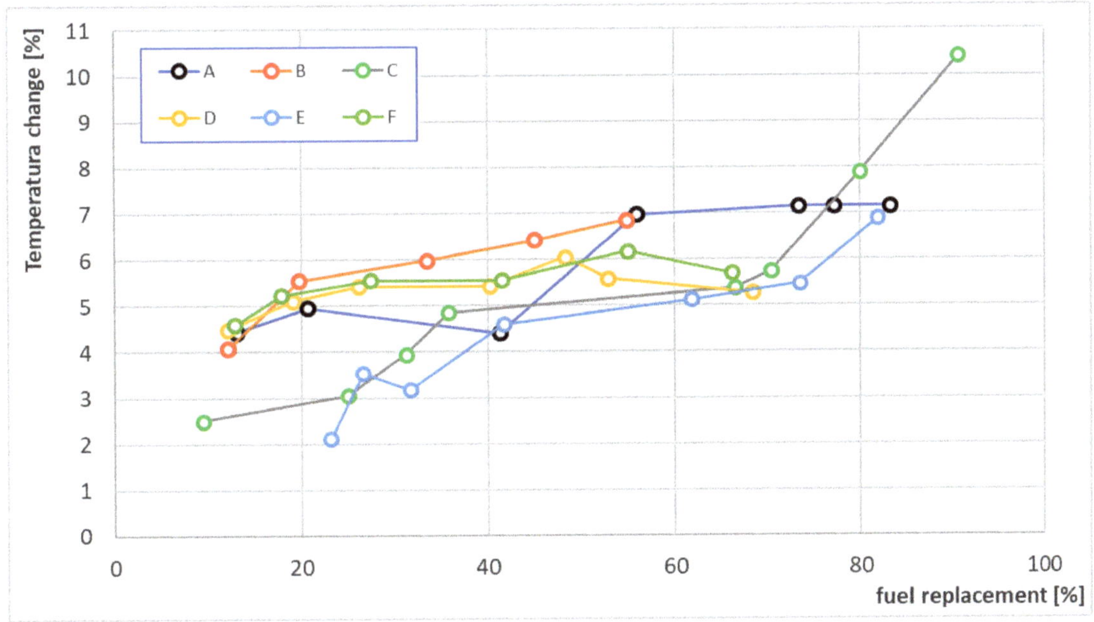

Figure 2. Curves from A to F showing the changes in the exhaust gas temperatures as the comparison of exhaust gas temperature noted only in single fuel (diesel) working mode for individual engine operating points, where the rotation speed, power, and torque were constant, as a dual-fuel mode (where engine uses CNG and diesel fuel) replacement function.

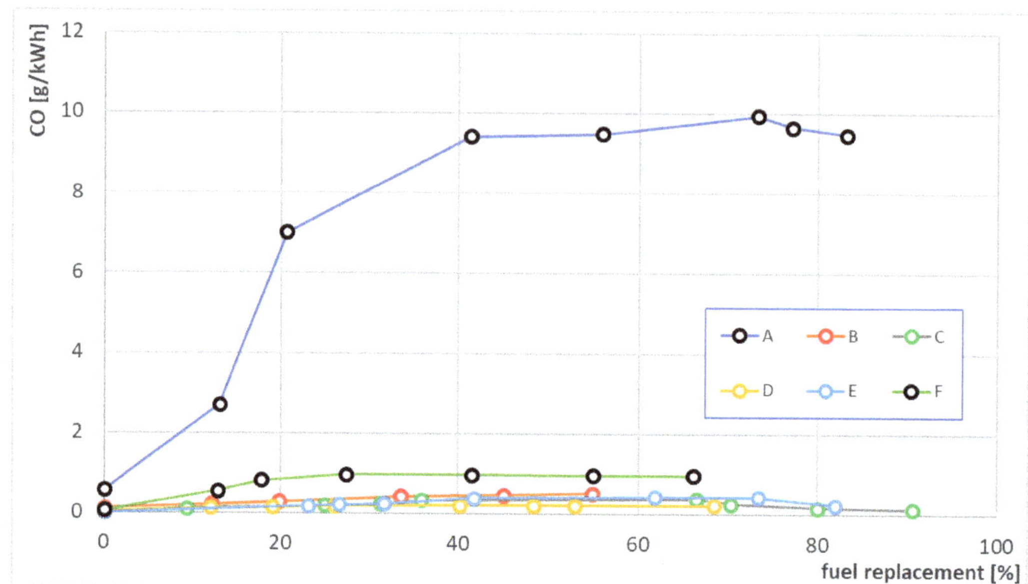

Figure 3. Curves from A to F showing engine's emission of CO (carbon monoxide) for individual engine operating points, where the rotation speed, power, and torque were constant, as a replacement function of diesel fuel replaced by CNG.

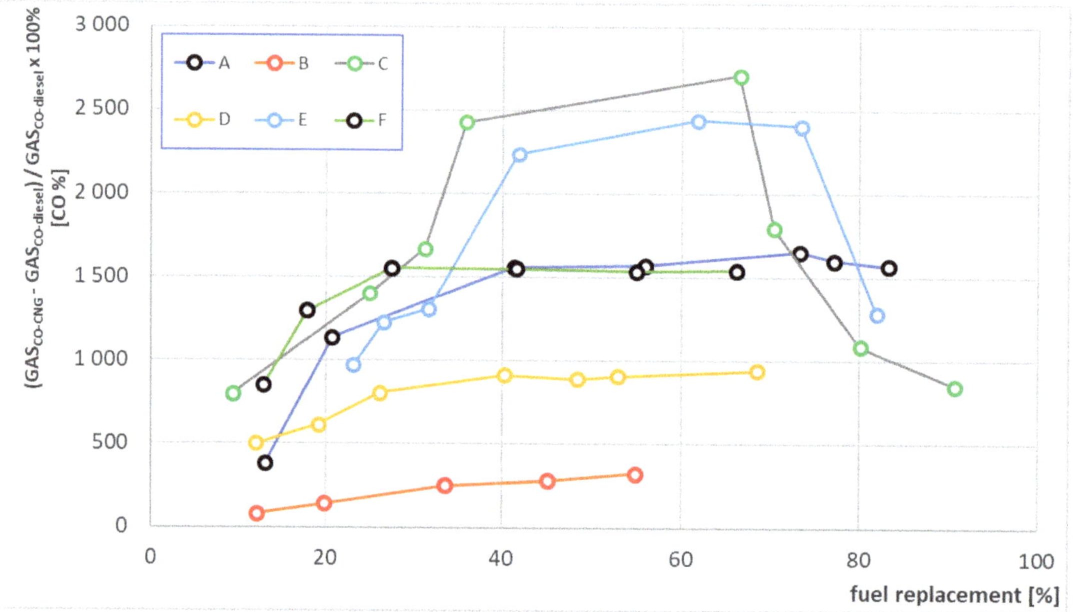

Figure 4. Emission changes of carbon monoxide in the exhaust gas for individual engine operating points (rotation speed, torque) in comparison to point zero of replacing diesel/CNG as a function of replacing diesel/CNG. Curves from A to F showing the changes of the emission of carbon monoxide as the comparison of exhaust gas temperature noted only in single fuel (diesel) working mode for individual engine operating points, where the rotation speed, power, and torque were constant, as a dual fuel mode (where engine uses CNG and diesel fuel) replacement function.

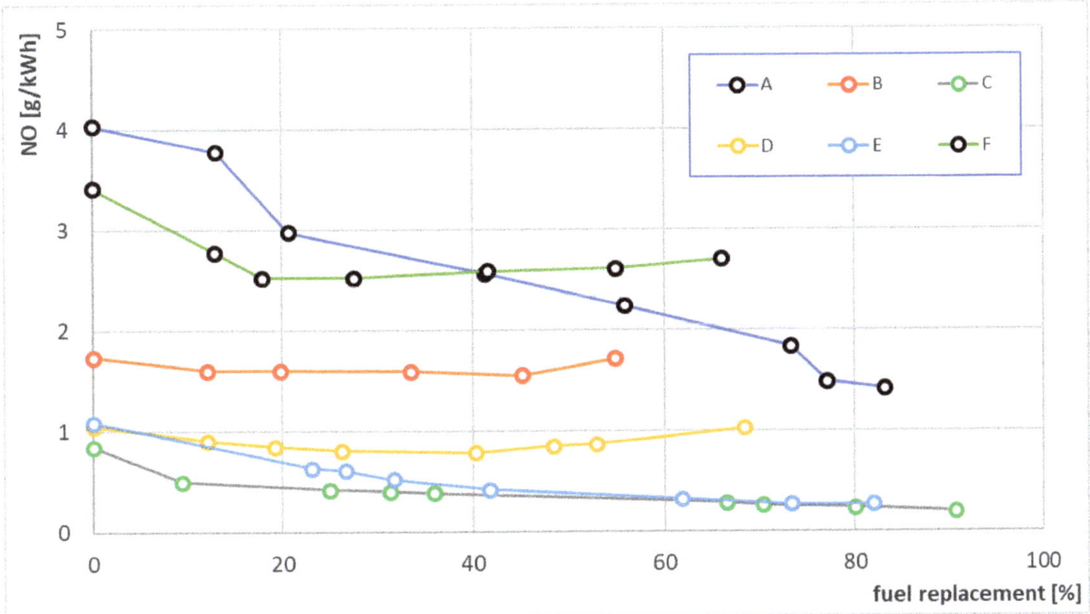

Figure 5. Curves from A to F showing engine's emission of NO (nitric oxide), for individual engine operating points, where the rotation speed, power, and torque were constant, as a replacement function of diesel fuel replaced by CNG.

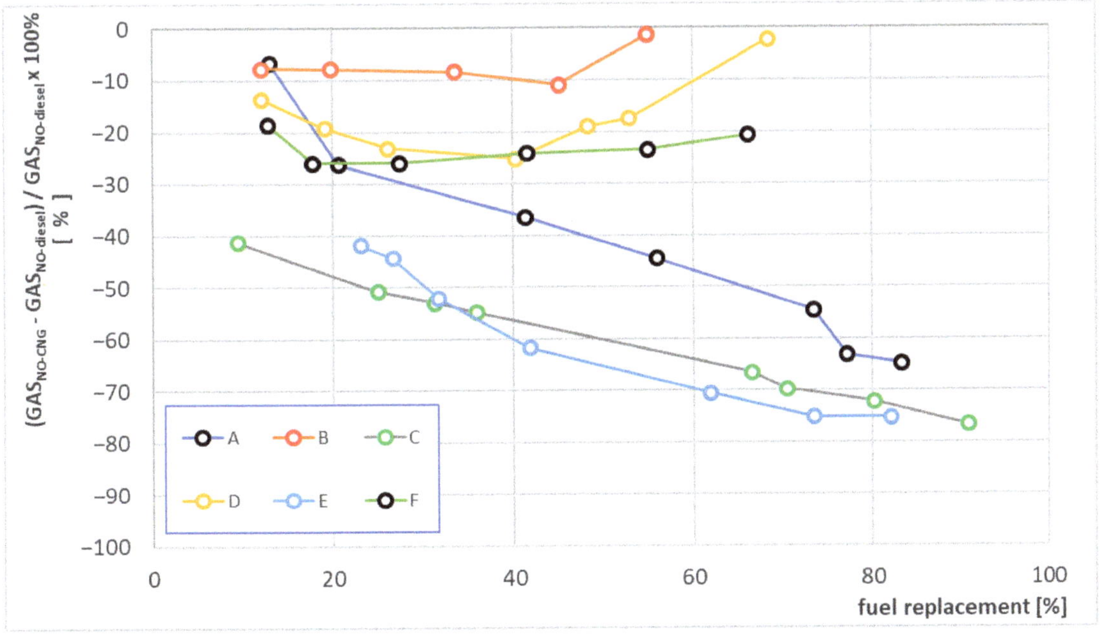

Figure 6. Curves from A to F showing the changes in the emission of nitric oxide as the comparison of exhaust gas temperature noted only in single fuel (diesel) working mode for individual engine operating points, where the rotation speed, power, and torque were constant, as a dual fuel mode (where engine uses CNG and diesel fuel) replacement function.

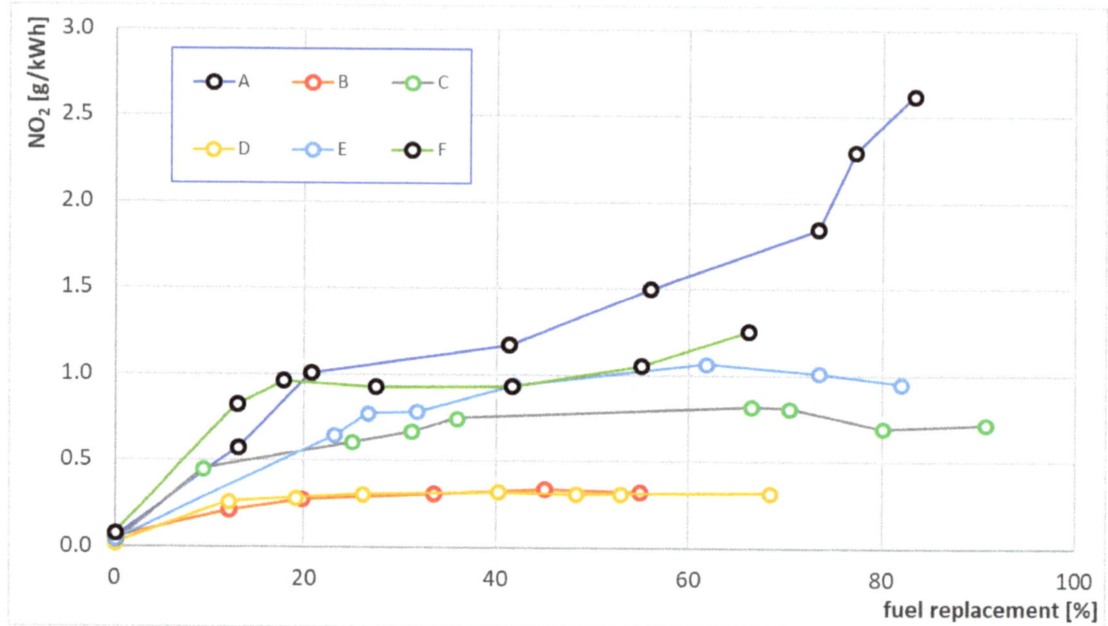

Figure 7. Curves from A to F showing engine's emission of NO_2 (nitrogen dioxide), for individual engine operating points, where the rotation speed, power, and torque were constant, as a replacement function of diesel fuel replaced by CNG.

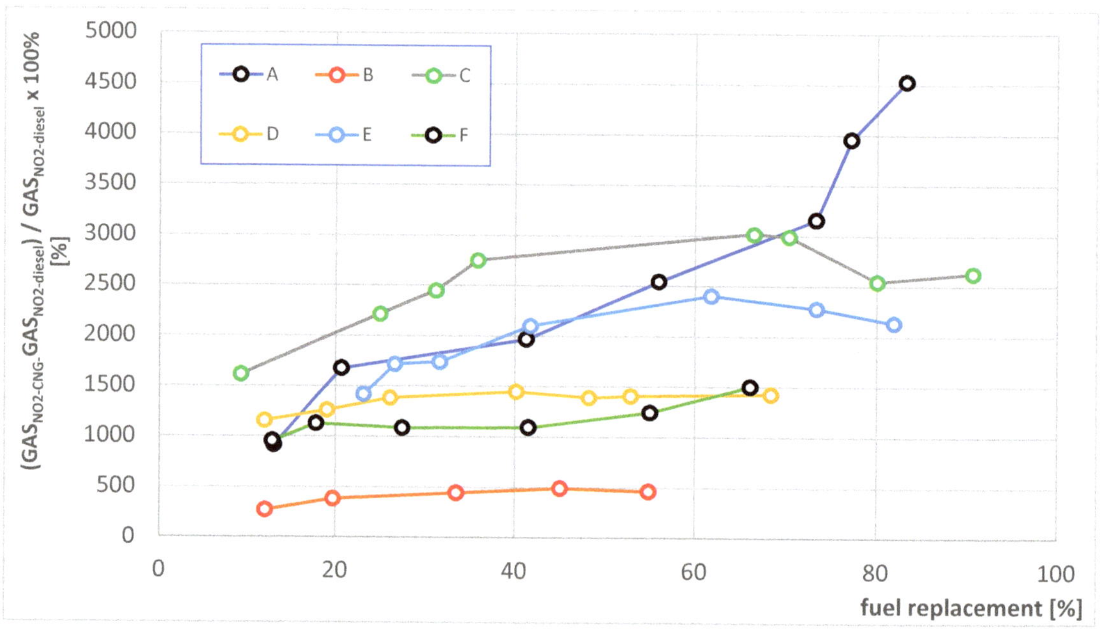

Figure 8. Curves from A to F showing the changes in the emission of nitrogen dioxide as the comparison of exhaust gas temperature noted only in single fuel (diesel) working mode for individual engine operating points, where the rotation speed, power, and torque were constant, as a dual-fuel mode (where engine uses CNG and diesel fuel) replacement function.

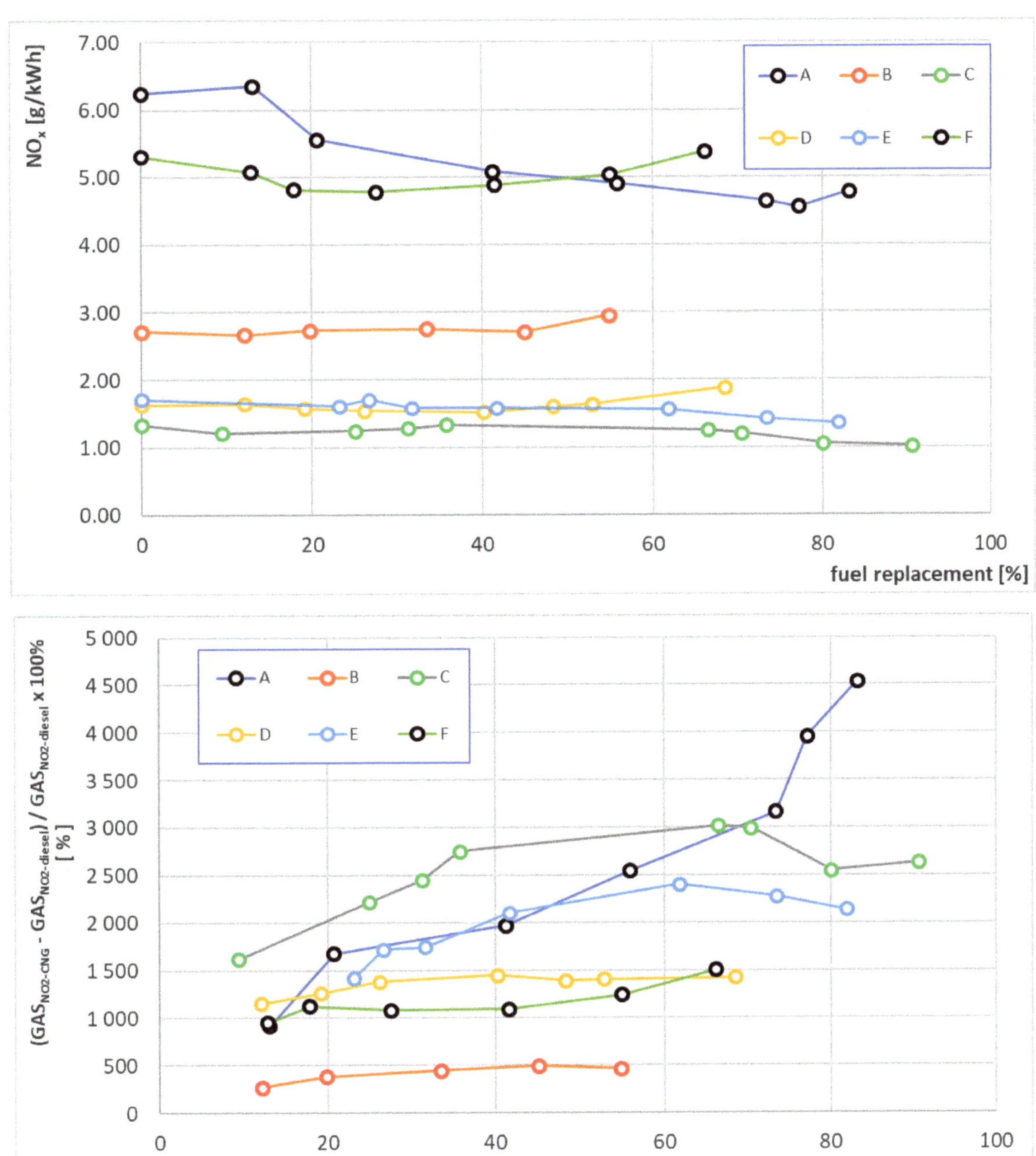

Figure 9. Curves from A to F showing engine's emission of NO_X (nitrogen oxides), for individual engine operating points, where the rotation speed, power, and torque were constant, as a replacement function of diesel fuel replaced by CNG.

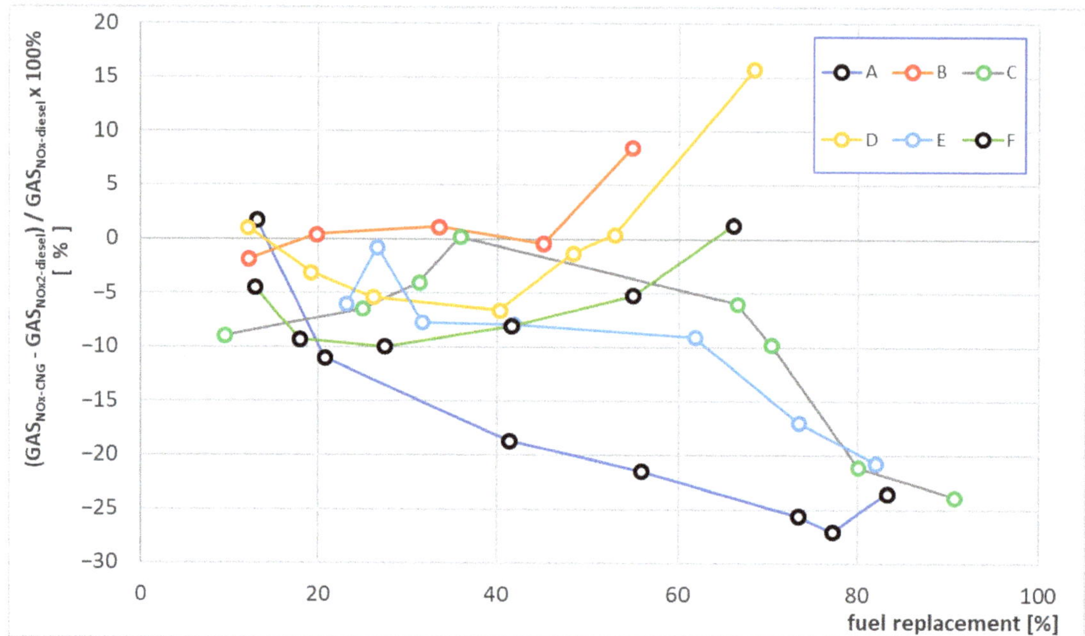

Figure 10. Curves from A to F showing the changes in the emission of nitrogen oxides as the comparison of exhaust gas temperature noted only in single fuel (diesel) working mode for individual engine operating points, where the rotation speed, power and torque were constant, as a dual-fuel mode (where engine uses CNG and diesel fuel) replacement function.

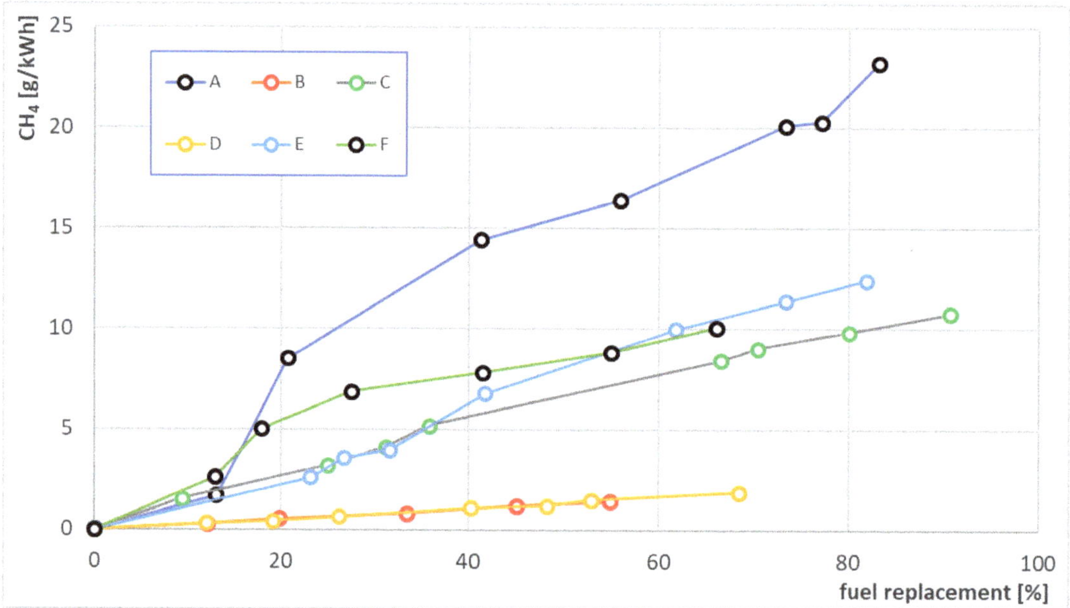

Figure 11. Curves from A to F showing engine's emission of CH_4 (methane), for individual engine operating points, where the rotation speed, power, and torque were constant, as a replacement function of diesel fuel replaced by CNG.

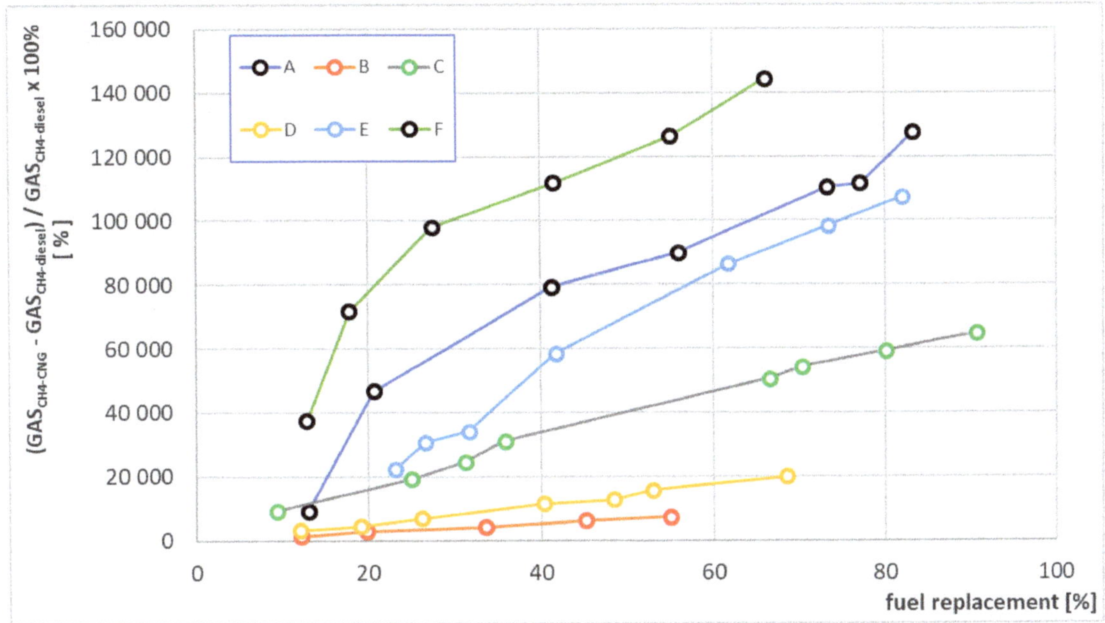

Figure 12. Curves from A to F showing the changes in the emission of methane as the comparison of exhaust gas temperature noted only in single fuel (diesel) working mode for individual engine operating points, where the rotation speed, power, and torque were constant, as a dual-fuel mode (where engine uses CNG and diesel fuel) replacement function.

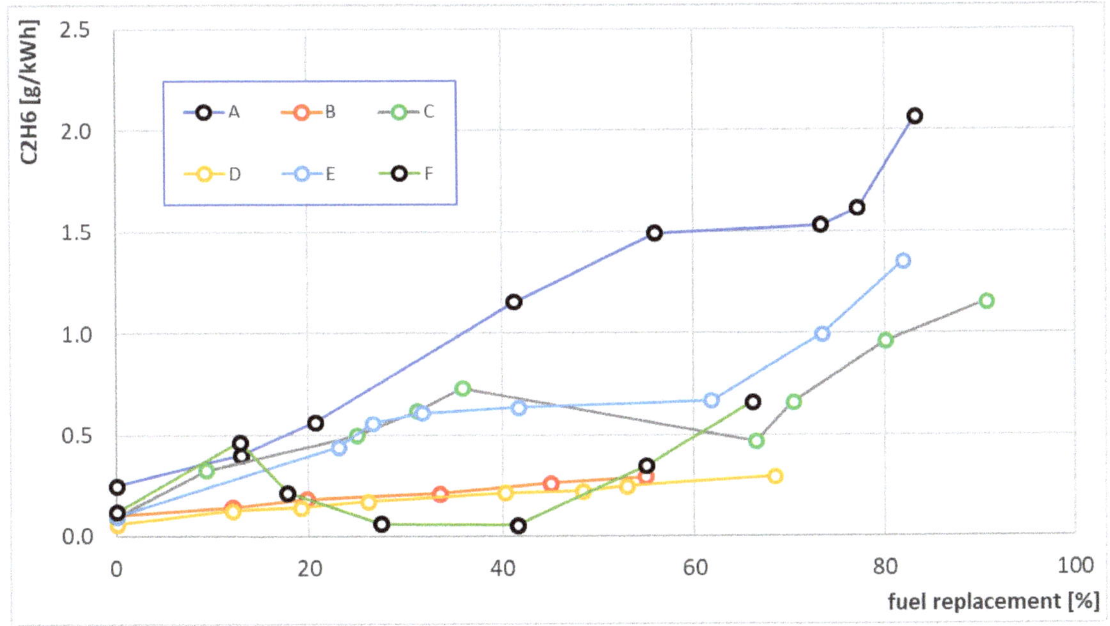

Figure 13. Curves from A to F showing the engine's emission of C_2H_6 (ethane), for individual engine operating points, where the rotation speed, power, and torque were constant, as a replacement function of diesel fuel replaced by CNG.

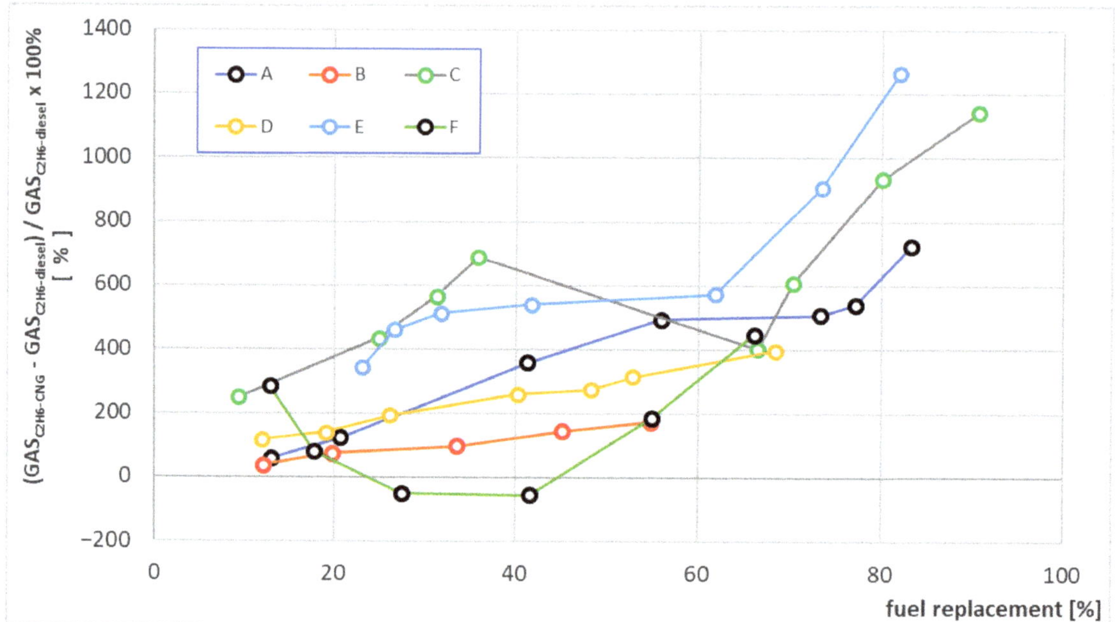

Figure 14. Curves from A to F showing the changes of the emission of ethene as the comparison of exhaust gas temperature noted only in single fuel (diesel) working mode for individual engine operating points, where the rotation speed, power, and torque were constant, as a dual-fuel mode (where engine uses CNG and diesel fuel) replacement function.

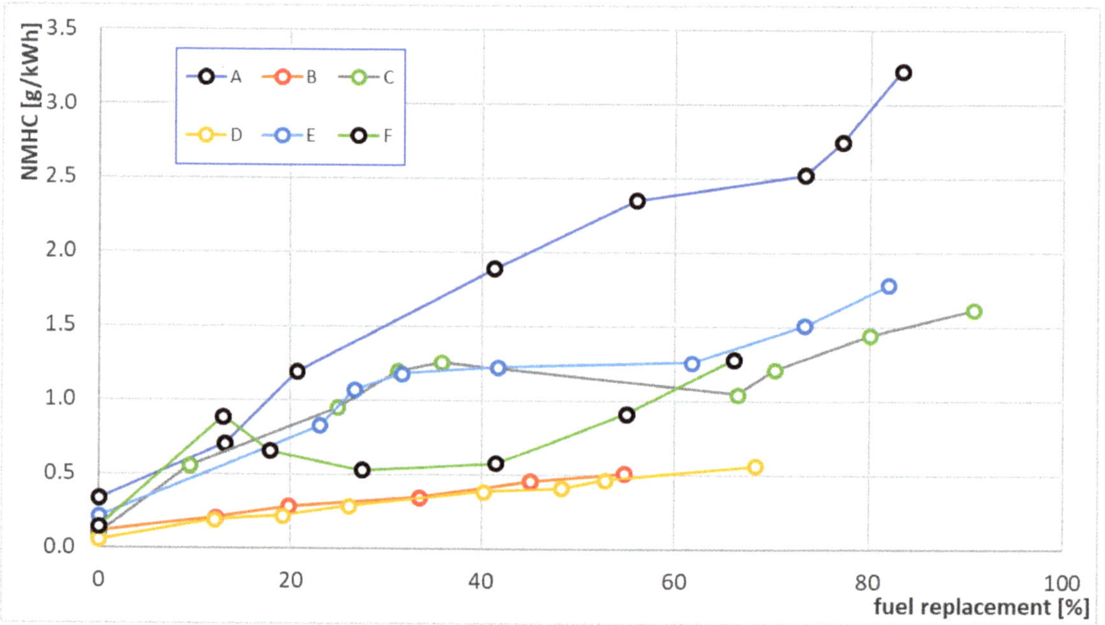

Figure 15. Curves from A to F showing engine's emission of NMOC (Non-Methane Organic Compounds), for individual engine operating points, where the rotation speed, power, and torque were constant, as a replacement function of diesel fuel replaced by CNG.

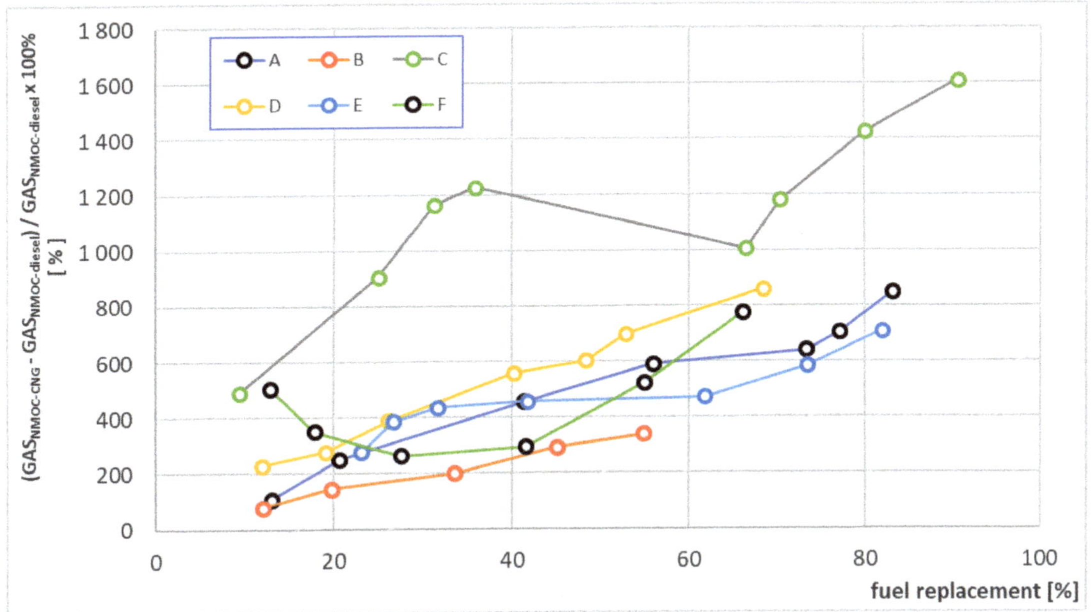

Figure 16. Curves from A to F showing the changes in the emission of NMOC as the comparison of exhaust gas temperature noted only in single fuel (diesel) working mode for individual engine operating points, where the rotation speed, power, and torque were constant, as a dual-fuel mode (where engine uses CNG and diesel fuel) replacement function.

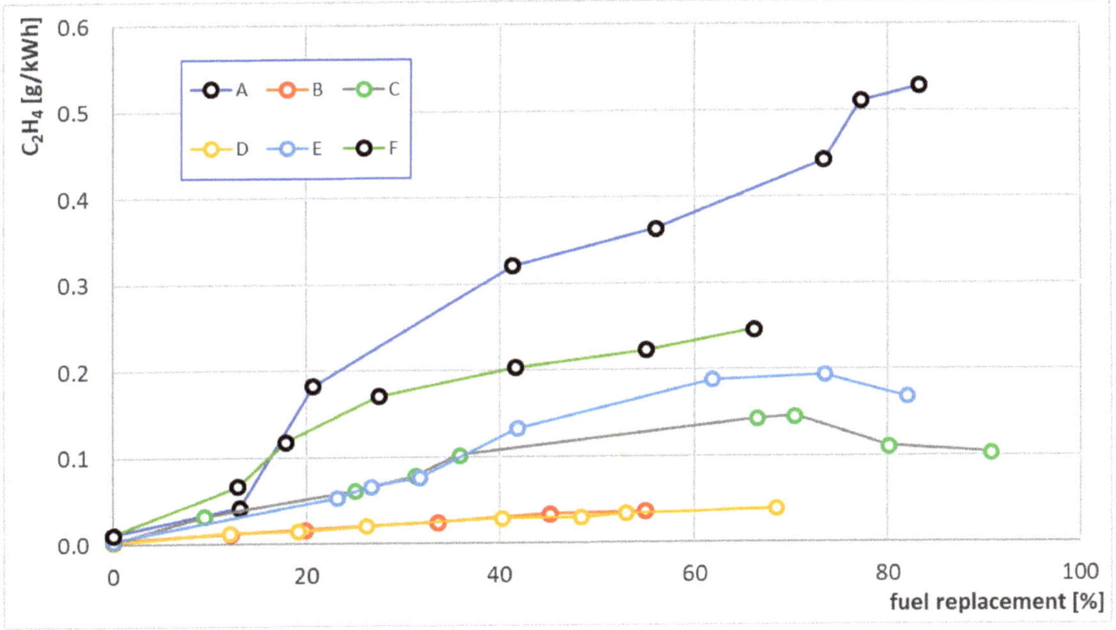

Figure 17. Curves from A to F showing engine's emission of C_2H_4 (ethene), for individual engine operating points, where the rotation speed, power, and torque were constant, as a replacement function of diesel fuel replaced by CNG.

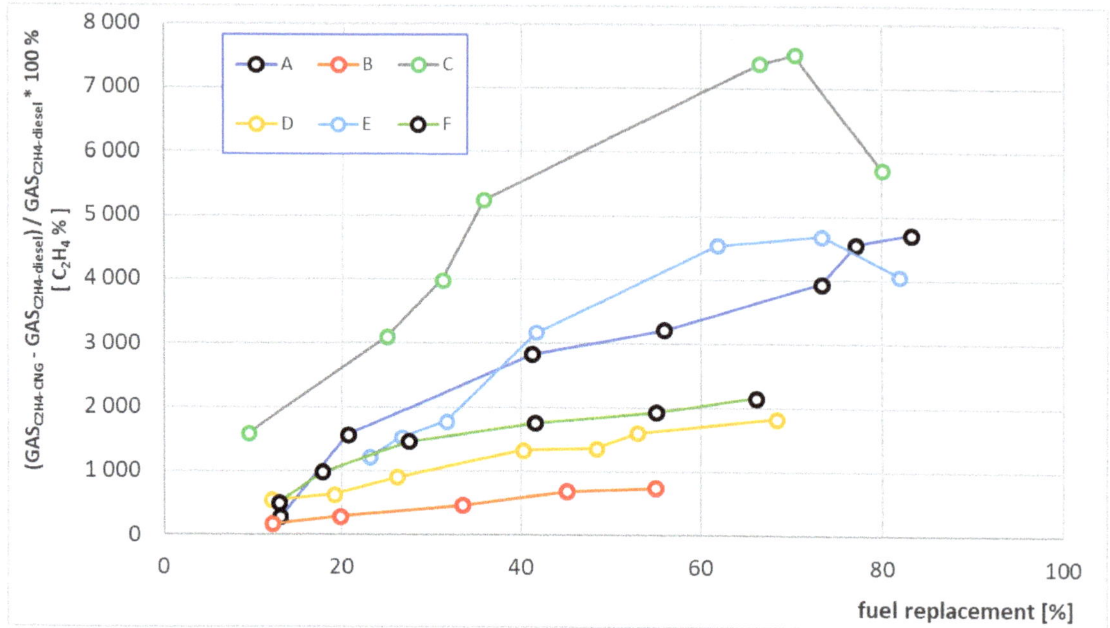

Figure 18. Curves from A to F showing the changes in the emission of ethene as the comparison of exhaust gas temperature noted only in single fuel (diesel) working mode for individual engine operating points, where the rotation speed, power, and torque were constant, as a dual-fuel mode (where engine uses CNG and diesel fuel) replacement function.

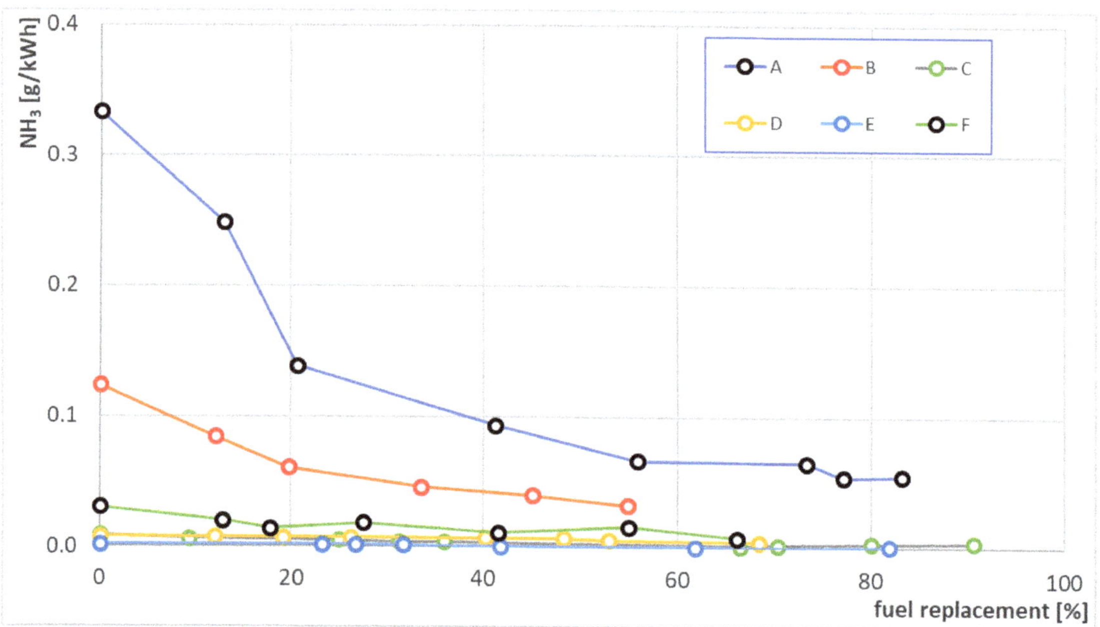

Figure 19. Curves from A to F showing engine's emission of NH_3 (ammonia), for individual engine operating points, where the rotation speed, power, and torque were constant, as a replacement function of diesel fuel replaced by CNG.

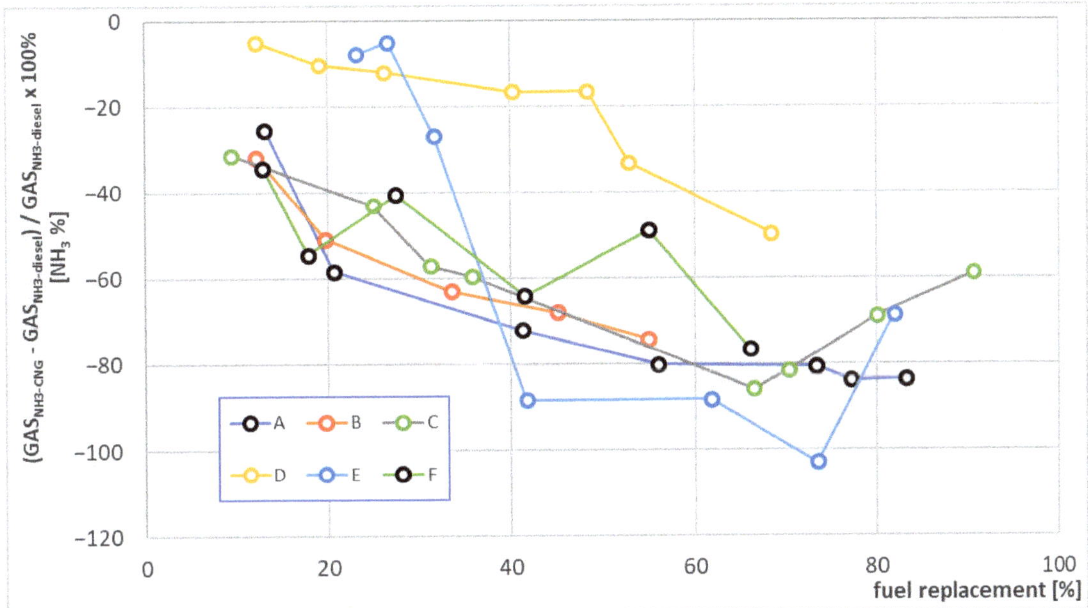

Figure 20. Curves from A to F showing the changes in the emission of ammonia as the comparison of exhaust gas temperature noted only in single fuel (diesel) working mode for individual engine operating points, where the rotation speed, power, and torque were constant, as a dual-fuel mode (where engine uses CNG and diesel fuel) replacement function.

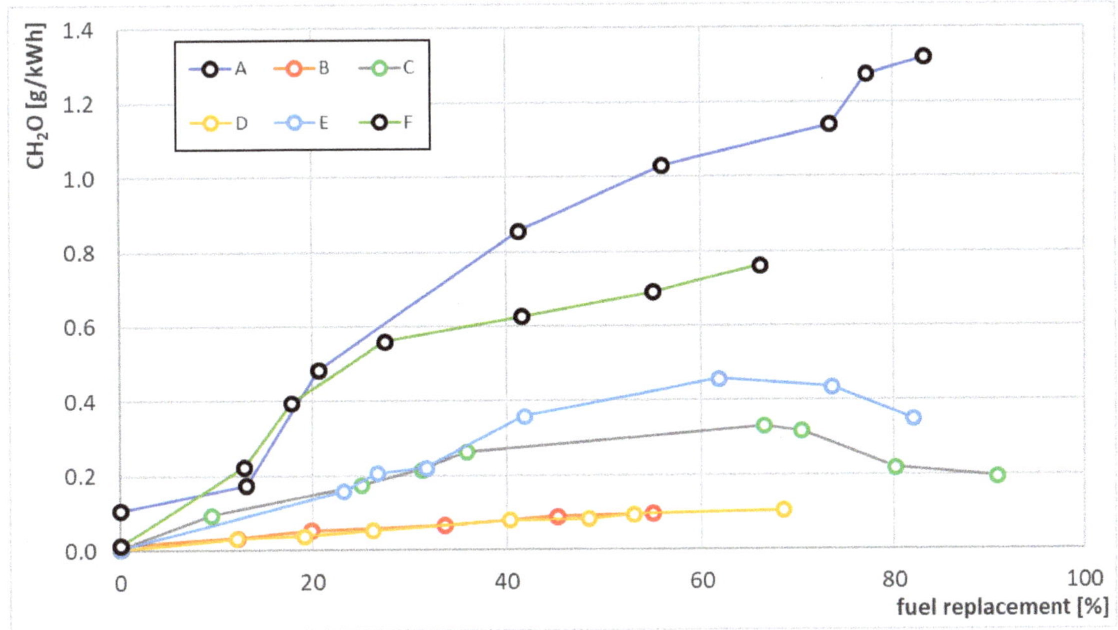

Figure 21. Curves from A to F showing engine's emission of CH_2O (formic formaldehyde), for individual engine operating points, where the rotation speed, power, and torque were constant, as a replacement function of diesel fuel replaced by CNG.

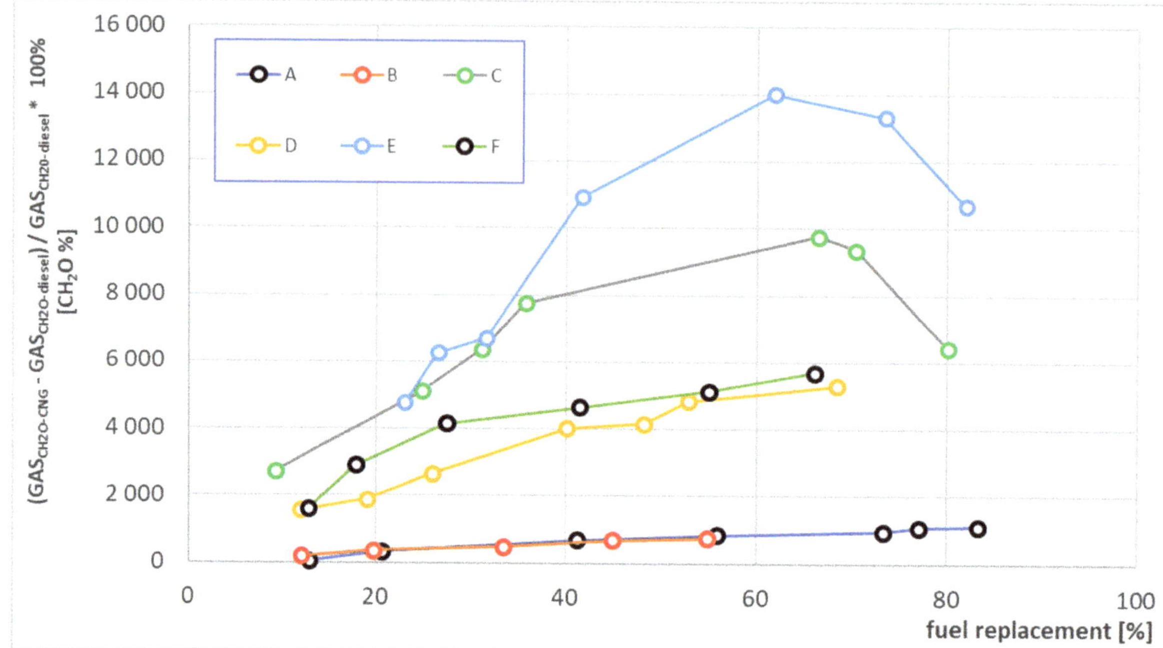

Figure 22. Curves from A to F showing the changes in the emission of formic formaldehyde as the comparison of exhaust gas temperature noted only in single fuel (diesel) working mode for individual engine operating points, where the rotation speed, power, and torque were constant, as a dual-fuel mode (where engine uses CNG and diesel fuel) replacement function.

Emissions of harmful exhaust constituents were calculated following EN ISO 8178-1 on the measurement of exhaust emissions from piston engines [35]. Changes in emissions were calculated as the ratio of the difference between the emissions of an engine operating in dual-fuel (DF) mode at a given operating point and the emissions of that engine operating in single-fuel (SF) mode at a given operating point to the emissions of that engine in single-fuel (SF) mode at that operating point, expressed as a percentage, as in the following formula:

$$y = \frac{(DF\ emission - SF\ emission)}{SF\ emission} \times 100\% \qquad (2)$$

where *DF* emission is the emissions of an engine operating in dual-fuel mode at a given point in its operation, and *SF* emission is the emissions of an engine operating in single-fuel mode at a given point in its operation.

An innovative, dedicated, diesel/CNG EG Diesel DI supply system in software version 1.7.2 was installed on a VOLVO D13C 460 semi truck's engine.

Key features of this installation include:

- Two injectors working in parallel-supplying one cylinder, mounted in a transition plate fitted between the head and the intake manifold;
- The operating pressure of the system: CNG-8 bar;
- Gas injection rail supply double-sided to avoid pressure drops;
- An achieved energy conversion rate of up to 95% for diesel/CNG, which is significantly better than installations generally available on the market;
- The installation is dedicated to a specific engine family/brand, when changing engines, it is re-optimized on a dynamometer bench;
- A bypass valve used in the intake system to regulate boost pressure when running on CNG to avoid knocking combustion.

It was possible to include the implementation of the following functions in the dedicated software:

-   Automatic changeover to dual-fuel operation when conditions are met—minimum reducer temperature, minimum engine crankshaft speed, and others;
-   Adaptation of the switching time from mono- to bi-fuel operation between cylinders to maintain engine stability;
-   Method of implementation of diesel to CNG changeover—fixed value or after map shaft rpm/diesel dose;
-   Minimum diesel pilot dose;
-   Factor which changes the gas dosage relative to the calculated diesel dosage x;
-   Reduction in the diesel dose in proportion to the preset replacement of diesel by CNG;
-   Possibility to program any coefficient of substitution of diesel by CNG over the whole field of work-speed-dose of basic fuel diesel;
-   Possibility of programming any gas injection start angle in relation to GMP in the whole working field-rotational-speed dose of the basic fuel, i.e., diesel;
-   Correction of gas dose in relation to the pressure in the engine intake system and the engine crankshaft speed—the function is necessary for the engine power correction-after switching to diesel/CNG supply;
-   Possibility of programming correction of the gas dosage in relation to the pressure in the engine intake system and to the engine crankshaft speed—a function necessary for correcting the engine power after switching to diesel/CNG supply;
-   Throttle control using bypass over the entire operating field-speed-base fuel dose diesel—after map or fixed value. This function is designed to reduce the pressure in the intake system by bleeding some of the air into the intake pipe before the charge air coolers.

The following optimization criteria, in order of importance, were adopted during the optimization studies previously carried out for the individual control maps:

-   No occurrence of detonation combustion during CNG fueling;
-   Maximum diesel substitution by CNG;
-   Obtaining the same useful power at a given accelerator pedal position and speed;
-   Limitation of the increase in exhaust gas temperature at the turbocharger outlet;

To develop optimum control maps for the diesel/CNG system, control maps had to be determined for the following controller functions:

-   Function-> Gas inj. end moment, end of injection angle—depending on engine speed— maximum engine power, and no knocking combustion were taken as criteria for the optimal value of the end of injection angle;
-   Diesel/CNG exchange/replacement ratio depending on engine speed with engine load identified as diesel output dose—the maximum value of the aforementioned ratio without the occurrence of knocking combustion was adopted as the criterion for the optimal diesel/CNG exchange/replacement ratio depending on engine speed and dose;
-   Correction of the gas dosage in relation to boost pressure—the function allows the generated power to be corrected after switching over to diesel/CNG. The correction allows the gas dosage to be increased or decreased to obtain the same effective power as diesel operation;
-   DSC function RPM/DIESEL DOSE bypass throttle opening map as a function of shaft speed and calculated ON dose as a criterion for optimum ON/CNG exchange/replacement ratio, were adopted as functions of crankshaft speed and liquid fuel dose. The maximum value of the abovementioned ratio was adopted without the occurrence of knocking combustion.

These maps were determined prior to the verification tests presented here.

Once the maps were determined for the individual controller parameters, they were entered into the controller software—the controller was considered to be programmed optimally.

3. Results

The obtained results of the tests and calculations are summarized graphically and tabulated as a summary of the results. The following graphs show each component's emissions and the relative change in emissions. Results are presented for three engine speeds, for each speed the engine was operated at low and high torque. The level of replacement of diesel by CNG shown on the x-axis of the graphs was calculated according to Formula (1).

As a result of our research, the emissions of individual harmful components of the exhaust gas depending on fuel replacement ratio of diesel by CNG were calculated. Based on these results, charts were developed showing the degree of increase or decrease in the emissions of individual exhaust constituents for each degree of substitution of diesel fuel by natural gas.

For the individual correlations of the fuel replacement ratio of diesel by natural gas with the engine load and its rotational speed, abbreviated designations were adopted to represent the given set of operating parameters maintained during the tests. This makes it easier to describe and present the obtained test results. They are summarized in Table 2.

Table 2. The definitions of research parameters in engine tested phases.

Abbreviation	Engine Torque		ENGINE Speed		Lines on Charts
[-]	Assumption	[kNm]	Assumption	[rpm]	[Colors]
A	Low	0.50	Low	1100	
B	High	1.15	Low	1100	
C	Low	0.50	Medium	1300	
D	High	1.00	Medium	1300	
E	Low	0.50	High	1500	
F	High	1.05	High	1500	

The degree of substitution, for which the results of the emission of unwelcome components of the exhaust gas were lowest, is presented in Table 3. Furthermore, the values of the absolute change (increase or decrease) in the emission of a given component in relation to its emission when running on diesel fuel have been presented using graphs, in the same way as in the previous study by the authors [1], which facilitates a comparison of the results obtained with the results of the concentrations of individual components of the exhaust gases.

Table 3. The replacement of diesel/CNG setting with the lowest emission of harmful components.

Rotational Speed [RPM]	Load [Nm]	CNG Replacement Setting [%] When There Is a Minimum Emission Share of:									
		CO	NO	NO_2	NO_X	NH_3	CH_4	C_2H_6	NMOC	C_2H_4	CH_2O
1100	520	ON	83.22	ON	77.12	77.12	ON	ON	ON	ON	ON
	1150	ON	45.05	ON	12.1	54.86	ON	ON	ON	ON	ON
1300	480	ON	90.68	ON	90.68	66.44	ON	ON	ON	ON	ON
	980	ON	40.2	ON	40.20	68.38	ON	ON	ON	ON	ON
1500	500	ON	81.91	ON	81.91	73.37	ON	ON	ON	ON	ON
	1040	ON	17.84	ON	27.49	66.08	ON	41.52	ON	ON	ON

Figure 1 shows the temperature of the engine exhaust gas, while the following diagrams illustrate the emission of unwanted exhaust gases elements and their changes at individual points of engine operation.

Figure 1 shows the effect of the diesel/CNG exchange ratio for six engine operating states with constant engine parameters differing in speed and load on the turbocharger output exhaust gas temperature. The graph illustrating the course of the exhaust gas temperature of the tested engine was already presented in the previous publication of the series [1]; however, it is important for the subsequent analysis of the test results, where the exhaust gas temperature has a significant impact on the course of combustion and the possibility of the occurrence of certain chemical reactions affecting the final emission of individual harmful components from the exhaust gas. The exhaust gas temperatures in the individual runs are directly correlated with the load under which the engine was operating—runs A, C, and E, for which the engine load was lower, also had an exhaust gas temperature slightly lower than in the case of runs B, D, and F, in which the load was more than twice as high. A clear increase in the absolute value of the exhaust gas temperatures in the diesel/CNG replacement coefficient of about 50–60 °C at the final point is visible. The changes in exhaust gas temperature can be seen in both Figure 1 and the following Figure 2, which show the relative values of the temperature changes. In the same way, the two graphs will show the emission levels of the harmful components of the exhaust gas, and the levels of change in emissions of the harmful components of the exhaust gas.

The temperatures in all runs increased with natural gas replacing diesel. According to the theory in [36], in a dual-fuel engine fueled with natural gas and diesel, the exhaust gas temperature should decrease. In the case of the tests carried out, the temperature rises, which is indicative of an insufficient homogeneous mixing of natural gas and air and the problem of post-combustion vaporization in the later phase of combustion due to insufficient areas in which the air required to burn the fuel was not available. Such poor areas are caused by a change in the way that additional fuel is burned, and this was thoroughly described in previous work by the authors [1]. It is also worth noting that the temperature changes are most pronounced at the lowest values of diesel-to-CNG substitution and the highest values of diesel-to-CNG substitution. This may be indicative of the extreme changes that occur in the engine in these cases, as at low replacement values these are changes from single-fuel to dual-fuel operation, and at high replacement values (runs B, C, E), the limit of knocking combustion may already appear, indicating extreme utilization of the amount of CNG that can be used. Figure 2 shows the relative changes in exhaust gas temperature calculated related to the exhaust gas temperature achieved with diesel combustion, according to [37] in reference to the Kelvin temperature scale.

Figure 2 shows the relative temperature changes, which clearly show an upward trend in temperature increments relative to the reference temperature reached, when the engine is fueled with diesel, as the degree of substitution of diesel for natural gas increases. This trend holds for all test runs—there are temporary drops at individual points, but the overall upward trend in temperature rise is visible. Irrespective of the temporary changes in exhaust gas temperature increases, the exhaust gas temperature invariably rises at all operating points relative to the exhaust gas temperatures of the single-fuel-powered engine, with the lowest increase achieved in run E being just over 2%. The highest temperature increase was recorded for test run C where it exceeded 10%. It is worth noting that this highest recorded increase was achieved with the highest degree of substitution achieved in the entire test run of over 90%. This demonstrates the clear, direct influence of the degree of diesel/CNG substitution on the combustion temperature in the combustion chamber.

Figure 3 shows that the CO emissions have a very strong upward trend for one of the engine modes (A) in which the increase already occurs sharply for the first point with the smallest substitution of diesel for natural gas. At this load (low load of 0.5 kNm) and speed (low revs, 1100 rpm), CO emissions were already noticeably higher at the lowest applied CNG dose to the engine. These values deviate significantly from the emission values for the other phases of engine operation, which may be indicative of serious fuel combustion problems in this area of engine operation. High CO emissions may be indicative of incomplete combustion of both diesel and CNG. This problem did not arise for any of the other test runs, and although run F is also characterized by noticeably higher CO emissions, all runs apart from a test run A are characterized by values so small that their cross-characterization misses the point, as the values are within the measurement error. These are operating modes with extremely opposite test assumptions, where for run A the speed and load were set to the lowest of the accepted values, while for run F the load and engine speed was set to the highest. It is therefore difficult to find a correlation between the two runs, but it can certainly be assumed that fuel combustion was incomplete in both cases, and this is a problem for this type of engine, which should meet modern emission standards.

The emission limit for CO in Euro VI is set at 4000 mg/kWh, while for Euro VII it is 1500, or 400 mg/kWh, respectively. This is well illustrated by the fact that for five out of the six test tests carried out, the recorded emissions are clearly below the Euro VI limit, and the more liberal Euro VII scenario. The problem, however, is the emissions for test A where there was a clear deterioration in combustion quality, significantly exceeding any acceptable emissions standard. This mileage (A) would not meet either the newer Euro VII or Euro VI emission standards, nor the Euro V emission standard to which the test engine was factory operated. This demonstrates the significant deterioration in combustion quality in this test run.

One important factor contributing to the increase in CO emissions is the low boost pressure associated with the low amount of exhaust gas generated. The low charge pressure does not favor adequate mixing of air and fuel, which promotes the post-mix in the cylinders with a spatially varying excess air ratio, which is an important factor affecting the creation of both CO and unburned HC.

Although Figure 4 shows that only sample run A appears to show a strong increase in CO emissions with increasing substitution of diesel by CNG, it is clear from Figure 5 that the increases are quite intense for most of the runs that were tested. It can be shown that runs B and D, generated at high engine load, are characterized by a greater increase in CO emissions with low substitution of diesel by CNG and a negligible further increase with higher substitution of diesel by CNG. The waveforms A and F, described in Figure 5 as having the highest CO emissions, after a very strong increase in emissions at low diesel substitution by CNG hardly increase any further when diesel substitution by CNG is increased above a value of 40%. The strongest increases in CO emissions were characterized by runs C and E, in which the strongest increase occurs in the range of diesel substitution by CNG from 30% to 40%. An important difference for these two runs is also that the most intense increases at high levels of diesel substitution by CNG are followed by an even more intense decrease in CO emissions above 50% liquid-fuel substitution. This indicates a significant deterioration in combustion quality at certain values of diesel substitution by CNG in certain areas of engine operation. This may be primarily due to the poor mixing of gaseous fuel with air and a failure to adapt the shape of the combustion chamber to the new engine operating conditions. It is worth noting that the increases for all ratios are high and range from several hundred to several thousand percent. The theoretical assumptions that describe dual-fuel engines present these engines as more efficient, cleaner, and more economical than classic CI engines, so the research carried out here clearly shows that this depends on other factors and will not always be true.

Figure 5 shows that NO emissions decrease in all test runs for most areas of operation, but these are not large decreases, and for some runs, after small decreases, above certain replacement values they start to increase. For runs B and D, the limit is the replacement of diesel by CNG at around 40%, while for run F the limit is already 20% fuel replacement.

NO emissions assimilated to NO_X emissions allowed under Euro VI or Euro VII standards exceed the permissible standards many times over. The limit for Euro VI is 460 mg/kWh, while for Euro VII it is 120 and 400 mg/kWh. Within the limits of the Euro VI emission standard, the engine was operating at high levels of diesel substitution by

CNG (above 60%) for runs C and E, where the engine load was low, but this was not a condition that allowed NO emissions to be reduced; as for run A, the NO emissions were cumulatively one of the two highest of all runs. The NO emission levels for four of the six test runs would allow the engine to meet the Euro V emission standard for which it was factory prepared. The emission limit for NO and NO_2 is 2000 mg/kWh in this standard, and was exceeded in two test runs, A and F, where the emissions were significantly higher; however, it is worth noting that this engine would also not have met the Euro V standard when running on diesel, and the emissions when running on diesel for run A were twice as high as that allowed by the standard. The decreasing trend in NO emissions with increasing substitution of diesel by natural gas is indicative of a trend towards potentially meeting the emissions standard that a depleted engine converted to dual fuel would no longer meet (in the case of mileage A, this could even be achieved with substitutions over 60%). This shows that simply converting a depleted diesel engine to run on CNG/diesel oil bi-fuel does not easily meet the newer emission standards, but may allow emissions of harmful exhaust constituents to return to the emission standard in force at the date of the engine's manufacture.

Figure 6 shows that increases in NO emissions only occurred in three cases for higher degrees of diesel substitution by CNG. This is the case for runs B, D, and F, and these are runs where the engine was running at a higher load. These increases follow earlier decreases and are small, starting above 40% CNG replacement of diesel. In contrast, test runs A, C, and E, in which the engine was under low load, are characterized by a continuous decrease of an approximately linear nature as the degree level of diesel replacement by natural gas increases.

These results do not show similarities with the temperature of the exhaust gases, local combustions with a higher temperature than the background could raise the average temperatures. This problem was described in previous work by the authors [1]. It is worth noting, however, that at lower engine loads, an increase in the proportion of CNG in the fuel supplied to the engine contributes to lower NO emissions, which may be related to the fact that the combustion temperature is lower the higher the CNG content is, and the combustion run itself is smoother. Increases in mileage B, D, and F where the engine load is high may be precisely due to the less stable combustion course. Decreases in NO concentration, on the other hand, most likely result from a reduced availability of methane used by the engine. There is a reverse correlation where there is an increase in the degree of fuel exchange and a decrease in the value of the excess air coefficient [38].

Figure 7 shows NO_2 emissions. The emissions of NO_2 for all test run increase steadily with an increasing substitution of diesel for CNG. The emission level for all runs starts from a similar level which corresponds to the emission level for an engine running only on diesel and is roughly similar. It then increases steeply for all runs to stabilize at around 0.3 g/kWh for two runs (B and D) with substitution values greater than 20%, and for the remaining runs the emissions increase further. The strongest increase can be observed, especially for high substitution rates of more than 60% for run A, where emissions exceed 2.5 g/kWh for the highest CNG/diesel substitution rate examined. The other runs at substitution rates above 40% stabilize at around 1 g/kWh, indicating an approximately similar combustion pattern for the three runs. Changes in NO_2 emissions are correlated with changes in exhaust gas temperature. An increase in flue gas temperature (Figure 2) results in an increase in NO_2 concentration in the flue gas [39].

The NO_2 emissions assimilated to the NO_X emissions allowed under the Euro VI or Euro VII standards exceed the permissible standards many times over. The limit for Euro VI is 460 mg/kWh, while for Euro VII it is 120 and 400 mg/kWh, and it is clear that these standards can only be met by this engine when running on pure diesel in mono-fuel mode. Once dual fuel is introduced, NO_2 emissions increase so strongly that it becomes impossible to meet today's emission standards. The Euro V NO_X emission limit would allow the achieved emission levels for five of the six test runs, in which NO_2 emissions did not exceed 1.5 g/kWh and the limit was set at 2000 mg/kWh (NO + NO_2). Run

A, when substituting close to, and exceeding, 80% natural gas for diesel, exceeded the 2 g/kWh emission level which would disqualify this run from being able to meet the Euro V standard.

Figure 8 shows that the increase in NO_2 emissions is most clearly visible, as in Figure 7, which shows that run A, where the increase is the most intense and, with the highest degree of substitution of diesel by CNG, reaches more than 4500% relative to the emission values of the diesel-only engine. All of the sample runs visible in the graph illustrate increases at levels of several hundred to several thousand percent, which is high. However, Figure 8 shows that the correlation between lines B, D, and F is more apparent, where the emission increases are linear with a small incremental increase in emission increases. These waveforms were produced when the engine was operated at a higher load, which may indicate a similar pattern of phenomena for these tests, and it is worth noting that for these waveforms the NO emission increases as higher replacement of diesel fuel with natural gasare growing even stronger. The differences between NO and NO_2 emissions may be due to the way in which these substances are formed. This has already been described previously in our study [1], where intensive growth with an initial small degree of replacement was noticed. Authors from [40,41] clearly described the mechanism of NO_2 formation in an internal combustion (IC) engine.

NO_X emissions represent the sum of NO and NO_2 emissions. For both of these components, the opposite correlation was shown to increase as a function of engine load during the tests. Figure 9 shows that the summed emissions of both these components assume a linear character, confirming previous observations regarding this correlation between both components. For runs B, D, and F, in which the engine load is higher, NO emissions increased more intensively than for NO_2 in Figure 10, showing that the absolute values of NO_X emissions are linear with little variation but take on varying emission values. It is noticeable that the emission values increase slightly in these runs for diesel substitution by CNG exceeding 40%, and this can also be seen in Figure 10. In contrast, for runs A, C, and E, a slight decrease in NO_X emissions can be seen as the degree level of diesel substitution by CNG increases, which can also be seen in the runs of Figure 10.

NO_X emissions in dual-fuel operations are no longer as fundamentally different from NO_X emissions for single-fuel operations as they are for NO_2 emissions. This is due to the much higher proportion of total NO in NOx than NO_2, the amount of which is proportionally lower.

Figure 10 shows that the changes in the NO_X emission values, although clearly visible, are in a very small range of a few tens of percent at most. This is due to the correlation between NO and NO_2 emissions described earlier and the fact that NO share of NOX emissions proportion is a much higher than NO_2 and changes with increasing substitution of diesel by CNG also reached a maximum of a few tens of percent.

NO_2 emissions, compared to NO_X emissions, allowed the limit to be exceeded many times over, under Euro VI or Euro VII standards. The limit for Euro VI is 460 mg/kWh, while for Euro VII it is 120 and 400 mg/kWh. It is therefore clear that these emission levels were not achieved during this test. This is not due to the specific operation of the engine in dual-fuel mode per se, as the NO_X emission levels in single-fuel and dual-fuel operations were similar, but rather due to the mileage and factory preparation of the engine used in the tests to meet the respective emission standard. This is because the engine had to comply with the Euro V emission standard, in which the NO_X emission limit was set at 2 g/kWh [22]. This limit was exceeded for half of the tests prepared for this paper, which may indicate the technical condition of the tested engine.

Figure 11 shows that CH_4 emissions, as the degree of substitution of diesel by CNG increases, also increase steadily for all engine loads and speeds. The increase in emissions is directly due to the increase in methane present in the combustion chamber and is produced by not burning all of the fuel supplied to the engine. Emission levels vary from one engine run to the next, and it is characteristic that for two runs (B and D), the emission levels and their variation are virtually identical, and this is the lowest emission level achieved in this study—not exceeding 2 g/kWh. The other trials reached levels of several g/kWh and, in the case of run F, close to 25 g/kWh.

The CH_4 emission limit for the Euro VII scenarios is 100 and 50 mg/kWh, respectively; for Euro VI, it was defined at 500 mg/kWh, but only for HDVs, as CH_4 emissions for LFVs were included in the total THC emissions and were not independently measured. However, the impact of CH_4 on the climate and the increasing use of natural gas in motor vehicles has resulted in CH_4 emissions becoming an important test point and being included in today's emissions standards. In the case of the tests carried out, however, it is clear that the Euro VI emission limits can only be met at very low exchange rates and only in two of the tests, while in the four other tests (A, C, E, and F) CH_4 emissions already rise above the limit value allowed by the standard at the lowest exchange rates. The Euro VII standard could not be met for any of the dual-fuel engine operating points; however, the engine running only on diesel met the Euro VII emission standard predicted in the scenarios. The Euro V standard was also not met in any of the test runs. The only operating points at which the Euro V CH_4 emission limit was not exceeded concern engine operation in test runs B and D with a maximum of 40% substitution of diesel by CNG.

Figure 12 shows that the changes in CH_4 emissions are roughly directly proportional to the amount of methane entering the combustion chamber systematically, but for some runs, the increase is more intense than for others. Runs B and D have the smoothest increase in CH_4 emissions, indicating a more complete combustion of the CNG supplied to the engine than for the other runs. The most intensive increase is seen for runs A, E, and F, but it is difficult to find a definite link between these relationships.

To reduce CH_4 emissions in dual-fuel engines, it is necessary to use a suitable combustion chamber and to prepare the mixture in such a way that as much fuel as possible is combusted.

Figure 13 shows that ethane emissions are a harmful component of the exhaust gas resulting from incomplete combustion of the fuel supplied to the engine—in this case, mostly from the heavy hydrocarbons contained in diesel fuel. Although its formation from unburned methane is not the main source of emissions (although CNG fuel may also contain some ethane) its emissions increase systematically for most tests as the degree level of substitution of diesel fuel by gaseous fuel increases. For run F, its emissions rise to the highest values, exceeding 2 g/kWh with a fuel change rate of more than 80%. For runs B and D, the increase in emissions was lowest and did not exceed 0.3 g/kWh. It is worth noting that minimal increases in CH_4 were recorded for the same runs. This means that for these engine operating areas, both fuels were burned more completely than for the other runs.

The allowable emission of this component is not precisely defined by the standard, as it is included in the total hydrocarbon emissions (THC), or the total hydrocarbon emissions group excluding CH_4–NMHC (Non-Methane Hydro Carbons). When comparing the C_2H_4 emission results obtained with this group, the emission limit values should be set at 50 and 25 mg/kWh for the Euro VII and 160 mg/kWh for the Euro VI emission standard scenarios, respectively. This engine can only comply with Euro VI with the mono-fuel operation because when a second fuel is added to the combustion chamber, the C_2H_6 emissions already increase sufficiently with small additions of C_2H_6 to not exceed the emission limits allowed by the current standards. Even the Euro V standard, for which the engine is homologized, could only be met with two test runs (B and D). In the other runs, only test run F met the Euro V emission standard for most of the operating points tested, exceeding it only at the maximum degree of substitution achieved in this test run.

Figure 14 shows the increases in C_2H_6 emissions range from several hundred to several thousand percent, which is a strong increase. It is worth noting, however, that for some runs the value of C_2H_6 emissions, after an initial increase, then decreases, which may be indicative of the inadequacy of both the combustion chamber shape and the matching of the correct ratio of oil used to natural gas supplied, as there may be areas inside the combustion chamber in which the amount of oxygen available for the combustion process is less than necessary for the proper combustion of the injected liquid fuel, leading to the formation of unburned hydrocarbons, such as C_2H_6.

The waveforms in which decreases in C_2H_6 emissions are observed as the amount of diesel replaced by natural gas increases are waveforms C and F, which were not correlated with each other under the research assumptions. The lowest increases were observed for run B, but it is difficult to find a noticeable correlation with the individual test samples.

NMOC (Non-Methane Organic Compounds) emissions are a key parameter characterizing the quantity of unburned hydrocarbons in the exhaust gas, excluding methane emissions, which would be overwhelmed by the use of a gaseous fuel, such as CNG. Figure 15 shows the emission of it. From the NMOC values it is possible to read, above all, the number of unburned particles of the base fuel, diesel. The main component of the measured NMOC emission values was propane and the heavier hydrocarbons that are produced by the incomplete combustion of diesel fuel. Propane emissions will not be presented directly, as they are included in NMOC emissions as one of the key components, and the other hydrocarbons included in NMOC emissions are presented in the other graphs of this article. NMOC emissions increase as the replacement rate of diesel to CNG increases. Emissions of this group of compounds are related to unburned compounds contained in diesel fuel. Natural gas may contain traces of propane and other heavy hydrocarbons, but the amount is negligible and depends on the chemical composition of the CNG used. Its emission is mainly related with uncombusted, to all the diesel fuel. The level of emissions varies widely from trial to trial, but in all cases increases with increasing substitution of diesel by CNG. For runs B and D, emissions ranged from 0.1 g/kWh to 0.6 g/kWh. In runs C, E, and F, emissions were less than 2 g/kWh, while in run F, emissions were already noticeably higher than in the previous five runs, with emissions reaching over 3 g/kWh. It is worth noting that NMOC emissions in run F were significantly higher than in the other tests, even when running on mono-fuel with diesel only, at 0.4 g/kWh, which is higher than the NMOC emissions from low levels of diesel substitution with natural gas in runs B and D.

The emission limit for NMOCs is not strictly defined in the emission standards, but they belong to an emission group referred to as non-methane hydrocarbons(NMHC) in which there are no methane emissions included. Their permissible emissions are also not defined for modern emission standards, so in order to compare the emission results achieved, reference can only be made to the NMHC emission limits contained in Euro VI and Euro VII of 160, 50, and 25 mg/kWh, respectively. These limits, corresponding to Euro VI, could only be met with single-fuel diesel for test runs B, C, D, and F. In the case of two runs (B and D), the engine would have met the Euro V emission standard, to which it was factory-adapted, only because tests were not continued for higher degrees of diesel substitution with natural gas. The other runs only met the Euro V standard with the mono-fuel operation. The Euro VII standard predicted by scenarios A and B would not have been met by the engine at any of the test points.

Figure 16 shows the trends in NMOC emission changes for all samples and are high, ranging from a few hundred to tens to over a thousand percent. Three groups of emission increases can be distinguished. The smallest increases can be observed for samples A and B, while the largest is for sample C. Runs A and B are jointly characterized by engine operation at the lowest of the tested speeds. The other runs are no longer correlated with each other, and it is difficult to find a common characteristic causing such increases in emissions.

The increase in NMOC emissions is due, as in the case of C_2H_6 increases, to unburned hydrocarbons remaining in the combustion chamber after combustion from the high reactive fuel injected into the combustion chamber. However, the increases in NMOCs are even higher than those of C_2H_6. This may support the thesis that there is too little oxygen available in the combustion chamber where natural gas is located where air should be the primary air in the engine. The combustion of organic hydrocarbon compounds requires more oxygen AND energy than CH4 and C_2H_6 alone, which therefore may result in the much higher increase in emissions of this group of substances in the exhaust gases. Another reason is that C_3H_8, which is the main component of measured NMOC emissions, is a more complex hydrocarbon than C_2H_4, so there is more of it in diesel fuel, whose composition is based primarily on heavy hydrocarbons.

Ethene–C_2H_4-is an unsaturated hydrocarbon, it is the simplest alkene—an organic chemical compound. Its emissions are related to the chemical transformations taking place in the combustion chamber. Figure 17 shows the emission of it. Ethane emissions increase as the substitution of diesel fuel for CNG increases, and their pattern is similar to that of propane emissions. During the B and D test runs, emissions were lowest—virtually the same as for propane emissions. These emissions did not exceed 0.05 g/kWh. Emissions for runs C and E increased more and reached almost 0.2 g/kWh at extreme operating points. Sample runs A and F grew significantly faster and reached the highest values—in the case of run A, over 0.5 g/kWh.

Emission limit values for ethene are not specified directly in the standards, but for hydrocarbon substances that are not methane or ethane the emission limit value can be compared to the NMNEHC emission limit value. For the standards currently in use, the NMHC emission limit value is known to be 160 mg/kWh for Euro VI, and 50 mg/kWh and 25 mg/kWh, respectively, for both Euro VII scenarios. The emission limit values for Euro VII scenario A and Euro VI are higher than those for B and D runs. The Euro VI emission limit is also higher than the maximum emission values recorded in sample run C. However, the remaining runs exceed the Euro VI and Euro VII emission limits. This indicates that the emissions of this harmful exhaust constituent are too high for the conversion of this engine to be considered sufficient to improve the quality of the harmful exhaust constituent emissions. It should also be remembered that the NMHC emission limit applies to emissions of all hydrocarbons in the exhaust gas except methane, and the ethene described here is only one of many that were emitted by this engine. In the case of run A, for the highest values of diesel substitution by CNG, even the Euro V standard for which the engine was factory prepared was exceeded at the operating points, but it should be remembered that an exhaust gas catalytic converter was not used in the tests.

Figure 18 shows the ethene emission changes. The variations in ethene emissions are highly variable, ranging from tens to many thousands of percent. Such large variations are indicative of widely varying combustion chamber conditions at different points in the engine's operation. Low oxygen availability in the combustion chamber can result in combustion residues remaining in the form of hydrocarbon bonds from unburned fuels—both diesel and CNG. Alkenes can also appear in exhaust gas as a result of chemical transformations occurring during the combustion of both fuels. The course of these transformations depends on many factors, the occurrence of which is difficult to assess based on the results obtained. However, when analyzing the results for ethene, it should be taken into account that the high variability in emissions is due to the fact that ethene emissions are very low when running on diesel.

However, in some runs (C and E), ethene emissions in the highest fuel changeover ranges start to decrease as the diesel-to-CNG changeover ratio increases, and in the case of run C, to an even level lower than at the previous two operating points. This may indicate not so much an improvement in the conditions for efficient combustion of the fuels used, but the non-existence of conditions for the conversion of more complex hydrocarbons into simpler hydrocarbons, such as ethene, while the systematic increase in the growth of ethene emissions indicates a deterioration in the quality of combustion and the entry into the exhaust gas of increasing amounts of unburned simple hydrocarbons.

Figure 19 shows that ammonia emissions decreased with increasing substitution of diesel by CNG for all test runs. For the four runs (C, D, E, and F), the absolute value of ammonia emissions did not change noticeably, but for emissions in trials A and B, the changes were noticeably higher, reaching 0.3 g/kWh absolute difference. The maximum emission values always occurred when the engine was mono fueled with diesel only, and their reduction through the use of natural gas is an interesting phenomenon that also coincides with the observations presented in the study in [1]. Ammonia emissions usually result from engine operation on stoichiometric or close to stoichiometric parameters of combustion [42]. In the case of our study, the opposite phenomenon occurs here, in which ammonia emissions decrease as the replacement of diesel by CNG increases. It is important to note that ammonia emissions are highest in runs A and B with the lowest engine speed. It is therefore possible to assume a thesis in which an increase in engine load by increasing engine speed induces a decrease in ammonia emissions.

The emission standards used today assume an acceptable limit for ammonia emissions. In the case of scenarios prepared for Euro VII, they assume a maximum permissible emission value of 20 mg/kWh and 10 mg/kWh. For Euro VI, it is around 40 mg/kWh (approximate, calculated value). It is therefore easy to see that, in the case of the engine tests, out of the six tests carried out, only four would meet the potentially acceptable Euro VI emission standards, while Euro VII in the scenario allowing maximum emissions of 10 mg/kWh would only be met by half of the tests (C, D, and E). The Euro V standard did not include a reduction in NH_3 emissions.

Figure 20 shows emission changes of ammonia, on which all test trials showed strong decreases in ammonia emissions with increasing substitution of diesel for natural gas. In the case of trials C, E, and F, there are isolated increases in emissions for the operating points preceding the operating point to which we refer. These increases are not high and are rather due to measurement uncertainties and changes associated with changes in emissions of other combustion substances and the average trend for all test runs.

This is because ammonia emissions are largely due to the occurrence of other substances in the combustion process, which in further chemical transformations result in the formation of ammonia and its persistence in the final flue gas composition. Its emission is therefore highly dependent on the emission of other substances, as described in the study [1]. To understand the processes that took place in the engine in this case, it would be important to find out about hydrogen and water emissions, which requires further study.

Figure 21 shows that formaldehyde emissions increase with increasing substitution of diesel by CNG. A correlation can be seen between formaldehyde emissions and those of other substances, the emission curves of which follow a very similar pattern. As in the case of hydrocarbon emissions, runs B and D increase slowly and reach low values of no more than 0.1 g/kWh, runs C and E reach higher values of no more than 0.5 g/kWh, while runs A and F reach the highest increases of up to 1.4 g/kWh. A correlation can be seen here with the emissions of hydrocarbons, such as ethene and butane.

Formaldehyde is formed during the incomplete combustion of substances containing the element carbon. Emissions are favored by elevated temperatures and high humidity. In the reaction of the oxidation process, formaldehyde is formed where the flame temperature has been reduced below 1000 K that could appear at the end of the combustion process or in the combustion chamber wall areas. This is when CH_2O as a combustion intermediate does not have sufficient energy to be converted into CO_2. In gas engines, several sources can be assumed to be responsible for the formation of formaldehyde [43]. The first is the inhomogeneity of the fuel–air mixture in the engine cylinder. This can result in the formation of zones with a λ-factor causing under-firing of the gaseous fuel and the formation of formaldehyde molecules. The second source is the walls of the combustion chamber, which reduce the combustion temperature in their vicinity by receiving heat. A similar effect is obtained in the case of the volumes between the piston and cylinder walls (crevice effect) where, due to the small distance between these surfaces, the flame is extinguished. Formaldehyde emissions are also significantly affected by the increase in temperature and pressure during the compression stroke, which can reach values that favor CH_2O formation.

Modern emission standards for harmful exhaust components do not limit the emission limits for formaldehyde. It can be compared similarly to how butane emissions were compared to the NMHC limit value. Somewhat similar to that case, only two test trials would meet any modern emission standards, and, in this case, they would only meet the Euro VI standard. For two of the six tests (A and F), not even the Euro V standard to which the test engine was adapted would be met; however, the tests did not take into account the catalytic reactor.

Figure 12 shows changes in formaldehyde emissions range from several hundred to several thousand percent. With an increase in the exchange of diesel to CNG, emissions increase in all runs, and they increase for most runs, except for runs C and E in which, for higher exchange values (from 70% diesel to CNG exchange and above), emission increases are already smaller than at previous engine operating points. The smallest changes were found for test runs A and B, i.e., for tests in which the engine speed was lowest. The largest emission changes occurred in runs C and E, and average changes (reaching 6000%) in D and F.

Summary of the Results

All emissions measured in test trials where diesel is substituted to any extent by CNG have changed relative to the single-fuel diesel-only engine. Some of these changes were large and had an impact on the potential failure of the dual-fuel engine to meet modern emission standards, and in some cases, the emissions changed to an extent that did not change the range of permissible emission standards under which the engine operated. A summary of the most significant changes and meeting acceptable emission standards is provided below.

- Summarizing all the above, the following analysis of the obtained results can be carried out: Studies that have been carried out indicate that emissions of certain harmful exhaust components must be reduced. Their emissions have increased too much, and the values must be corrected if the engine is to be used today.
- CO emissions at one run are much higher than the other runs.
- NO_X emissions, including both NO and NO_2 in some test runs, increase with the degree of diesel replacement with CNG, and decrease in others.

The overall emissions of this substance did not change significantly, affecting the ability of the test engine to meet modern emission standards. From the theoretical assumptions [44–46], NO_X emissions should be decreasing—however, this is only the advantage of the RCCI engine. At the same time, these emissions have not increased significantly.

In the analysis [1], a table summarizing the results was presented. The same operation but with the lowest emission level was carried out and is shown in Table 3 below:

The results presented in Table 3 clearly show in which areas the engine converted to bi-fuel did not reduce emissions of harmful exhaust components. Evident deficiencies in any emission reductions for certain substances include CO, NO_2, CH_4, C_2H_4, and CH_2O. Decreases recorded only for post-single runs (a correlation with the high engine speed used is noticeable here) were recorded for C_2H_6 for a test run E. For two substances and one substance group (NO + NO_2), decreases were recorded for all test runs in which the lowest emissions were most often (more than 2/3 of cases) achieved at higher (more than 50%) substitution rates. These substances were NO, NH_3, and a group of compounds: NO_X. In the Section 4, a discussion is held on the results obtained as seen in Table 3.

4. Discussion

The emission standards that the motor industry is trying to introduce today are very strict. To meet them, engines that have already been manufactured and adapted to meet the older emission standards are trying to use procedures such as the use of alternative fuels, or the addition of a second fuel to an existing fuel system. Such conversions are becoming increasingly popular, but as can be seen from the research carried out for this article, simply replacing part of the high cetane fuel with CNG does not allow the vehicle to meet the newer emission standards.

The appearance of higher unburned hydrocarbon emissions proved to be a major problem. This was to be expected, due to the unpreparedness of the combustion chamber for the stoichiometric gaseous fuel supplied and the lack of a catalytic reactor in the system from which the exhaust gases were drawn. As their emissions increase, they can be dealt with by carrying out ground-truthing changes to the design of the combustion chamber and the way fuel is delivered to the combustion chamber. Hydrocarbon emissions are also efficiently corrected by a catalytic reactor, which can post-combust unburned hydrocarbons. A more significant problem in meeting today's emission standards is nitrogen oxide emissions. The scenarios envisaged for Euro VII are very restrictive in this respect, as illustrated by the graph below showing the average contemporary NO_X emission levels in the RDE test trials as a function of the speed of the vehicle under test and the limits to be imposed in the latest standard (red lines).

As can be seen, Figure 23, based on work from [47] shows that the anticipated limitations of the Euro VII standard are very difficult to meet, even for contemporary engines. Several studies [48–51] predict the possibility of meeting the new standards through the use of dual-fuel natural gas in engines. In the case of the engine used in this study, no noticeable changes in NO_X emissions could be made. This indicates, as in the case of the increase in hydrocarbon emissions, a mismatch between the combustion chamber and the conditions required for the efficient combustion of natural gas.

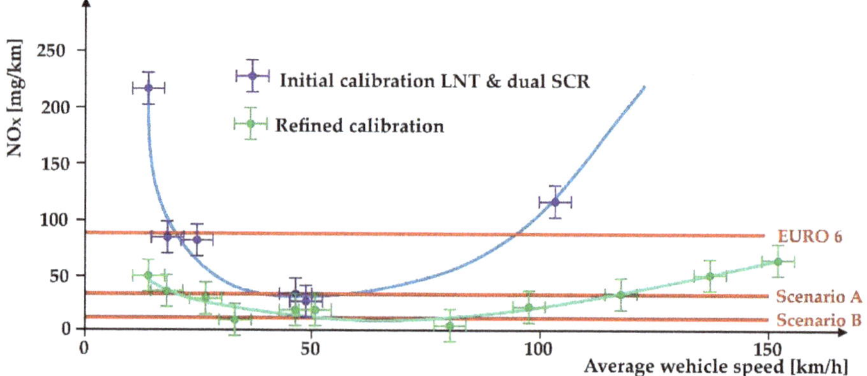

Figure 23. Predicted reductions in toxic exhaust emissions according to alternative scenarios projected for Euro 7, based on charts from [47].

On the other hand, it may also indicate a real problem with the ability to reduce emissions of this substance, which may negate the possibility of reducing nitrogen oxide emissions with this method. However, this requires, in the authors' opinion, more tests preferably using a different engine and types of gas installations adapted for the engine under study.

The summary quoted in Table 3 clearly shows the deterioration of engine performance in the area of harmful hydrocarbon emissions, which arise in the case of incomplete combustion of the used fuels. This issue is most easily addressed by the use of a catalytic reactor in the exhaust system to post-combust these substances, of which the largest increases in emissions are related to methane. The remaining hydrocarbon compounds appearing in the exhaust gas are indicative of a deterioration in the quality of combustion processes, which points to the need to modify the shape of the combustion chamber in the future. Adapting the combustion chamber to run on two fuels will be necessary to reduce unburned hydrocarbon emissions without the use of a catalytic reactor and thereby improve the thermal efficiency of this engine. Additional evidence of the deterioration in combustion quality is the increase in CO emissions for all test runs. The appearance of this substance is mainly due to insufficient oxygen in the areas of formation. Reduction of its emissions can be achieved by both methods described for hydrocarbons.

The reduction in NO_X emissions is linked to a reduction in NO emissions, which account for a larger proportion of this group of substances than NO_2. The decrease in NO emissions is in line with the most common assertions in the scientific literature about NO_X emission decreases in dual-fuel engines. Usually, detailed data on independent NO and NO_2 emissions are not cited, so the commonly found decreases in NO_X emissions in dual-fuel engines may translate mainly into decreases in NO emissions, and the focus should be on the phenomena of the simultaneous increase in NO_2 emissions. Ammonia emissions, which were also able to be reduced in all test runs, are due to the appearance in the combustion chamber of other substances whose presence has a significant effect on the formation of ammonia, formed by a series of chemical transformations. The reduction in NO_X emissions can also affect ammonia emissions, given that its natural reactants, nitrogen oxides, absorb fewer ammonia molecules, which naturally decompose NO_X molecules into water and oxygen.

5. Conclusions

1. The tests carried out to make it possible to state unequivocally that the engine modification carried out did not make it possible to limit the emission of all harmful components of exhaust gases whose emission is limited by contemporary EU regulations.
2. The emission of hydrocarbons has increased in such a way as to make it impossible to comply with contemporary exhaust emission standards. Their emission is related to the incomplete combustion of fuels supplied to the combustion chamber, and the largest increases relate to methane emissions, which is the main and basic component of CNG fuel.
3. Emissions of nitrogen oxides did not change with the increase in the substitution of diesel for CNG in a way that significantly affects the engine's ability to meet modern emission standards. This is evidenced by the similar rate and temperature of the combustion process for both mono- and bi-fuel fueling, and the change in the amount of CNG used does not significantly affect emissions.
4. NO_X emissions did not change significantly as the degree of diesel-to-CNG replacement increased. This indicates that there was no significant change in the combustion of fuels in the engine under study where the combustion temperature should have decreased, resulting in lower NO_X emissions.
5. The failure to adapt the combustion chamber to burn gaseous fuel in a fuel–air mixture close to stoichiometric results in increased hydrocarbon emissions. Emissions of all hydrocarbon compounds in most test runs increased with the degree level of substitution of diesel by CNG, indicating an insufficient availability of oxidant in the combustion process, and therefore insufficient mixing of fuel and air, and the existence of areas in the combustion chamber that are too rich and potentially too poor.
6. Simple conversion of an engine to so-called "gas-diesel" does not make it possible to reduce the emission of harmful components by the engine to a level that would allow the engine to meet the higher emission levels envisaged in the production of a bi-fuel converted engine.

The way for a dual-fuel engine fueled with a high reactive fuel, such as diesel, and low reactive fuel, such as CNG, to meet higher exhaust emission standards than those for which the engine was designed is through greater intervention in both the design of the engine combustion chamber and the timing gear main construction and the operation of the engine's power supply system, which should be carried out as part of further research work. It may also be necessary to change the exhaust after-treatment system used, which was expected to work more effectively with different concentrations of harmful exhaust components than under dual-fuel operation.

The authors' expertise in this area allows them to set out clear guidelines for the way forward on the topic:

-   Carry out tests in which additional flue gas components such as H_2, H_2O, SO_2, N_2O, N_2O_5, and other NO_Y- and NO_Z-forming substances will be measured.
-   Check the emissions of the various components of the exhaust gas after being routed through the original catalytic reactor of the engine under test.
-   Repeat the tests carried out, as the determination of emission values with a single measurement is an unreliable source of information and, when comparing changes in

emission values, it is important to establish a trend based on a number of the same tests in order to identify differences and similarities indicating whether a result has been achieved.

- Compare the results obtained when using a different generation of gas system on the same engine
- Carry out a full modification of the engine, including the rebuilding of the combustion chamber and valve train in order to compare emission test results on the engine thus modified.

Author Contributions: Conceptualization, M.K. and G.S.; methodology, M.K.; software, M.K.; validation, M.K.; formal analysis, M.K.; investigation, M.K.; resources, M.K.; data curation, M.K.; writing—original draft preparation, G.S.; writing—review and editing, G.S.; visualization, M.K.; supervision, M.K.; project administration, M.K.; funding acquisition, M.K. All authors have read and agreed to the published version of the manuscript.

Funding: This work was financed by the Military University of Technology under research project UGB 758/2022.

Data Availability Statement: Data are contained within the article.

Conflicts of Interest: The authors declare no conflict of interest.

References

1. Karczewski, M.; Szamrej, G.; Chojnowski, J. Experimental Assessment of the Impact of Replacing Diesel Fuel with CNG on the Concentration of Harmful Substances in Exhaust Gases in a Dual Fuel Diesel Engine. *Energies* **2022**, *15*, 4563. [CrossRef]
2. Reitz, R.D.; Duraisamy, G.D. Review of high efficiency and clean reactivity controlled compression ignition (RCCI) combustion in internal combustion engines. *Prog. Energy Combust. Sci.* **2015**, *46*, 12–71. [CrossRef]
3. Chojnowski, J.; Karczewski, M.; Szamrej, G. The phenomenon of knocking combustion and the impact on the fuel exchange and the output parameters of the diesel engine operating in the dual-fuel mode (Diesel-CNG). *NAŠE MORE* **2021**, *17*, 30.
4. Karim, G.A. *Dual-Fuel Diesel Engines*; CRC Press: New York, NY, USA, 2015.
5. Li, Y.; Jia, M.; Xu, L.; Bai, X.-S. Multiple-Objective Optimization of Methanol/Diesel Dual-Fuel Engine at Low Loads: A Comparison of Reactivity Controlled Compression Ignition (RCCI) and Direct Dual Fuel Stratification (DDFS) Strategies. *Fuel* **2020**, *262*, 116673. [CrossRef]
6. Available online: https://www.ac.com.pl/products-autogas/controllers-lpg-cng-diesel/stag-diesel/338/ (accessed on 20 October 2022).
7. Available online: https://www.europegas.pl/en/eg-injecto-duo/ (accessed on 21 October 2022).
8. Available online: https://polandasia.com/cz%C5%82onkostwo/nasi-partnerzy/dual-fuel-systems/ (accessed on 21 October 2022).
9. Available online: https://fuelfusion.pl/firma/ (accessed on 21 October 2022).
10. Al-Saadi, A.A.A.; Bin Aris, I. CNG-diesel dual fuel engine: A review on emissions and alternative fuels. In Proceedings of the 2015 10th Asian Control Conference (ASCC), Kota Kinabalu, Malaysia, 31 May–3 June 2015; pp. 1–4. [CrossRef]
11. Smallwood, J.S. Investigation of the Dual-Fuel Conversion of a Direct Injection Diesel Engine. Master's Thesis, Department of Mechanical and Aerospace Engineering, Morgantown, WV, USA, 2013; p. 3410.
12. Andrych-Zalewska, M.; Chlopek, Z.; Merkisz, J.; Pielecha, J. Comparison of Gasoline Engine Exhaust Emissions of a Passenger Car through the WLTC and RDE Type Approval Tests. *Energies* **2022**, *15*, 8157. [CrossRef]
13. Stelmasiak, Z.; Larisch, J.; Pietras, D. Wpływ dodatku gazu ziemnego na wybrane parametry pracy silnika Fiat 1.3 MultiJet zasilanego dwupaliwowo. *Combust. Engines* **2015**, *54*, 672–682.
14. Jamrozik, A.; Tutak, W.; Grab-Rogaliński, K. An Experimental Study on the Performance and Emission of the diesel/CNG Dual-Fuel Combustion Mode in a Stationary CI Engine. *Energies* **2019**, *12*, 3857. [CrossRef]
15. Hanna, R.N.; Mustafa, A.M.; Zefaan, H.; Sameh, M.M. Aftermarket exhaust gas emissions analysis of a direct injection CNG-Diesel dual fuel engine. *Int. J. Curr. Res.* **2016**, *8*, 29321–29329.
16. Singh, R.; Maji, S. Performance and exhaust gas emissions analysis of direct injection cng-diesel dual fuel engine. *Int. J. Eng. Sci. Technol.* **2012**, *4*, 833–846.
17. Reitz, R.D.; Foster, D.; Ghandhi, J.; Rothamer, D.; Rutland, C.; Trujillo, M.; Sanders, S. Optimization of Advanced Diesel Engine Combustion Strategies. DOE Merit Review, ACE020. 2011. Available online: https://www.energy.gov/sites/prod/files/2014/03/f10/ace020_reitz_2012_o.pdf (accessed on 24 October 2022).
18. Reitz, R. Fuel Reactivity Controlled Compression Ignition (RCCI)—A Practical Path to High-Efficiency, Ultra-Low Emission. *Int. J. Engine Res.* **2011**, *12*, 209–226.
19. Dziubak, T. Experimental Studies of Dust Suction Irregularity from Multi-Cyclone Dust Collector of Two-Stage Air Filter. *Energies* **2021**, *14*, 3577. [CrossRef]

20. Dziubak, T.; Bąkała, L. Computational and Experimental Analysis of Axial Flow Cyclone Used for Intake Air Filtration in Internal Combustion Engines. *Energies* **2021**, *14*, 2285. [CrossRef]

21. Zheng, J.; Wang, J.; Zhao, Z.; Wang, D.; Huang, Z. Effect of equivalence ratio on combustion and emissions of a dual-fuel natural gas engine ignited with diesel. *Appl. Therm. Eng.* **2019**, *146*, 738–751. [CrossRef]

22. Pham, V.C.; Choi, J.-H.; Rho, B.-S.; Kim, J.-S.; Park, K.; Park, S.-K.; Le, V.V.; Lee, W.-J. A Numerical Study on the Combustion Process and Emission Characteristics of a Natural Gas-Diesel Dual-Fuel Marine Engine at Full Load. *Energies* **2021**, *14*, 1342. [CrossRef]

23. Benajes, J.; García, A.; Monsalve-Serrano, J.; Boronat, V. Dual-Fuel Combustion for Future Clean and Efficient Compression Ignition Engines. *Appl. Sci.* **2017**, *7*, 36. [CrossRef]

24. Regulamin EKG ONZ nr 49 Europejskiej Komisji Gospodarczej Organizacji Narodów Zjednoczonych (EKG/ONZ). Available online: https://eur-lex.europa.eu/LexUriServ/LexUriServ.do?uri=OJ:L:2011:180:0053:0069:PL:PDF (accessed on 15 October 2022).

25. Kuiken, K. Book I: Principles. In *Gas- and Dual-Fuel Engines for Ship Propulsion, Power Plants, and Cogeneration*; Target Global Energy Training: Onnen, The Netherlands, 2016; pp. 1–488.

26. Karczewski, M.; Chojnowski, J.; Szamrej, G. A Review of Low-CO_2 Emission Fuels for a Dual-Fuel RCCI Engine. *Energies* **2021**, *14*, 5067. [CrossRef]

27. DieselNet Emission Standards: EU: Cars and Light Trucks. Available online: https://dieselnet.com/standards/eu/ld.php (accessed on 30 October 2022).

28. DieselNet Emission Standards: EU: Heavy-Duty Truck and Bus Engines. Available online: https://dieselnet.com/standards/eu/hd.php (accessed on 30 October 2022).

29. Preliminary Findings on Possible Euro 7 Emission Limits for LD and HD Vehicles. Available online: https://circabc.europa.eu/sd/a/fdd70a2d-b50a-4d0b-a92a-e64d41d0e947/CLOVE%20test%20limits%20AGVES%202020-10-27%20final%20vs2.pdf (accessed on 30 October 2021).

30. DieselNet Emission Standards: EU World Harmonized Transient Cycle (WHTC). Available online: https://dieselnet.com/standards/cycles/whtc.php (accessed on 30 October 2021).

31. Mulholland, E.; Miller, J.; Bernard, Y.; Lee, K.; Rodríguez, F. The role of NOx emission reductions in Euro 7/VII vehicle emission standards to reduce adverse health impacts in the EU27 through 2050. *Transp. Eng.* **2022**, *9*, 100133. [CrossRef]

32. Karczewski, M.; Wieczorek, M. Assessment of the Impact of Applying a Non-Factory Dual-Fuel (Diesel/Natural Gas) Installation on the Traction Properties and Emissions of Selected Exhaust Components of a Road Semi-Trailer Truck Unit. *Energies* **2021**, *14*, 8001. [CrossRef]

33. Wajand, J.A. *Silniki o Zapłonie Samoczynnym*; WNT: Warsaw, Poland, 1988; pp. 123–130.

34. *Norm PN-ISO 15550:2009*; Reciprocating Internal Combustion Engines—Determination and Method of Measuring Engine Power—General Requirements. Polski Komitet Normalizacyjny PKN: Warsaw, Poland, 2009.

35. *Norma PN-EN ISO 8178-1*; Pomiar Emisji Spalin dla Tłokowych Silników Spalinowych. Polski Komitet Normalizacyjny PKN: Warsaw, Poland, 1999.

36. Kokjohn, S.L.; Hanson, R.M.; Splitter, D.A.; Reitz, R.D. Fuel reactivity-controlled compression ignition (RCCI): A pathway to controlled high-efficiency clean combustion. *Int. J. Engine Res.* **2011**, *12*, 209–226. [CrossRef]

37. Wiśniewski, S. *Pomiary Temperatury w Badaniach Silników i Urządzeń Cieplnych*; WNT: Warsaw, Poland, 1983.

38. Muscala, W. Tworzenie i Destrukcja Tlenków Azotu w Procesach Energetycznego Spalania Paliw. Available online: https://docplayer.pl/5623726-Tworzenie-i-destrukcja-tlenkow-azotu-w-procesach-energetycznego-spalania-paliw.html (accessed on 28 October 2022).

39. Heywood, J.B. *Internal Combustion Engine Fundamentals*, 2nd ed.; McGraw-Hill Education: New York, NY, USA, 2018.

40. Nova, I.; Tronconi, E. *UREA SCR Technology for deNOx after Treatment of Diesel Exhausts*; Springer Science+Buisness Media: New York, NY, USA, 2014.

41. Kowalewicz, A. *Podstawy Procesów Spalania*; WNT: Warsaw, Poland, 2000; pp. 248–268.

42. Żółtowski, A.; Gis, W. Ammonia Emissions in SI Engines Fueled with LPG. *Energies* **2021**, *14*, 691. [CrossRef]

43. Mitchell, C.E.; Olsen, D.B. Formaldehyde Formation in Large Bore Natural Gas Engines Part 1: Formation Mechanisms. *J. Eng. Gas Turbines Power* **2000**, *122*, 603–610. [CrossRef]

44. Kuiken, K. *Gas- and Dual-Fuel Engines for Ship Propulsion, Power Plants and Cogeneration. Book II: Engine Systems and Environment*; Target Global Energy Training: Onnen, The Netherlands, 2016; pp. 1–544.

45. Wei, L.; Geng, P. A Review on natural gas/diesel dual fuel combustion, emissions, and performance. *Fuel Process. Technol.* **2016**, *142*, 264–278. [CrossRef]

46. Engine Management Systems for CPBC and RCCI Engines. Available online: https://www.arenared.nl/cpbc+~{}+rcci (accessed on 30 October 2021).

47. DieselNet ACEA releases Euro 7/VII Position Paper. Available online: https://dieselnet.com/news/2020/12acea.php (accessed on 30 October 2021).

48. Adzuan, M.A.; Yusop, A.F. A review of dual fuel RCCI engine fundamental and effect on the performance, combustion and emission characteristic. *AIP Conf. Proc.* **2022**, *2465*, 040001. [CrossRef]

49. García, A.; Monsalve-Serrano, J.; Villalta, D.; Guzmán-Mendoza, M. Methanol and OMEx as fuel candidates to fulfill the potential EURO VII emissions regulation under dual-mode dual-fuel combustion. *Fuel* **2020**, *287*, 119548. [CrossRef]

50. Garcia, A.; Monsalve-Serrano, J.; Villalta, D.; Guzmán Mendoza, M. "OMEx Fuel and RCCI Combustion to Reach Engine-Out Emissions Beyond the Current EURO VI Legislation," SAE Technical Paper 2021-24-0043, 2021. Available online: https://doi.org/10.4271/2021-24-0043 (accessed on 21 October 2022).
51. Dahodwala, M.; Joshi, S.; Koehler, E.; Franke, M.; Tomazic, D.; Naber, J. *Investigation of Diesel-CNG RCCI Combustion at Multiple Engine Operating Conditions*; SAE International: Warrendale, PA, USA, 2020.

Article

Experimental Investigation of Possibilities to Improve Filtration Efficiency of Tangential Inlet Return Cyclones by Modification of Their Design

Tadeusz Dziubak

Faculty of Mechanical Engineering, Military University of Technology, 2 Gen. Sylwestra Kaliskiego St., 00-908 Warsaw, Poland; tadeusz.dziubak@wat.edu.pl; Tel.: +48-261-837-121

Citation: Dziubak, T. Experimental Investigation of Possibilities to Improve Filtration Efficiency of Tangential Inlet Return Cyclones by Modification of Their Design. *Energies* **2022**, *15*, 3871. https://doi.org/10.3390/en15113871

Academic Editor: Theodoros Zannis

Received: 5 May 2022
Accepted: 23 May 2022
Published: 24 May 2022

Publisher's Note: MDPI stays neutral with regard to jurisdictional claims in published maps and institutional affiliations.

Abstract: It has been shown that tangential inlet return cyclones are commonly used for inlet air filtration of off-road vehicle engines. The wear of the engine elements, and thus their durability, is determined by the efficiency and accuracy of the inlet air filtration. It has been shown that the possibilities of increasing the separation efficiency or decreasing the pressure drop of a cyclone by changing the main dimensions of the cyclone are limited, because any arbitrary change in one of the dimensions of an already operating cyclone may cause the opposite effect. A literature analysis of the possibility of increasing the filtration efficiency of cyclones by modifying the design of selected cyclone components was conducted. In this paper, three modifications of the cyclone design with a tangential inlet of the inlet air filter of a military tracked vehicle were proposed and performed. The symmetrical inlet of the cyclone was replaced with an asymmetrical inlet. The cylindrical outlet tube was replaced with a conical tube, and the edges of the inlet opening were given an additional streamlined shape. The modification process was carried out on three specimens of the reversible cyclone with a tangential inlet. After each modification, an experimental evaluation of the modifications was carried out. The influence of the modifications on the cyclone's efficiency characteristics and pressure drop was examined. Subsequent modifications of the cyclone were performed on the same specimen without removing the previous modifications. Tests were performed in the air flow range Q_G = 5–30 m^3/h. Polydisperse "fine" test dust with grain size d_{pmax} = 80 µm was used for testing. The dust concentration at the cyclone inlet was set at 1 g/m^2. The performed modifications caused a slight (about 1%) increase in separation efficiency in the range of small (up to Q_G = 22 m^3/h) flux values and about 30% decrease in pressure drop in the whole range of the Q_G flux, which positively influences the increase in engine filling and its power. There was a noticeable increase in filtration accuracy in the range of low and high values of Q_G flux, which results in a decrease in the wear of engine components, especially the piston-piston ring-cylinder (P-PR-C) association, and an increase in their durability.

Keywords: air filter; tracked vehicle; return cyclone with tangential inlet; cyclone design modifications; separation efficiency; pressure drop; experimental studies

1. Introduction

Trucks and special vehicles, including tracked military vehicles, as well as agricultural machines and working machines, due to their application, are most often used in roadless, unpaved, sandy and desert terrain, where air dust concentration is significant and often exceeds the value of 1 g/m^3 [1–5]. The main air pollutant sucked in by the engines of these vehicles is road dust, the dominant component of which is mineral dust carried from the ground to a considerable height by moving vehicles or by the wind.

The residence time of mineral dust grains in the air depends on the speed of their fall, which is determined by their size and density and is a result of the mutual relation of the aerodynamic drag force and gravity force. The falling velocity increases significantly with increasing grain diameter. For example, SiO_2 silica grains with sizes of 10 μm, 50 μm and 100 μm (density 2650 kg/m^3) fall at speeds of 0.08 m/s, 0.19 m/s and 0.7 m/s, respectively [6]. Dust grains with diameters in the order of 1 μm, due to their very small mass, are characterized by a low velocity of descent. In addition, dust grains with a size of 1 μm and smaller perform Brownian motions in a gaseous medium. These are disordered movements of solid particles suspended in a liquid or gas resulting from collisions of these particles with molecules (particles) of the surrounding medium. Very small particles move in suspension along irregular paths. The particles move independently of each other, their motion does not diminish in time, but becomes more intense when the temperature of the medium increases. As a result, dust particles of small size do not fall. This means that only dust grains with diameters of 1 μm and less can float in the still air for an almost indefinite period of time. Their contribution to the total mass of dust is usually small.

Due to their low velocity of descent, dust grains with a size of d_p = 2–10 μm remain in the air for a long time and are sucked in together with the intake air to the engine cylinders. On the other hand, airborne dust grains with a size of d_p = 10–50 μm are sucked in with the engine intake air when the vehicle is operated in conditions of high dust concentration in the air. In raised dust clouds, dust grains with diameters not exceeding d_p = 50–80 μm account for 80–100% of the total dust mass. Dust grains larger than 50 μm are found in the air during the use of vehicles on training grounds, in quarries, on construction sites and during take-off or landing of a helicopter on an adventure landing pad [7].

The main components of mineral dust are silica SiO_2 and alumina Al_2O_3. The share of these two components in the total mass of dust reaches up to 95%. The residue (trace amounts) is oxides of various metals including Fe_2O_3, MgO, CaO, K_2O, Na_2O and SO_3 (Figure 1). The chemical composition of dust depends strictly on the composition and type of substrate, the location and altitude above the ground and climatic factors, such as winds, rains, snow, frost and droughts, as well as dust precipitation that has entered the atmosphere as a result of forest fires, landfills, industrial activities and volcanoes [8–11].

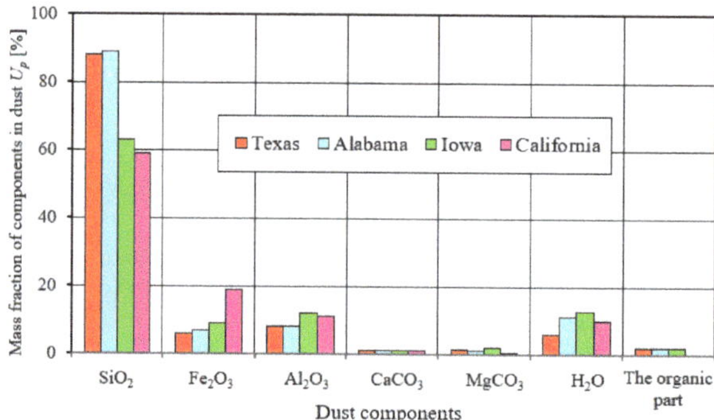

Figure 1. Chemical composition of selected dusts from U.S. substrates. Figure made by author based on data from paper [11].

Mineral dust grains are very irregularly shaped lumps with sharp edges and are characterized by high hardness. On the 10-grade Mohs scale, where 10 corresponds to the hardness of diamond, silica has a hardness of 7 and corundum has a hardness of 9. Mineral dust has strong abrasive properties. When it enters the engine cylinders together with the air, it is the most frequent cause of accelerated wear of two frictionally mating parts, such as the P-PR-C (piston—piston rings—cylinder wall) mating parts of an engine. Dust that is drawn in with the air enters above the piston and causes the most wear on the upper part of the cylinder and piston as well as the upper piston rings. The abrasive wear of engine components is caused mainly by particles of 1–40 μm, with dust grains of 1–20 μm being the most dangerous [9,11–13]. The authors of [8] state that about 30% of the pollutants entering the engine may escape from the cylinders to the exhaust system unchanged together with the exhaust gases, thus increasing the emission of particulate matter (PM) from the engine. Only 10–20% of the dust that enters the engine with the air through the intake system settles on the cylinder liner walls. This part of the dust settles on the cylinder liner, where it forms a kind of abrasive paste together with the oil, which causes abrasive wear when it comes into contact with the surfaces of engine combinations, for example the piston-ring-cylinder (P-PR-C) combination.

The air filter is responsible for supplying air of sufficient quality (purity) to the engine cylinders to minimize wear on engine components. For filtration of inlet air of modern passenger car engines, due to low values of dust concentration in the air (Table 1), single-stage filters with a panel insert made mainly of pleated filter paper, which is characterized by the mass of dust retained per unit filtration area, are used. This property is determined by the dust absorption coefficient k_m of the filter material, which is defined as the quotient of the total dust mass m_{cw} retained by the filter insert until the value of the permissible resistance Δp_{fdop} is reached and the active area of the filter insert paper A_c.

$$k_m = \frac{m_{cw}}{A_c} [g/m^2] \tag{1}$$

For typical cellulose-based filter materials and standard dusts whose grain size does not usually exceed 100 μm, the dust absorption coefficient reaches a value in the range k_m = 200–250 g/m^2 [14,15].

Fibrous filter baffles, as a result of dust accumulation, increase the pressure drop during operation, which causes a decrease in engine power. When the paper absorption capacity is used up, the filter cartridges must be changed periodically, the more often the vehicle is used in higher concentrations of dust in the air. Therefore trucks, working machines and military vehicles (mainly tanks, infantry fighting vehicles and armored personnel carriers), which are used in conditions of high dust concentration in the air (Table 1), are equipped with two-stage filters [16–18]. The first stage of filtration is then an inertia filter, a multicyclone or a monocyclone with a swirl guide, and the second stage is a porous baffle arranged in series in the form of a cylindrical insert made of pleated filter paper with a suitably selected surface area. The multicyclone used in engine inlet air filtration technology is a set of several dozens of identical return or through cyclones arranged side by side whose internal diameter does not exceed D = 40 mm [19–21].

The concentration of dust in the air around a moving vehicle is a variable quantity and depends mainly on the type and condition of the ground, precipitation, wind direction, conditions of vehicle movement (speed, single vehicle or column) and the type of running gear (wheeled or tracked). Therefore, the concentration of dust around a moving vehicle assumes very different values, as seen in Table 1.

Table 1. Particulate matter concentrations in the air for different vehicle operating conditions.

Author	Ambient Conditions	Value [g/m^3]
[1,2]	Clean rural environments	from 0.01 mg/m^3
	Movement of a column of tracked vehicles in desert conditions	about 20
[3]	Dusty environments	0.001–10
[22]	On highways	0.0004–0.1
[22]	When driving a column of vehicles on sandy terrain	up to 0.03–8
[23]	During takeoff or landing of a helicopter on an adventure landing site at helicopter propeller tip height—0.5 m above ground	3.33
[24]	Limited visibility	0.6–0.7
	Zero visibility	about 1.5

Table 1. *Cont.*

Author	Ambient Conditions	Value [g/m³]
[4]	At a distance of several meters behind a column traveling at 30 km/h: • tanks, • armored personnel carriers, • trucks.	1.17 0.62 0.18
[5]	A tracked vehicle driving on sandy terrain at 18 km/h: • at a distance of about 80 mm from the side surface of the armor, • at the air filter inlet.	2.1–3.8 0.8–1
[11]	A few meters away from the sand road on which the terrain vehicles were moving	0.05–10
[1,3,11]	At the inlet to the combustion engine intake system of a vehicle	not more than 2.5

Other sources of pollutant emissions also influence the concentration of pollutants in the air, including road transport emitting particles from tire wear [25], heating systems and thermal power plants emitting particulates and toxic gases [26], cement industry [27], field work (plowing, harvesting), biomass transport [28] and open pit coal mines emitting dust and gases [29].

The idea of using two-stage air filters is to initially separate large masses of dust from the polluted air in the multicyclone, without increasing the pressure drop, but with not very high efficiency (87–95%) and accuracy ($d_p > 15$–35 μm), as seen in Figure 2. The remaining small mass of dust is then directed to the filter cartridge made of pleated paper, where the dust grains are retained with high efficiency (above 99.5%) and accuracy (above $d_p = 2$–5 μm) [30–32]. As a result, air of the required purity is supplied to the engine cylinders and the operation time of the filtration system is extended, and thus the service interval is extended, which is limited by reaching a certain value of the pressure drop—the permissible resistance Δp_{fdop}.

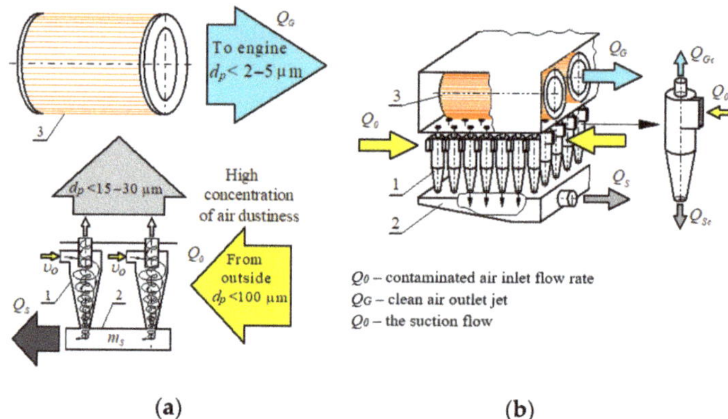

Figure 2. Two-stage (multicyclone, filter bed) filtration system for the inlet air of an off-road vehicle engine: (**a**) essence of the two-stage filtration system; (**b**) functional diagram of the two-stage filter, 1—multicyclone (first filtration stage), 2—dust settler and 3—paper filter bed (second filtration stage).

Multicyclones can be constructed of tangential inlet or axial inlet return cyclones or pass-through cyclones. Because of their simple and robust design, lack of moving parts, high operational safety and stable and low pressure drops, as well as their comparatively low investment and operating costs, cyclones have been a very efficient and popular device for separating solids from the air stream for over 100 years. In addition, cyclones can operate under extreme conditions, including large gas flow volumes and high solids concentrations, temperatures and pressures, which is their advantage over other solid-to-gas separation technologies. Since the appearance of the first industrial cyclone in 1886, the cyclone has been widely used in aerosol filtration processes in the chemical, coal, petroleum, metallurgical, power generation and other industries [33].

A cyclone works by making the gas stream rotate so that the particles acquire centrifugal force. The rotational movement of the gas stream can be achieved either tangentially into the cylindrical part (Figure 3a–c) or axially (Figure 3d) by flowing through stationary blades with a helical outline (swirler) at the inlet of the cylindrical part.

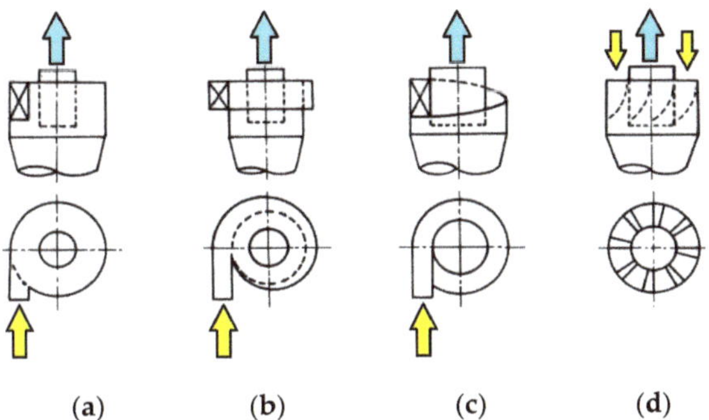

(a)   (b)   (c)   (d)

Figure 3. Methods of gas supply to the cylindrical part of a return cyclone with inlet: (**a–c**) tangential, (**d**) axial. Figure made by the author based on data from the work [34].

Due to the principle of operation, cyclones can be divided into reverse, when the aerosol changes the direction of flow by 180° (Figure 4a,b), clean gas and collected particles leave the device on opposite sides and through, when the aerosol does not change the direction of flow, and then the gas and particles pass through the cyclone in one direction only, leaving from the same end of the device (Figure 4c).

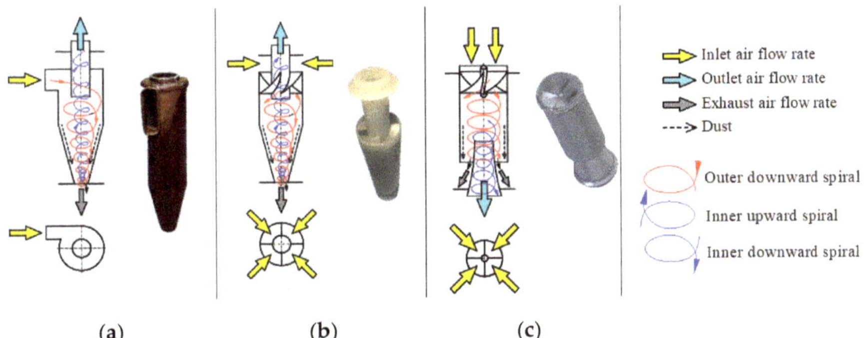

(a)   (b)   (c)

Figure 4. Types of cyclones used for filtering the intake air in motor vehicles: (**a**) reverse-flow cyclone, (**b**) return with axial inlet and (**c**) axial flow (vortex tube separators).

A conventional tangential inlet reciprocating cyclone comprises a cylindrical part, which then tapers downward in the form of a cone terminated by a discharge opening and a dust collector. A tangential inlet is attached to the cylindrical part of the cyclone, which causes the gas containing the dust to swirl, resulting in a spiral motion that first descends along the outer spiral, then ascends through the inner spiral and finally leaves the cyclone through the outlet pipe at the top of the cyclone (Figure 4a). The centrifugal force created by the swirling motion causes the higher mass particulates in the gas to move (overcoming the drag force of the air) toward the outer wall of the cyclone, where they are braked and stopped, and then fall by gravity into the dust container. Gas flowing spirally downwards, due to the shrinking diameter of the circular cone and the attraction of the low-pressure zone in the middle of the cyclone, changes its direction of movement spirally upwards at the bottom of the cone. The gas purified from larger dust grains is discharged from the cyclone through a cylindrical outlet tube located in the upper part of the cyclone. The cylindrical part of the cyclone is connected to the outlet pipe by a tight cover. The cyclone design is simple, but the gas flow inside is a complex three-dimensional vortex flow [35].

Cyclones differ in construction, principle of operation and separation efficiency, which is described by the characteristics of separation efficiency and pressure drop. Examples of characteristics of cyclones, which are elements of air filters of off-road vehicles, are shown in Figure 5. The diagrams show a close relation between the separation efficiency and pressure drop of the cyclone, indicating that the greater the filtration efficiency, the greater the value of the pressure drop and vice versa. A high value of filter pressure drop adversely affects engine operation, causing a decrease in engine cylinder filling and effective power, hence the desire to reduce them. High separation efficiency is important to minimize wear and tear on engine components, which increases engine durability and vehicle mileage. Air filter performance is a technical compromise between its pressure drop, separation efficiency and accuracy, as well as engine wear and durability and vehicle reliability [1–3,7,13].

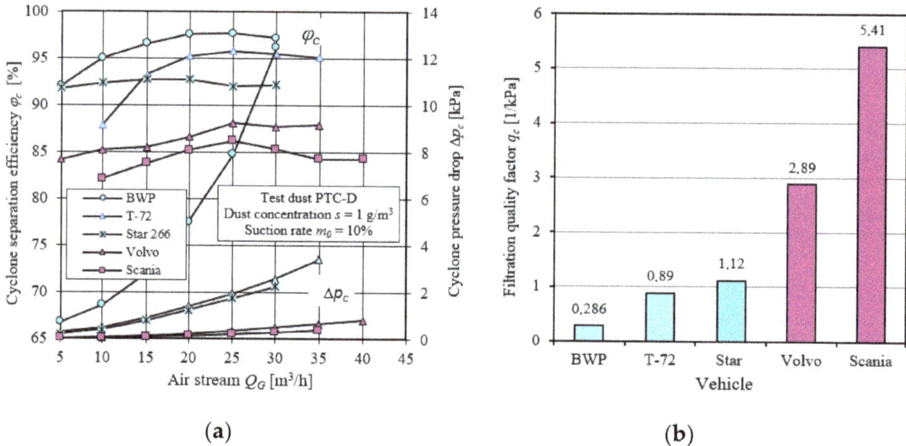

(**a**)   (**b**)

Figure 5. Filtration properties of cyclones as elements of off-road vehicle air filters: (**a**) characteristics of separation efficiency $\varphi_c = f(Q_G)$ and pressure drop $\Delta p_c = f(Q_G)$; (**b**) filtration quality factor.

At the maximum value of the air stream $Q_G = 30$ m³/h, the tangential inlet reversible cyclone BWP reaches a pressure drop of the order of 12 kPa, exceeding by several times the pressure drop of the tangential inlet cyclones. At a slightly higher value of air stream $Q_G = 35$ m³/h, the pressure drop of another tangential inlet reverse cyclone T-72 reaches the value of 3.4 kPa. The filtration efficiencies of the tangential inlet return cyclones take the values 96.9% and 95.6%, respectively. Scania and Volvo tangential inlet backflow cyclones have much lower pressure drop values at 0.36 kPa and 0.58 kPa, and much lower values of filtration efficiency at 87.7% and 85.2%, respectively [36].

A comparison between cyclones of different filtration properties can be made with the use of the conventional filtration quality factor qc, which relates the separation efficiency and pressure drop with the following relation the quality factor [37]:

$$q_c = \frac{-\ln(1 - \varphi_0)}{\Delta p}\,[1/\text{kPa}] \tag{2}$$

where φ_0 is the cyclone filtration efficiency (–) and Δp is the pressure drop for nominal air stream (kPa).

The higher the value of the q_c coefficient, the more favorable the cyclone efficiency. It should be noted that the filtration quality factor q_c defined in this way optimizes the cyclone only due to separation efficiency and pressure drop. The q_c coefficient does not take into account the filtration accuracy of the cyclone, which clearly increases as the air velocity through the cyclone increases, just as the pressure drop increases.

The values of filtration quality coefficients of cyclones, whose characteristics are shown in Figure 4a, are shown in Figure 4b. The pass-through cyclones take values of the filtration quality factor q several times higher than the tangential inlet return cyclones.

The continuous pursuit to increase the durability of truck and special vehicle engines, which are operated in conditions of high values of dust concentrations in the air, sets more and more stringent requirements for two-stage air filters in terms of separation efficiency and accuracy, as well as pressure drop. For this reason, it is reasonable to take measures to use return cyclones, which are characterized by high values of filtration efficiency, for the filtration of inlet air in these vehicles. This will reduce the mass of dust directed to the second filtration stage, thereby extending the vehicle's service interval. However, due to the high pressure drop, measures are needed to reduce it, which will minimize the decrease in engine power.

High cyclone operating efficiency, determined by separation efficiency and pressure drop p, is obtained by using appropriate air flow velocities and proportions of its main dimensions, also called geometrical parameters or dimensionless numbers. For tangential inlet reciprocating cyclones, such proportions in the form of quotients of the main dimensions have been established as a result of research work conducted over a number of years [38–54].

These research works have shown a mutual dependence between the main dimensions of a cyclone and have led to the determination of such proportions of these dimensions (quotients of main dimensions), for which a cyclone obtains optimal working conditions in the scope of the maximum separation efficiency and minimum pressure drop. The main dimensions of a tangential inlet return cyclone are shown in Figure 6.

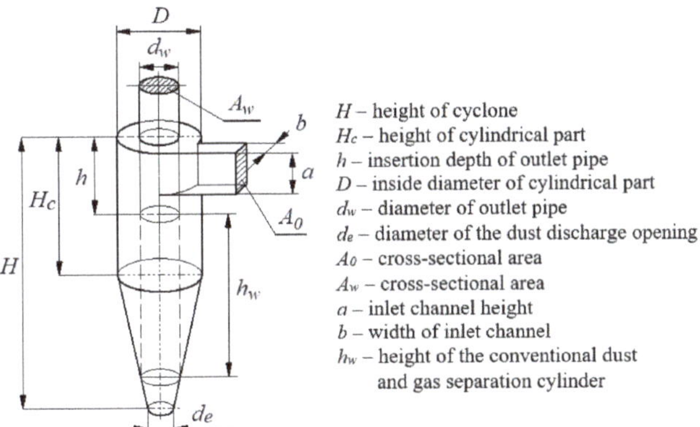

Figure 6. Characteristic dimensions of a tangential inlet return cyclone.

For example, the author of [55] combined the variables determining the cyclone separation efficiency (cyclone main dimensions, air and dust flow parameters) and presented the cyclone efficiency as a function:

$$\eta = f\left(\frac{d_w}{D}, \frac{a}{D}, \frac{b}{D}, \frac{h}{D}, \frac{H}{D}, \frac{H_c}{D}, \frac{d_e}{D}, Re, Fr, St_k, D_{en}\right). \tag{3}$$

The quotients d_w/D to d_e/D are dimensionless numbers for the cyclone configuration, Re and Fr are dimensionless numbers for the gas flow, called Reynolds number and Froude number and can be defined as follows, respectively:

$$Re = \frac{\rho_g v_i D}{\mu},$$

(4)

$$Fr = \frac{v_i}{gD^{1/2}}.$$

(5)

S_{tk} is a dimensionless number that determines particle dynamics, which is called the Stokes number and can be defined as follows:

$$S_{tk} = \left(\frac{C_c \rho_p d_p^2}{18\mu} \right)$$

(6)

D_{en} is a density number that can be defined as the quotient of the gas density ρ_g and the dust density ρ_p:

$$D_{en} = \frac{\rho_g}{\rho_p}.$$

(7)

From the presented (Figure 7) example changes of separation efficiency c and pressure drop coefficient c as a function of the quotient A_w/A_0 and H/d_w, it can be seen that the optimal range of these parameters is in the range of 1.5–2 and 7–9, respectively [39,40]. The use of these data during the design work of cyclones makes it possible to obtain a relatively high separation efficiency of the inlet air to internal combustion engines with respect to micron particles. Any arbitrary change in one of the cyclone dimensions can cause the opposite effect.

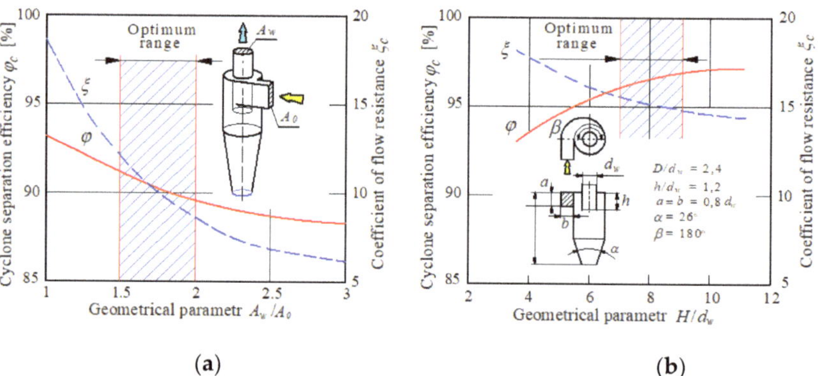

Figure 7. Separation efficiency φ_c and coefficient of pressure drop ξ_c of cyclones as a function of: (a) A_w/A_0 and (b) H/d_w parameter. Figure made by author based on data from paper [39].

The pressure drop coefficient c is defined by the relation:

$$\xi_c = \frac{2\Delta p_c}{\rho_p \cdot v_0^2},$$

(8)

where ρ_g is the gas density, Δp_c is the air pressure drop through the cyclone and v_0 is the outlet velocity from the cyclone.

The gas velocity in the cyclone is represented by the average velocity at the inlet port (inlet velocity) v_0, which is defined by the relation:

$$v_0 = \frac{Q_0}{A_0} [\text{m/s}],$$

(9)

where Q_0 is the flow of air entering the cyclone and A_0 is the cross-sectional area of the inlet spigot at its narrowest point.

The range of inlet velocities 0, within the limits of which the changes in the purification efficiency are relatively small, should be as large as possible because it makes it possible to use the best separating capabilities of the cyclone at a widely varying air stream Q_G as a result of changes in the engine rotational speed n. The range of inlet velocities 0 is at the level of $v_0 = 12$–20 m/s, and exceptionally reaches the value of 25 m/s [56–59]. The use of high velocities generates high pressure drop. According to the study of the author of [60], the cyclone inlet velocity reaches $v_0 = 35$ m/s, but the pressure drop also reaches a large (about 4 kPa) value.

Therefore, work is constantly conducted to increase the efficiency of aerosol particle separation in cyclones and to reduce their pressure drop. This goal is achieved by modifying the cyclone design but keeping the cyclone's main dimension quotients. The available literature describes various ways to improve the efficiency of cyclones of different diameters over a wide range. Each cyclone is designed to perform a specific task and is associated with unique flow physics, which is strongly influenced by its geometry.

This paper aims to analysis the available design solutions of tangential inlet return cyclones in terms of their applicability to modify a return cyclone with a cylindrical part diameter of 40 mm, which is a component of an off-road vehicle air filter multicyclone. Proposed and performed modifications of cyclones will be verified during experimental tests on a laboratory bench.

2. Analysis of Options for Increasing the Separation Efficiency of Tangential Inlet Return Cyclones

Ongoing research on tangential inlet return cyclones is aimed at improving their performance by increasing the efficiency of aerosol particle separation in cyclones and reducing air pressure drop. This effect can be achieved by modifying the cyclone design, but without interfering with its basic geometric dimensions. To date, many such experimental and numerical studies of tangential inlet return cyclones have been conducted, where the researchers have focused on the individual effects that a single geometric factor has on the separation efficiency and pressure drop of the cyclone. The studies included modification of three cyclone components:

(1) The cyclone inlet (shaping the inlet tube or changing the way the inlet stream is fed into the cylindrical part of the cyclone) [61–79];
(2) The cyclone outlet tube (shaping the outlet tube and using special steering wheels (deswirlers) inside the outlet tube) [80–86];
(3) The dumping hole and dust container [87–94].

In tangential inlet return cyclones, a single symmetrical inlet with rounded edges in the form of a tube with a rectangular cross-section with sides a and b is commonly used (Figure 1), with the longer side a aligned parallel to the cyclone's main axis. It has been found that the optimal ratio of inlet width to inlet height b/a is in the range of 0.25–0.5 [53,54], although the authors of [59] report a different range from 0.5 to 0.7

Therefore, a lot of work has been undertaken on the organization of the aerosol inlet into the cyclone. The authors of the work [51–64] believe that replacing one inlet channel with several inlet channels will improve the cyclone efficiency. Reference [61] presents a comparative analysis of the original cyclone (one inlet channel) with the designs of two cyclones that are equipped with four equal inlet channels. In cyclone one, the inlet channels were set symmetrically around the circumference of the cylindrical part (every 90°) at the same level as the inlet channel in the original cyclone. In cyclone two, the inlet channels were set symmetrically and arranged as stairs along the axial direction of the cyclone. The total inlet area of the four channels in the new designs remained the same. Therefore, the height and width of the inlet channel in the two new designs are equal to half the height and width of the inlet channel in the original cyclone. The cyclone design with four inlets achieved, in the inlet velocity range of 10–25 m/s, better filtration performance than the original cyclone.

In Reference [62], a numerical study of cyclones of new design with one, two and three tangential inlet channels symmetrically arranged in the cylindrical part at the same level was carried out. This cyclone differs from the conventional cyclone in the separation space. Instead of the conical part of the cyclone, a cylindrical part and a vortex limiter were used. The inlet channels have the same design. It was found that the cyclone with three inlets has higher separation efficiency and lower pressure drop, which is beneficial to cyclone performance. The cyclone with three inlets has a more favorable distribution of turbulent kinetic energy compared to cyclones with one and two inlets, which may lead to less noise in the cyclones and their pipelines.

Numerical results of cyclones of a new design with one, two and three tangential inlet channels symmetrically arranged in the cylindrical part at the same level are presented in [63]. The total inlet area of two and three channels in the new designs remained the same. The results show that the modification proposed in this paper increases the separation efficiency of particles smaller than 3 μm to 50% and particles larger than 6 μm to 15% compared to the conventional design. Increasing the inlet velocity from 14 to 20 m/s increases the separation efficiency for all particle sizes, but is associated with a significant increase (doubling) of the pressure drop. The proposed modification simultaneously reduces the erosion rate by 54%.

The object of study of the authors of [64] was a cyclone with two inlet channels (main and auxiliary), where the auxiliary inlet channel with a square cross-section was much smaller than the main channel. The inlet channels were rotated 180° with respect to each other. Tests were conducted for three variables: the flow rate through the secondary inlet channel, the cross-sectional area of the secondary channel and the position of the top of the main inlet channel relative to the cyclone lid. The optimum design obtained from the tests provides better performance than conventional cyclones.

El-Emam et al. [65] fabricated and optimized the geometry of a tangential inlet return cyclone by varying the position and dimensions of the inlet channel and the length of the cylindrical part of the cyclone at a constant air inlet velocity of v_0 = 30 m/s. Five cyclones with different sizes and dimensional proportions of the inlet channels and the position of the inlet relative to the cyclone top cover, were modeled computationally. The inlet channel width and the diameter of the cylindrical inlet channel were assumed constant. It was shown that cyclone modification (III), in which the inlet channel has twice the height of the original cyclone, provided the highest (98.6%) filtration efficiency. Increasing or decreasing the height of the inlet channel more than in cyclone (III) has a negative effect on cyclone performance. Additionally, decreasing the length of the cylindrical part and changing the original position of the cyclone inlet channel gives poor cyclone performance.

In the work [66], a tangential inlet reciprocating cyclone (without cylindrical part) with two equal symmetrical inlets attached to the conical section at the top of the section was numerically investigated. In contrast, Lim et al. [67] installed additional inlets in the conical section of a tangential cyclone separator, numerically investigated and then experimentally verified their effect on filtration efficiency and pressure drop. When aerosol was metered through the additional inlet channels and clean air was fed into the original inlet channel, the boundary grain diameter decreased compared to when aerosol was metered only through the original inlet channel. It is shown that there is an optimum flow rate through the additional inlet ducts that increases the separation efficiency of the cyclone. Dosing the aerosol simultaneously through the additional inlet channels and through the original inlet channel resulted in a larger aerosol stream being filtered, while showing similar levels of efficiency and pressure drop compared to the original cyclone.

Wang et al. [68] studied the effect of the inlet channel angle of a feedback cyclone on its filtration efficiency. The angular inclination (−30°, −15°, 0°), the downward inlet channel with respect to the horizontal and the upward channel angular inclination (0°, 15°, 30°)

were analyzed. It is shown that the inlet channel inclination angle has a direct effect on the gas introduction direction, resulting in changes in pressure drop and filtration efficiency.

The pressure drop initially decreases, reaches a minimum value at an inlet angle of $0°$, and then increases as the inlet angle increases (Figure 8). A negative inlet angle results in a greater pressure drop compared to a positive inlet angle of the same value.

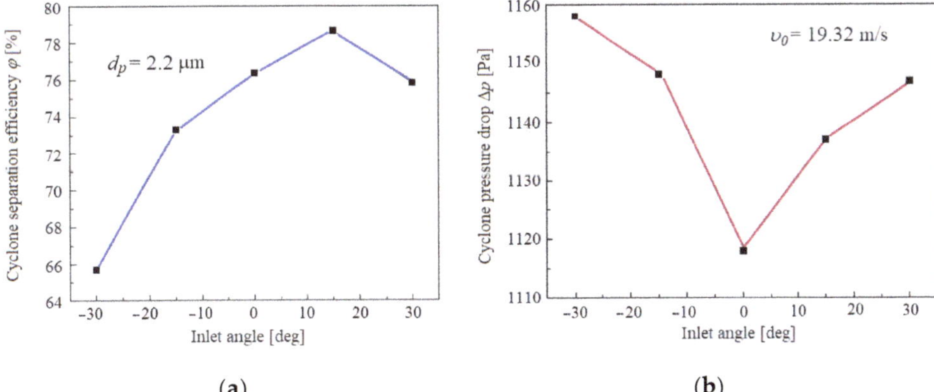

Figure 8. Effect of inlet channel angle of the return cyclone on: (**a**) filtration efficiency and (**b**) cyclone pressure drop. Figure made by the author based on data from paper [68].

The separation efficiency first increases, reaches a maximum value at an inlet (channel) angle of $15°$ and then decreases as the inlet angle changes from negative to positive. High agreement was obtained between the numerical simulation results and the experimental results.

A similar problem was addressed by the authors of [69], who studied the effect of changing the inlet channel angle contained in the vertical plane α and additionally in the horizontal plane β (Figure 9) on the changes in pressure drop and filtration efficiency. The lowest static pressure was recorded in the basic model, and for the other models, the value increased with the increasing channel angle on both sides.

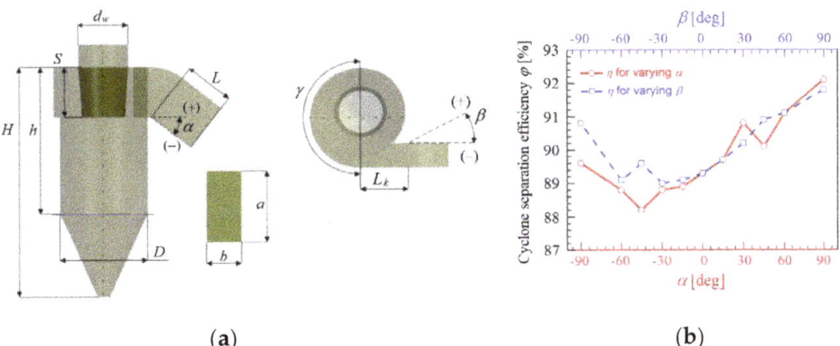

Figure 9. Effect of changing the angle of the cyclone inlet channel: (**a**) cyclone inlet channel angle in vertical plane α and horizontal plane β; (**b**) effect of angles α and β on filtration efficiency. Figure made by the author based on data from the paper [69].

For changes in angle in the vertical plane α, feeding the solids from the top (angle α positive) provided higher values of separation efficiency due to better penetration of both phases into the lower areas of the cyclone. The separation efficiency increased with increasing angle (except for angle $\alpha = 45°$), with the cyclone reaching its maximum value for channel angle $\alpha = 90°$ (Figure 9). There was a 3.1% increase in the separation efficiency

rical inlet with an asymmetrical inlet (Figure 12b), or the use of "K" vanes immediately before such an inlet (Figure 12c), causes a significant mass of dust to be directed (already at the inlet) towards the outer wall of the cylindrical part of the cyclone, which naturally facilitates its separation from the air and thus increases separation efficiency.

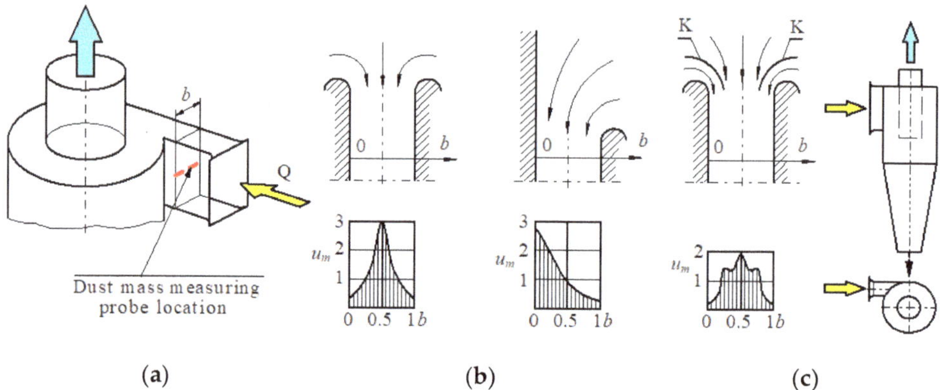

Figure 12. Cross-sectional dust particle size distribution—flat inlet: (**a**) symmetrical, (**b**) asymmetrical and (**c**) symmetrical with guide vanes u_m—relative dust fraction in the inlet air stream [79].

From the presented analysis, it can be seen that the organization of the aerosol inlet into the cyclone is being addressed by many researchers. The ongoing work is directed towards the use of a larger number (up to four) of inlet channels arranged symmetrically in the cylindrical part at the same level or shifted with respect to each other towards the conical part. Channels of the same or changed shape and cross-sectional area are used. It is proposed to construct cyclones with a different angle of inclination of the inlet channel in the vertical and horizontal plane. Numerical and experimental tests of modified cyclone constructions showed an increase in separation efficiency and a decrease in pressure drop, which confirms the purposefulness of conducting such tests.

All previous studies show that the largest share, estimated at 75–90%, of the total gas pressure losses in the cyclone are the losses incurred during flow through the cyclone inner tube (outlet tube) [40]. The pressure losses are a result of the resistance at the inlet to the outlet tube and the friction resistance during the gas flow through the tube. In addition, gas flows through the outlet tube and makes a helical motion towards the cyclone outlet. Hence, there have been numerous searches for ways to reduce these losses, primarily by giving the outlet tube more favorable shapes (conical) and giving streamlined shapes to the edges of the inlet opening, installing internal vanes and adding reversible heads to convert the kinetic energy of spiral motion of the gas inside the outlet tube into pressure energy and the kinetic energy of linear motion [80–86].

Lim et al. [80] performed an experimental analysis of the effect of the outlet tube shape on the separation efficiency and pressure drop. The experiments were performed for two air flow rates (30 dm^3/min and 50 dm^3/min) of four cyclone models with cylindrical outlet tubes with diameters of 15, 11 and 7 mm and six cyclone models with cone-shaped outlet tubes. The cone-shaped tubes had different cone lengths (10, 25 and 45 mm), with diameters of 7 and 15 mm at both ends of the cone (Figure 13). The pressure drop and separation efficiency of the conical tubes were greater than that of the 15 mm diameter cylindrical tube and less than that of the 7 mm diameter cylindrical tube. In addition, the pressure drop per unit flow rate ($\Delta p / Q$) for the cone-shaped outlet tube design (7-B, 7-C and 7-D) is less compared to the cylindrical-shaped design (7-A).

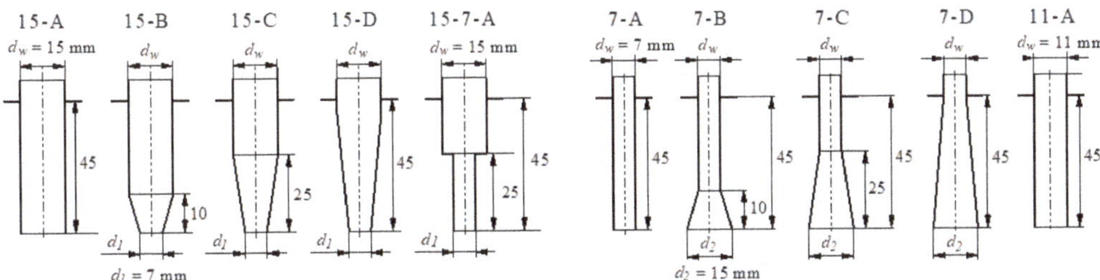

Figure 13. Geometry and dimensions of the cyclone outlet tube. Drawing made by author based on data from work [80].

The study presented in [81] aimed to obtain detailed flow information using CFD simulations in cyclones tested previously by Lim et al. [80]. Figure 14 shows the pressure contours at a flow rate of 70/min in a cyclone with different outlet tube inlet shapes. By decreasing the divergence angle of the outlet tube, the low-pressure zone in the center of the cyclones expands. More particles are trapped in this zone and transported to the cyclone outlet. Thus, the efficiency of the cyclone is lower.

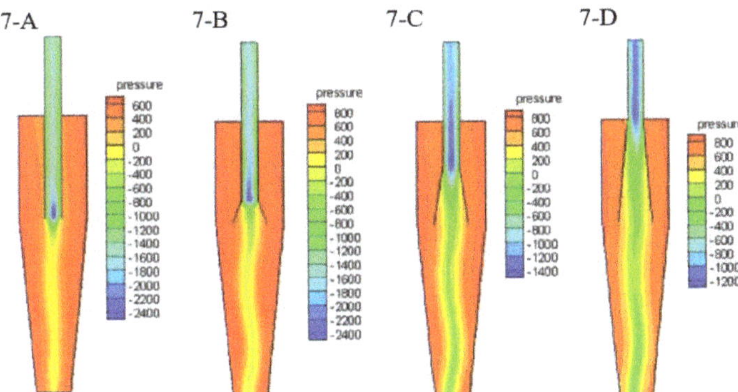

Figure 14. Pressure contours (Pa) at a flow rate of 70/min in the cyclone. Figure made by author based on data from paper [81].

Another way to reduce the pressure drop of the cyclone is to use special guides (deswirlers) inside the outlet tube to convert the kinetic energy of spiral motion, which the gas stream moves inside the outlet tube, into kinetic energy of linear motion.

Misiulia et al. [82] presented the results of optimizing the geometry of a deswirler located in the outlet tube of a feedback cyclone with a spiral inlet channel (Figure 15) using the pressure drop of the cyclone as a criterion. For this purpose, 70 different configurations of the deswirler (also known as the unscrewing or straightening device) were investigated, which were obtained by varying the range of its geometrical parameters: $d_c/D_v = 0.4–0.8$; $n = 1–9$; $h/D_v = 0.4–2$; $\alpha = 10–50°$. The optimization results indicated that the most significant geometric parameters of the deswirer are the number of blades, blade angle and blade height. It was shown that the new optimized deswirler significantly (by about 56%) reduces the static pressure on the cyclone wall and about 20% in the core. It reduces the pressure loss in the outlet tube by 95.67%, leading to a 43.17% reduction in total cyclone pressure drop. It slightly reduces the separation efficiency for some particle diameters, increasing the cyclone boundary grain diameter from 1.5 to 1.72 μm.

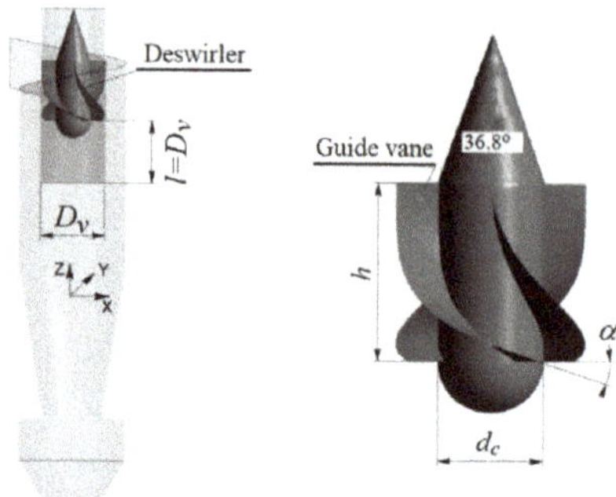

Figure 15. Deswirler geometry and dimensions: D_v—vortex finder diameter, l—mounting height in the vortex finder, d_c—deswirler core diameter, h—height of guide vanes and α—leading edge angle. Figure made by author based on data from paper [82].

In [83], based on the analysis of available flow straightening (unstraining) devices, a new design of flow straightening device using vanes was proposed, which reduces the energy consumption of the most commonly used cyclones by 25–29%, respectively, without compromising the separation efficiency. Figure 16 shows the components (proposed by LG Electronics) mounted on the inlet of the outlet tube, which reduce the pressure loss in the cyclone.

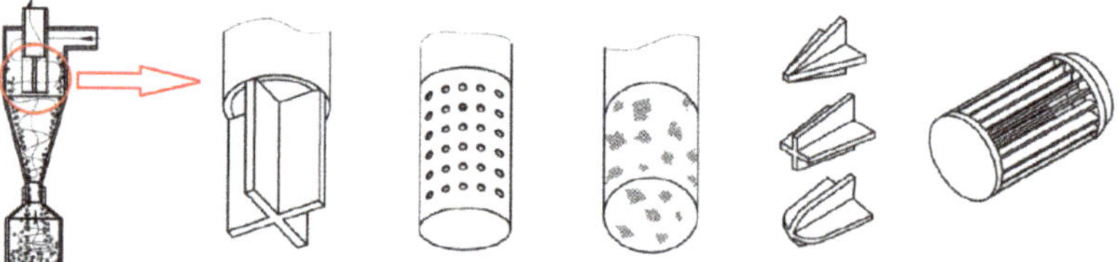

Figure 16. Devices proposed by LG Electronics to reduce pressure loss in cyclones. Figure made by author based on data from paper [83].

Figure 17 shows the components used to straighten the flow inside the outlet tube by Samsung Gwangju Electronics Co. [83], resulting in a reduction of pressure loss. From Figure 17, it can be seen that the pressure loss in the cyclone is reduced mainly by using the F-type rectifying element. Using it in the cyclone outlet tube reduced the power loss by about 11% [8].

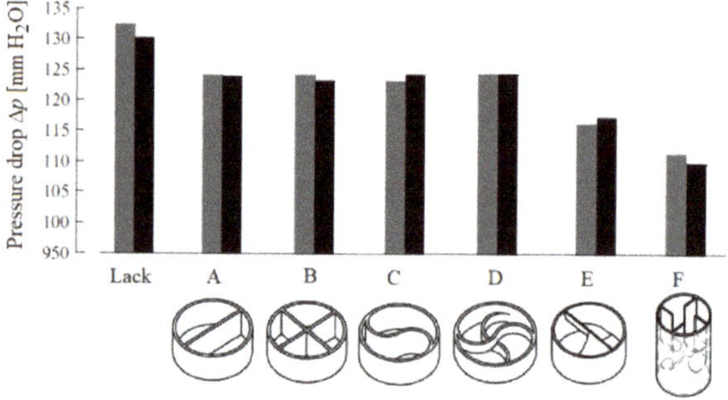

Figure 17. Pressure drop values for individual components used to straighten the flow inside the cyclone outlet tube of Samsung Gwangju Electronics Co. Figure made by author based on data from paper [83].

Reference [84] presents a numerical and experimental study of four reciprocating cyclone models with tangential inlet. In the three cyclone models, oblique (at an angle of $\alpha = 45°$) bevels (notches) were made at the inlet of the outlet tube, which were deviated from the cyclone inlet by the following angles: $\alpha = 45°$, $135°$ and $225°$. The simulation results show that the separation efficiency of the cyclones that used outlet tubes with notched cutoffs increased significantly compared to the separation efficiency of the original cyclone. The largest increase in efficiency was obtained for the cyclone in which the bevel of the outlet tube was deflected from the cyclone inlet by an angle $\alpha = 225$. The experimental results show that this cyclone model can remove particles with a diameter greater than 3.5 µm, while the original cyclone can only remove particles with a diameter greater than 3.5 µm. The separation efficiency of the original cyclone is 73% and that of the cyclone with the modified outlet tube is 76%.

Wang et al. [85] conducted an experimental study of two cyclone models with single and double inlet duct and with cylindrical and inverted cone outlet tube. The cyclone with cylindrical outlet tube has higher separation efficiency and pressure drop for both single and dual inlet channel cyclone. For dual inlet cyclones, the smaller the diameter of the outlet tube (cylindrical or inverted cone), the higher the separation efficiency and pressure drop.

The author of [40] presented the possibilities of modifying the outlet tube in a tangential inlet return cyclone, which can reduce the pressure drop (Figure 18).

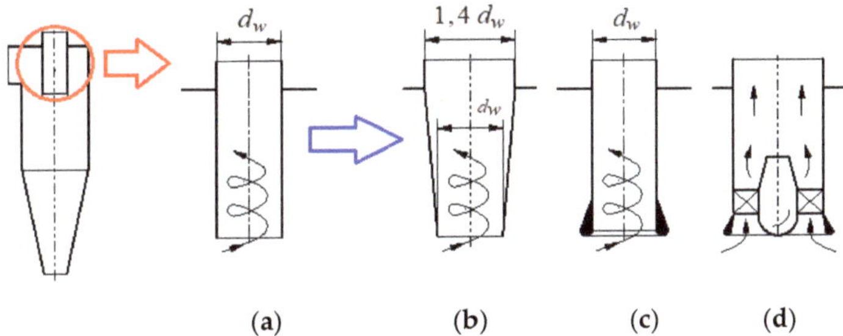

Figure 18. Available shapes of an inner cyclone tube: (**a**) cylindrical tube—standard, (**b**) conical tube, (**c**) cylindrical tube with streamlined inlet port and (**d**) tube with guide vanes. Figure made by author based on data from paper [40].

The analysis presented here shows that there are a number of design options for modifying the outlet tube to reduce the flow resistance through this element. Replacing the traditional cylindrical outlet tube with a conical one. and introducing a streamlined shape of the inlet opening of the outlet tube. can significantly reduce the flow resistance of the cyclone without causing a decrease in filtration efficiency. The gas stream flowing out of the cyclone through the outlet tube still performs an upward helical motion, resulting in pressure losses. These losses depend on the geometry of the cyclone and can be up to 50% of the total pressure drop in the cyclone. This remaining energy in the swirling stream can be recovered or used to increase the cyclone's filtration efficiency by installing a special element, the deswirler (also called a de-swirler or straightening device), in the cyclone's outlet tube. This can significantly reduce the pressure loss when flowing through the outlet tube. In the available literature, one does not encounter works analyzing the aerosol flow through the outlet tube with the streamlined shape of the inlet opening edge (Figure 18c).

In order to prevent the transfer of the gas vortex motion to the cyclone dust settling tank and lifting separated and already retained dust from it, a special container (settling tank) is attached to the dust discharge opening—dipleg [87–96]. This method is recommended for use in those cyclones, where extracting separated dust from the settling tank was not applied, and in "short" cyclones, i.e., with a small ratio of the cyclone height H to its diameter D. Problems of dust extraction from the settling tank of a multicyclone are presented in other works [97,98].

Elsayed et al. [87] studying the influence of the dimensions of the cylindrical dust settling tank (height and diameter) on the cyclone performance showed that the optimal design of the settling tank and the dipleg (an extended part of the lower part of the cyclone, connecting the cyclone with the dust settling tank) leads to better separation efficiency than a conventional cyclone. The authors of [88], using the CFD program, investigated the effect of the dipleg shape (cylindrical, inverted cone, conical and diamond) on flow characteristics and cyclone performance, as shown in Figure 19.They found that the dipleg geometry significantly affects the pressure drop of the cyclone. The cylindrical dipleg generated the highest pressure drop, while the lowest pressure drop was observed for the cyclone without dipleg. Among the geometries tested, the highest separation efficiency was found for the cyclone with a conical dipleg, while the diamond dipleg led to the lowest efficiency.

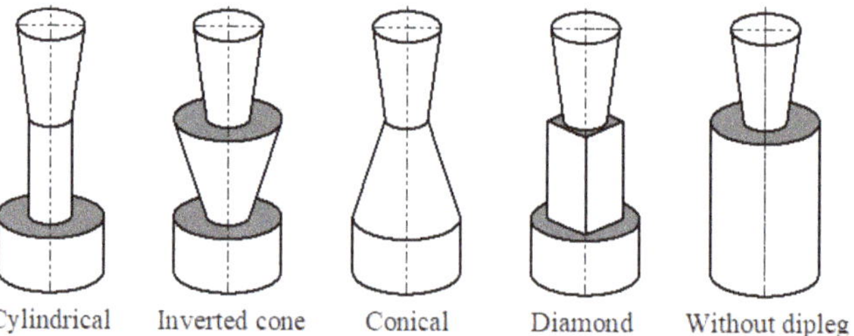

Figure 19. Different shapes of dust reservoirs (dipleg) of cyclones. Figure made by author based on data from paper [83].

Kaya et al. [89] numerically investigated the particle filtration process inside two cyclones whose conical part was extended by a cylindrical dipleg with a diameter equal to that of the outlet opening. They found that the length of the dipleg significantly affects the cyclone separation efficiency by providing a larger separation space and minimizing the re-lift of separated particles. The calculated cyclone filtration efficiencies for different inlet velocities in the range $v_0 = 7$–16 m/s and for particles with diameters of 1.5 μm are shown in Figure 20a. A trend of increasing separation efficiency up to a dipleg length of $L_d = 38.5$ mm can be seen. However, the cyclone separation efficiency begins to decrease for inlet velocities less than 10 m/s. For higher inlet velocities, the cyclone efficiency assumes significant values that remain almost constant. The optimum dipleg length, in terms of maximum separation efficiency, is obtained with a parameter value of $L_d/L = 0.5$, where L is the total length of the cyclone.

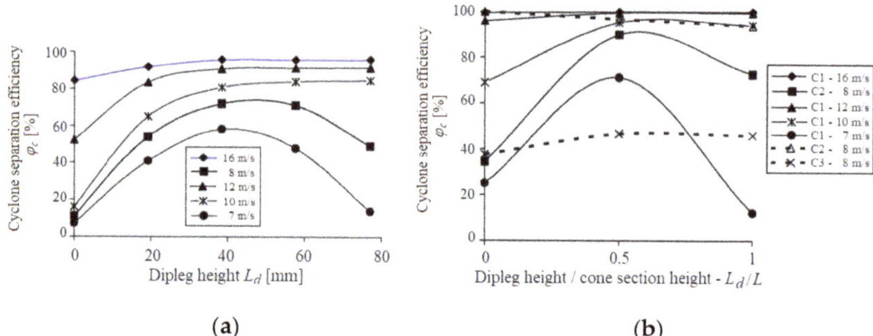

Figure 20. Cyclone separation efficiency for different inlet velocities depending on (**a**) dipleg length L_d and (**b**) L_d/L parameter value. Figure made by the author based on data from the work [89].

In [90], the results of the separation efficiency of three cyclones with different lengths of cylindrical diplegs and with an invariant dust settling tank are presented. It was shown that the tangential velocity, axial velocity and turbulent kinetic energy in the settling tank decreased significantly when the dipleg length was extended. This means that the dipleg dust settler can effectively prevent the re-entrainment of already separated dust. In addition, the extended part of the dipleg increases the dust separation space, and thus the separation efficiency increases, at the cost of a slightly increased pressure drop.

One of the methods of increasing separation efficiency of a cyclone is mounting an additional element (cone) on the outlet of dust from the cyclone to the settling tank. The task of this element, also called an apex cone, is to limit secondary circulation of separated dust particles. It is very important to precisely select the diameter of the cone, the apex angle, and its proper location. Incorrect selection of these parameters may block the bottom outlet of the cyclone and hinder dust falling into the container, and thus reduce the cyclone separation efficiency.

Therefore, in [91], using two research methods—computational fluid dynamics (CFD) and experimental studies—the effect of 15 variants of geometric configurations of the counter cone on the cyclone efficiency was analyzed. The geometry of the counter cone was determined by the following parameters (Figure 16): D_s—diameter of the counter cone (368, 436 and 520 mm), α—tip angle (85°, 95° and 105°), B_h—distance from the base of the counter cone to the bottom outlet of the cyclone of diameter (0.1B, 0.15B and 0.35B).

It was found that the use of a counter cone in each of the geometric variants tested led to an increase in cyclone efficiency. The results show that the counter cone should be located above the bottom outlet of the cyclone (Figure 21a,c). The optimal position of the counter cone (assuming maximization of separation efficiency) can be obtained when the distance of the counter cone base from the bottom outlet is $B_h = 0.15B$ (82.5 mm). The apex angle of the counter cone should $\alpha = 85°$. An improvement in separation efficiency was observed for particles smaller than 60 μm. This is due to the fact that the phenomenon of particle re-entrainment concerns fine particles in particular, particles up to 15 μm in diameter. The use of an apex angle with a value above 100° may be associated with the risk of excessive agglomeration of particles in the counter cone area.

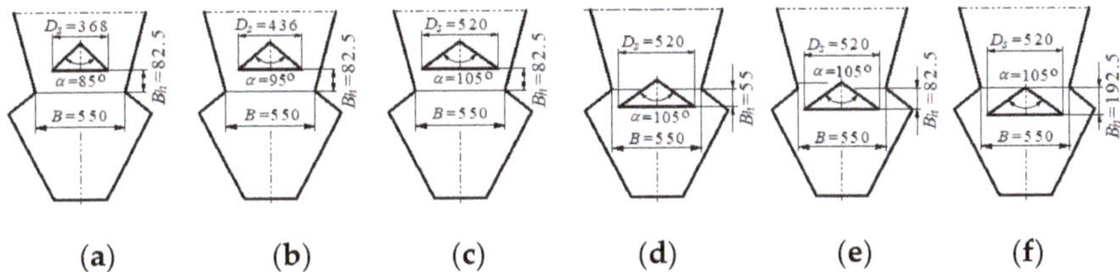

Figure 21. Examples of variants of analyzed geometric configurations of the counter cone [86]. (**a–c**)—cones with different values of angle α at the same height B_h above the diameter of the lower (dust) outlet of the cyclone, (**d–f**)—cones with the same value of angle α at different heights B_h below the diameter of the lower (dust) outlet of the cyclone.

The authors of [92–94] studied the apex angle of the counter cone at the dust outlet of the cyclone in the range $\alpha = 40–80°$. They found that the optimal value of the angle due to maximum separation efficiency is $\alpha = 70°$. Furthermore, they observed that increasing the gap between the cone and the wall of the conical part of the cyclone increased the number of small particles that enter the dust container.

The author of [95] compared two suggested locations of the cone (below the cyclone outlet and in the conical part of the cyclone) using CFD. He concluded that placing the cone below the conical part of the cyclone is more beneficial as it helps to increase the particle separation efficiency.

Obermair et al. [96] experimentally studied cyclones with three different cyclone dust outlet geometries: a cyclone with a dust trap with a cone at the cyclone outlet, and a cyclone with a cylindrical dipleg. They found that in the case of a cyclone with a dust settler only, the rotating flow goes from the cyclone cone to the settler, which causes the collected particles to be entrained again. For the cyclone with an apex cone, an improvement in cyclone efficiency was achieved. Two design variants of the counter cone were analyzed. In the first variant, the counter cone (apex angle $\alpha = 90°$) was located under the bottom outlet of the cyclone. In the second variant, the counter cone ($\alpha = 120°$) was located above the lower cyclone outlet. The latter solution led to an increase in separation efficiency by 2% and an increase in pressure drop from about 1.2 kPa to 1.4 kPa.

The cited examples indicate that the selection of appropriate dimensions of the dust settling tank and mounting an additional element (cone) on the outlet of the dust cyclone have a very significant impact on the performance of a single cyclone. However, the cyclone modifications presented in the literature have been performed on single cyclones and tested mainly for industrial use. Optimization of the size of the settling tank does not take into account the condition of dust extraction.

The above analysis shows that:

(1) It is possible to improve cyclone performance in terms of reducing pressure drop or increasing separation efficiency by modifying its design without changing its main dimensions.
(2) The modifications of cyclones presented in the literature have been performed in single cyclones and tested mainly towards their use for industrial purposes.
(3) There are no proposals for modifications of cyclones applicable to motor vehicle air filters.
(4) The following solutions can be used to modify cyclones that are elements of a multi-cyclone constituting the first stage of a motor vehicle air filter:

   - Changing the cross-sectional geometry and shape of the inlet stub;
   - Changing the shape of the outlet duct (outlet tube);
   - Use of a streamlined shape of the inlet opening to the outlet duct.

3. Experimental Research on the Influence of the Modification of the Design of the Tangential Inlet Reversible Cyclone on the Improvement of Separation Efficiency

3.1. Object of Research

The subject of the design modification was a tangential inlet return cyclone (*D*-40), which is a component of an air filter multicyclone (Figure 22) for the inlet of tracked vehicle engines operated under conditions of high dust concentration in the air. The parameters of the cyclones are shown in Figure 23. The multicyclone is constructed of 96 *D*-40 cyclones, which are arranged vertically and grouped in rows and columns. Dust removal holes are connected to a common, for all cyclones, settling tank of separated dust. Dust from the settling tank is removed ejectively by air stream Q_{SF} generated by an ejector located in the stream of outgoing engine exhaust gases. Air to the cyclones Q_{0F} flows in from the surroundings through inlet ducts of a rectangular cross-section [97].

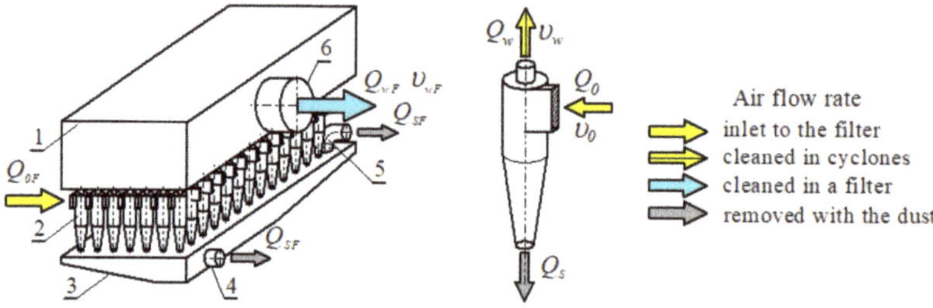

Figure 22. Tracked vehicle air filter: 1—porous baffle, 2—cyclone *D*-40 multicyclone, 3—dust trap, 4/5—right/left) suction port and 6—cleaned air port.

On the full engine power range, the averaged air flow velocities in the characteristic cross-sections of the air filter and single cyclone are the inlet velocity to the cyclone $\upsilon_0 = 32$ m/s, outlet velocity of cleaned air from the cyclone $\upsilon_w = 22.7$ m/s and velocity in the air filter outlet port $\upsilon_{wF} = 60$ m/s, respectively.

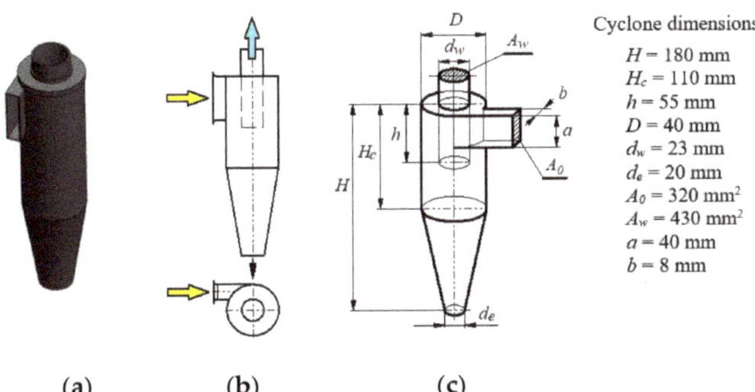

Figure 23. Return cyclone with tangential inlet D-40: (**a**) general view, (**b**) functional diagram and (**c**) characteristic dimensions.

The construction of this D-40 cyclone is simple (Figure 23). In the upper part of the cylindrical hull, with internal diameter $D = 40$ mm, there is attached a tangentially symmetrical inlet spigot with a rectangular cross-section with sides $a = 40$ mm and $b = 8$ mm. The cylindrical outlet tube with internal diameter d_w is situated coaxially in relation to the cylindrical part of the hull and sunk into it to a depth of h. The cyclone outlet tubes are connected by a common plate, which is the lower wall of the baffle filter housing. The cylindrical cyclone hull is terminated by a cone, the opening of which, with a smaller diameter d_e, provides a connection to the dust settling tank, which is a common dust container for all cyclones.

3.2. Proposal for Modification of D-40 Cyclone Design

The tangential inlet return cyclone D-40, due to its simple design, is susceptible to design modifications, as a result of which an improvement in performance can be expected in the form of a reduction in pressure drop or an increase in separation efficiency. However, design modifications should not compromise the basic dimensions of the cyclone, as this can have the opposite effect.

From the information resulting from the analysis conducted in Chapter 2, it appears that there are many opportunities to improve the performance of a tangential inlet return cyclone without changing its basic geometric dimensions, which are shown in Figure 24.

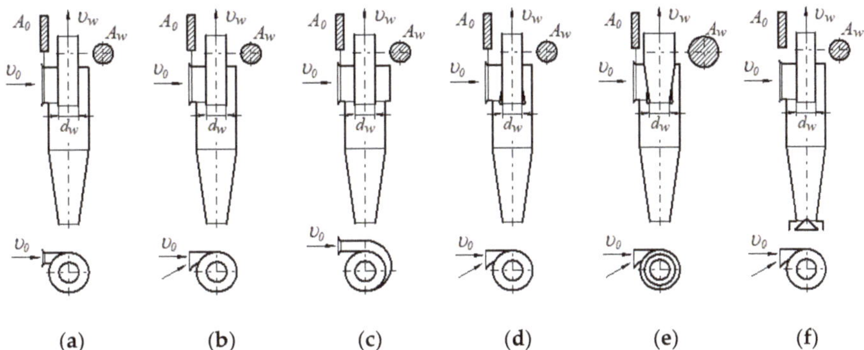

Figure 24. Possible changes in the design of cyclone D-40: (**a**) original version, (**b**) changing the shape of the inlet from symmetrical to asymmetrical, (**c**) using a spiral inlet described on the cyclone's hull, (**d**) giving a streamlined shape to the edges of the inlet opening of the outlet pipe, (**e**) changing the cylindrical outlet pipe into a conical one and (**f**) using a special protective cone in the lower part of the cyclone.

The design modifications proposed in *b* and *f* seem to be technologically too complicated and expensive in practical application. The use of special "protective" cones in the lower part of the cyclone (Figure 24f) prevents the entrainment of dust already separated there from the settling tank, and thus may increase the efficiency of air filtration in the cyclone. This method is recommended for use in those cyclones where extracting separated dust from the settling chamber is not applied, and in "short" cyclones, i.e., with a small ratio of the cyclone height *H* to its diameter *D*. Cyclone *D*-40 is characterized by a small value of the *H*/*D* parameter, but dust is extracted from the settling chamber in a continuous manner. Structural changes of the cyclone presented in points d and e are aimed at reducing pressure losses during air flow through the outlet pipe. Their share in the total flow losses in the cyclone is estimated at 75–90%. Therefore, in order to modify the design of cyclone *D*-40, the following solutions were selected and applied:

A   An asymmetrical shape of the inlet spigot was used instead of a symmetrical one, while maintaining the previous size and shape of its cross-section at the inlet to the cylindrical part (Figure 24);

B   A streamlined shape was given to the edges of the inlet opening of the outlet tube (Figure 25);

C   A tapered outlet tube was used (Figure 26).

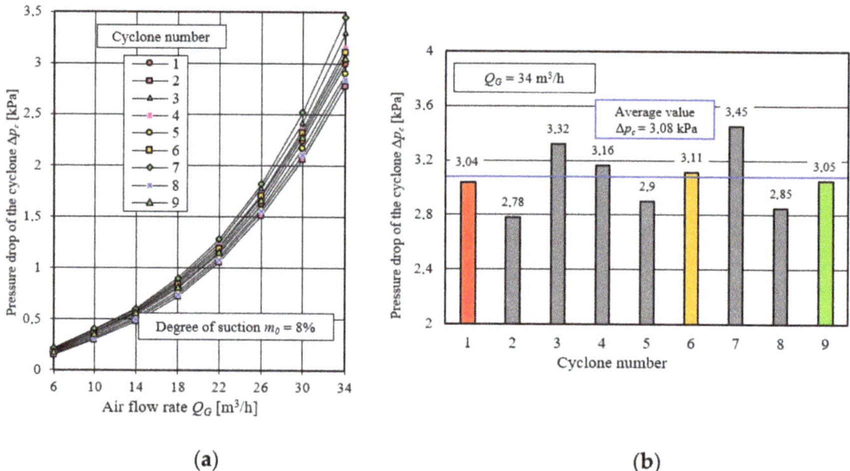

(a)   (b)

Figure 25. Preliminary test results of tracked vehicle air filter cyclones: (**a**) cyclone flow characteristics and (**b**) nominal cyclone pressure drop values.

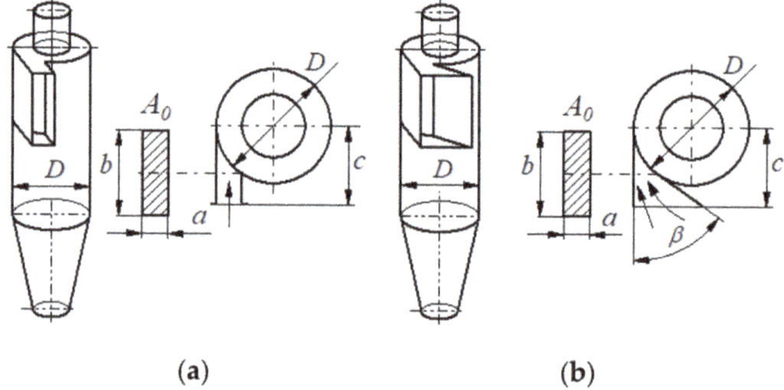

(a)   (b)

Figure 26. Modification A of cyclone *D*-40 design: (**a**) original (symmetrical) inlet channel shape and (**b**) asymmetrical inlet channel—Version W1.

The modifications were performed successively (without eliminating the previous ones) on one specimen of the cyclone. As a result, three versions of the modified cyclone were obtained:

- Version W1—modification element A;
- Version W2—elements of modification A + B;
- Version W3—elements of modification A + B + C.

After the implementation of the next elements of modification A, B and C, tests were carried out on the efficiency and accuracy of filtration and pressure drop.

Three pieces of cyclones *D*-40 were selected to perform the proposed modifications of the design. For this purpose, preliminary tests were carried out; flow characteristics of any nine pieces of cyclones *D*-40 in the air flow range Q_G = 6–34 m^3/h were performed (Figure 25). Three copies of cyclones No. 1, No. 6 and No. 9, whose pressure drops are close to the average value of the pressure drop of the nine tested cyclones, were selected to perform modifications and further tests (Figure 25b).

The design of the asymmetrical inlet consisted in rotating by an angle β = 60° the side (inner) wall of the inlet channel while maintaining the height *b* of the channel and its width *a* at the inlet to the cylindrical part of the cyclone as well as the length c of the channel (Figure 26). This causes a significant amount of dust mass to be directed (already at the inlet) towards the outer wall of the cylindrical part of the cyclone, which naturally facilitates the concentration of dust on the outer wall of the cyclone. This may cause, as proved by the research results presented in [79], an increase in the cyclone separation efficiency.

Modification *B* of cyclone *D*-40 consisted of making the edges of the inlet opening of the two outlet tubes streamlined (Figure 27). The streamlined shape of the inlet opening of the outlet tube was given by applying a ring tapering in the upward direction of the cyclone on the surface of the outlet tube from the external side. The wall of the outlet tube with a thickness of g = 1 mm and the superimposed ring was rounded with a radius of R = 2 mm, maintaining the internal diameter of the outlet tube d_w. The resulting narrowing of "z" between the internal wall of the cyclone and the external surface of the ring (Figure 27) causes an increase in gas flow velocity at this point, but this should not have a significant effect on the increase in pressure drop of the cyclone.

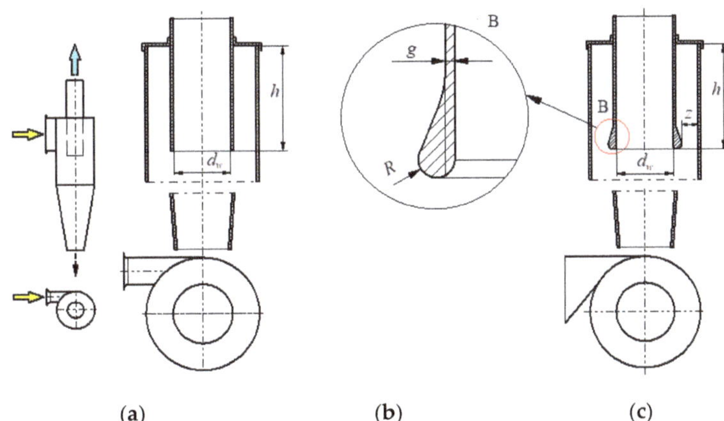

Figure 27. Modification of *D*-40 cyclone design—Version W2: (**a**) original cyclone version, (**b**) modification component B and (**c**) modification A + B.

This eliminated the contraction phenomenon that occurs when air enters the inlet tube opening with sharp edges. The contraction phenomenon caused narrowing of the cross-sectional area of the stream A_k, as a result of which it is smaller than the cross-sectional area of the inlet opening A_w (Figure 28). There is an increase in flow velocity at this location and consequently an increase in pressure drop.

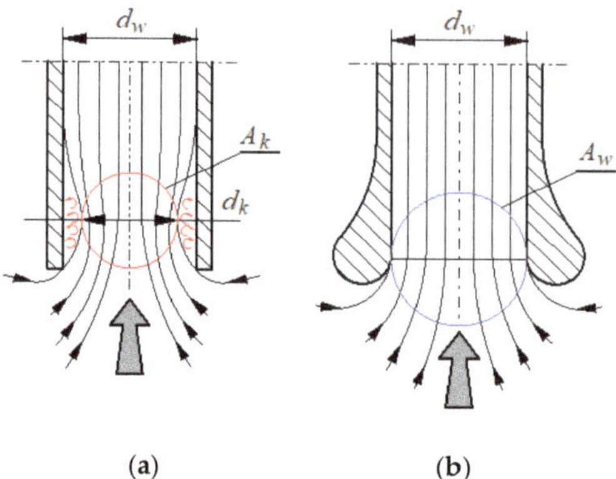

Figure 28. Air inflow into the cylindrical tube: (**a**) with sharp inlet opening edges and (**b**) with streamlined inlet opening shape.

Modification C of cyclone *D*-40 consisted of replacing the existing cylindrical outlet tube of inner diameter d_w with a tube of truncated cone shape, the smaller diameter of which equals the inner diameter of the outlet tube d_w = 23 mm (Figure 29). The conical outlet tube was made as a diffuser with an opening angle of α_s = 7°. For this value of the angle, there is no detachment of the jet from the walls or formation of vortices, and the pressure drop coefficient has the smallest value [98]. Therefore, in the duct after the cyclone, there is a decrease in the velocity of the flow in the air and thus a decrease in the pressure drop of the cyclone represented by the relation:

$$\Delta p = \xi \frac{\rho_p \cdot v_w^2}{2},\tag{10}$$

where ρ_p is the density of air and ξ is the resistance coefficient of air flow through cyclone.

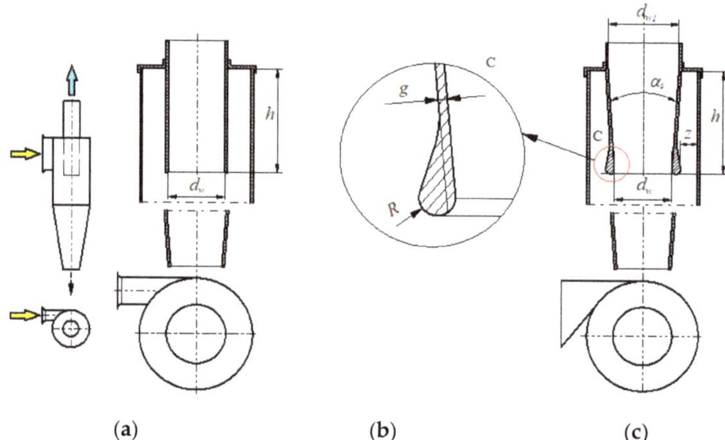

Figure 29. Modification of *D*-40 cyclone design—Version W3: (**a**) original cyclone version, (**b**) element C of modification and (**c**) modification A + B + C.

3.3. Experimental Testing of Cyclones

The specific aim of the research was to determine the effect of performed construction modifications of three (No. 1, No. 6 and No. 9) copies of cyclones *D*-40 on their performance. After each performed modification of the cyclone construction, its influence on the characteristics was evaluated:

- Separation efficiency $\varphi_c = f(Q_G)$;
- Pressure drop $\Delta p_c = f(Q_G)$.

After performing the last element of cyclone modification (element C), cyclone filtration accuracy was additionally determined by measuring the number and size of dust grains in the air stream Q_G behind the cyclone:

- Filtration accuracy $d_{zmax} = f(Q_G)$.

The above characteristics were also performed for three (No. 1, No. 6 and No. 9) copies of cyclones D-40 before their modification.

3.3.1. Methodology and Test Conditions

Testing of the D-40 cyclones was performed on a special stand, the diagram of which is shown in Figure 30. Testing was performed for an air stream within the range Q_{Gmin}–Q_{Gmax} resulting from the number of cyclones used in the filter multicyclone and the demand of air by the engine (within the rotational speed range n_{min}–n_{max}), for which the filter is intended. Thus, the cyclone outlet air flow rate was changed in the range $Q_G = 6$–34 m^3/h, which corresponds to the velocity in the cyclone outlet duct $v_w = 4$–22.7 m/s. Tests were conducted with the assumed constant concentration of dust in the air $s = 1$ g/m^3 with test dust and degree of ejective dust extraction from the cyclone settling chamber $m_0 = 8\%$.

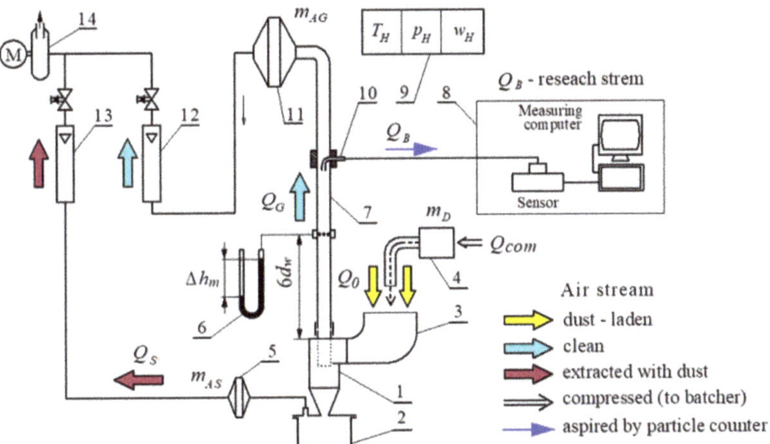

Figure 30. Schematic of cyclone test stand: 1—cyclone; 2—dust settler; 3—dust chamber; 4—dust dispenser; 5—absolute filter of dust extraction duct; 6—liquid manometer U-tube; 7—cyclone pressure drop measuring duct; 8—particle counter; 9—ambient air temperature, pressure and humidity measuring set; 10—measuring probe; 11—absolute filter of main duct; 12, 13—rotameters of main and extraction flux measuring, respectively; 14—fan producing air flux.

The stand was equipped with a particle counter enabling determination of the number and size of dust grains in the air stream behind the cyclone from the range of 0.7–100 m, in $i = 32$ measurement intervals, limited by the diameters (d_{pimin}–d_{pimax}). The widths of the measuring compartments could be freely programmed. The sensor of the device is adapted to work at the maximum concentration of pollution particles in the air up to 1000 pieces/mL, which corresponds to the mass concentration of air dust $s \approx 0.25$ g/m^3.

The measuring line was terminated with an absolute filter, which is used to determine the mass of m_{AG} dust flowing through the cyclone and, consequently, to determine the separation efficiency c of the cyclone. At the same time, the absolute filter prevented dust from entering the rotameter. The duration of one measurement (time of uniform dosing and distribution of test dust at air dust concentration $s = 1$ g/m^3) was assumed $t_{pom} = 3$ min. The test dust used was PTC-D, which is a domestic substitute for AC-fine dust, whose granulometric and chemical composition is given in Figure 31 [99].

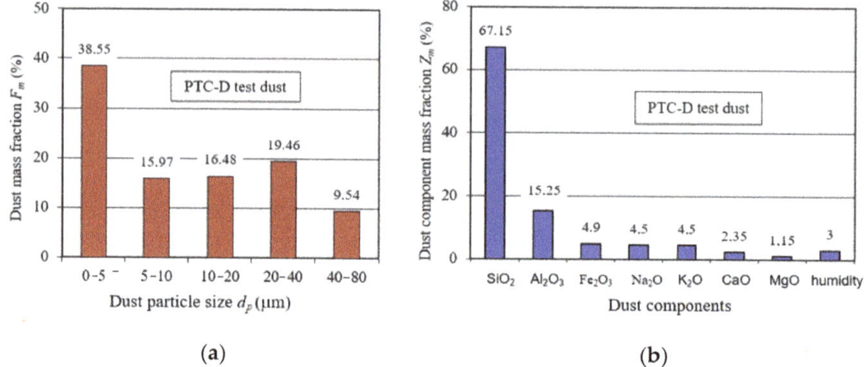

Figure 31. PTC-D test dust: (**a**) size distribution and (**b**) chemical composition [99].

The separation efficiency c of the cyclone was determined by the mass method for successively determined values of the air stream within the range Q_G = 6–34 m^3/h (in equal intervals, every 4 m^3/h) and for corresponding values of the suction flows Q_S for the assumed degree of ejection of dust from the cyclone settling pond m_0 = 8%.

For each measurement point p = I, II, III … k, starting from Q_G = 6 m^3/h, j = 5 measurements were made successively, calculating for each measurement the separation efficiency φ_{cj} of the cyclone from the relation:

$$\varphi_{cj} = \frac{m_{zcj}}{m_{Dcj}} \cdot 100\%,\qquad(11)$$

where m_{Zcj} is the dust mass retained by the cyclone and m_{Dc} is the dust mass dosed to the cyclone at the assumed time t_{pom}.

The mass of dust dosed to the cyclone m_{Dcj} was determined after each measurement as the difference of the mass of the doser before and after the measurement. The mass of dust retained by the cyclone was determined after each measurement from the relationship:

$$m_{zcj} = m_{Dcj} - m_{AGj},\qquad(12)$$

where m_{AGj} is the mass of dust retained by the absolute filter of the main duct determined as the difference of the mass of the absolute filter after and before the measurement.

The mass of the dust dispenser container and the mass of dust retained by the absolute filter were determined with an analytical balance with a measuring range of 220 g and an accuracy of 0.1 mg.

The pressure drop Δp_c of the cyclone was determined after each measurement (completion of dust dosing) based on the measured static pressure Δh_m (mm H$_2$O) at a distance of 6 d_w from the front plane of the cyclone hull, where d_w is the internal diameter of the cyclone outlet duct.

The dust concentration in the cyclone inlet air was determined after each measurement using the relation:

$$s_j = \frac{m_{Dcj}}{Q_{0cj} \cdot t_{pom}} = \frac{m_{Dcj}}{(Q_{Gcj} + Q_{Scj})},\qquad(13)$$

where t_{pom} is the time at which the dust mass m_{Dcj} was dosed uniformly into the cyclone inlet air stream Q_0.

After a cycle of j = 5 measurements at a given measurement point p, the average values of separation efficiency, pressure drop and dust concentration in the air were calculated.

During the measurement, at time $t_{cz} = \frac{1}{2} t_{pom}$, the procedure of measuring the number and size of dust grains in the cyclone outlet air was started in the particle counter. The U_{pi} contribution of dust grains from each measurement interval $(d_{pimin}-d_{pimax})$ to the total number of dust grains from all measurement intervals i was calculated.

$$U_{pj} = \frac{N_i}{\sum_{i=1}^{32} N_i}.$$

(14)

The largest dust grain size d_{pimax} in the cyclone exhaust air was read from each measurement interval (d_{pimin}–d_{pimax}), which is the filtration accuracy of the cyclone under these conditions.

3.3.2. Analysis of Test Results of Original Version Cyclones

The characteristics of separation efficiency $\varphi_c = f(Q_G)$ of three (No. 1, 6 and 9) cyclones D-40 in the original version differ slightly as to the value (Figure 32), but their course is similar and consistent with the information given by many authors of research papers. With the increase in the value of the air stream Q_G, the separation efficiency φ_c of the tested cyclones increases rapidly and, in the range $Q_G = 18$–26 m^3/h ($v_w = 12$–14.7 m/s), reaches the maximum value $\varphi_{cmax} \cong 94.8\%$, after which it decreases slightly, and at the value of the stream $Q_G = 34$ m^3/h, it reaches varied values of the separation efficiency in the range 93.8–94.2%.

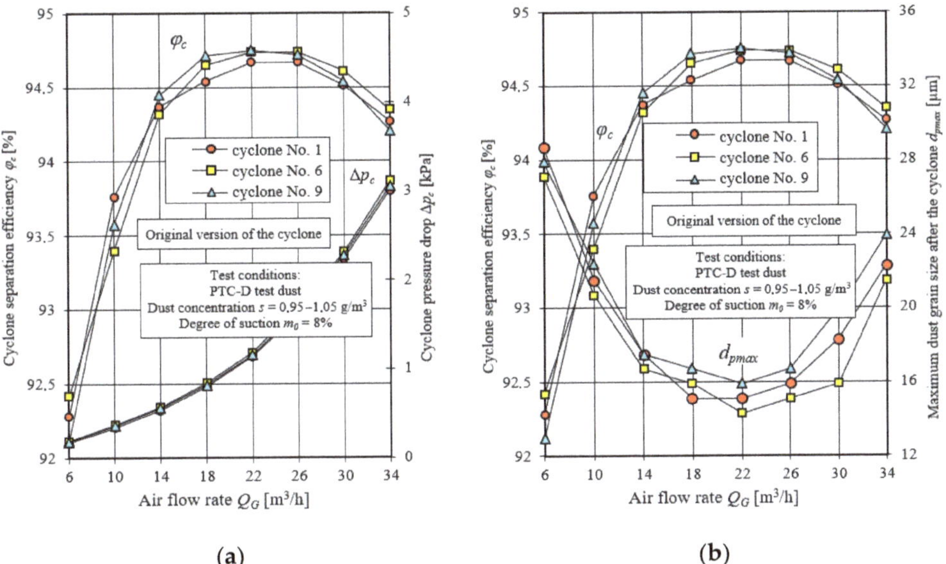

Figure 32. Characteristics of the original D-40 cyclones: (**a**) separation efficiency $\varphi_c = f(Q_G)$ and pressure drop $\Delta p_c = f(Q_G)$; (**b**) separation efficiency and accuracies $\varphi_c = f(Q_G)$, $d_{pmax} = f(Q_G)$.

Such a course of changes in the separation efficiency $\varphi_c = f(Q_G)$ is mainly the result of mutual relations between the inertia force P_d acting on the dust grain and the aerodynamic force P_R of the gas stream interaction. Along with an increase in the stream of air flowing through the cyclone (flow velocity), the values of both forces are increasing, and for higher values of the stream (in the tested cyclone, after exceeding $Q_G = 18$ m^3/h), the increase in the aerodynamic force P_R begins to be faster than the inertia force P_d, which causes braking the movement of grains; thus, their secondary entrainment by the stream (vortex) of air outlet from the cyclone may occur. As a consequence, the separation efficiency of the cyclone decreases. On the other hand, according to the authors of [79], the decrease in effectiveness is also caused by the fact that large dust grains hitting the cyclone wall at high speed are reflected, and then they are again entrained by the air stream and carried with it to the cyclone outlet.

The pressure drop of the selected cyclones changes parabolically in the range of the tested flux and, for the maximum value of $Q_{Gmax} = 34$ m^3/h, is in the range of 3.01–3.11 kPa.

The change in separation efficiency φ_c of the cyclones and the simultaneous change in the size of the maximum d_{pmax} of the dust grains found in the exhaust air from the tested cyclones in the original version is shown in Figure 32. A clear relationship between the two parameters can be seen. An increase in the air flow rate in the range Q_G = 6–22 m³/h causes an intensive increase in the separation efficiency φ_c of the cyclones and a decrease in the size of the maximum dust grains d_{pmax}, which for Q_G = 6 m³/h, take values in the range d_{pmax} = 27.1–28.9 µm. With a further increase in the air flow rate Q_G, there is a slight decrease in the separation efficiency c of the cyclones and an increase in the size of the maximum dust grains d_{pmax}. For Q_G = 34 m³/h, the sizes of dust grains d_{pmax} = 18.3–23.9 µm.

The presence of large dust grains in the air behind the cyclone at higher air flow rates may be due to the fact that the dust grains, while performing a swirling motion, hit the cyclone wall at high speed, bounce off it and then are again entrained by the air flow and lifted to the cyclone outlet. These are individual grains, but of large size and mass. Therefore, the dust grain with the largest size $d_p = d_{pmax}$ found in the air behind the cyclone was used as a criterion for evaluating the accuracy of air filtration in the cyclone.

The PTC-D test dust used for testing, which is a representation of road dust sucked together with the intake air for motor vehicle engines, is a polydisperse dust of varied chemical composition (Figure 30) with a dominant share of SiO_2 silica. The chemical composition of the dust makes the grains have different densities, which for the main components of the dust have a value in g/cm³: for silica SiO_2—2.65, for alumina Al_2O_3—3.99, for iron oxide Fe_2O_3—5.24 and for calcium oxide CaO—3.40. Therefore, the dust grains with the same volume have a different mass, which determines the force of inertia. In addition, the dust grains have an irregular shape, which affects the value of the aerodynamic force PR. For these reasons, the separation in the cyclone from the air containing road dust grains of grains only above a certain diameter d_{pgr} is impossible, as shown in Figures 33 and 34.

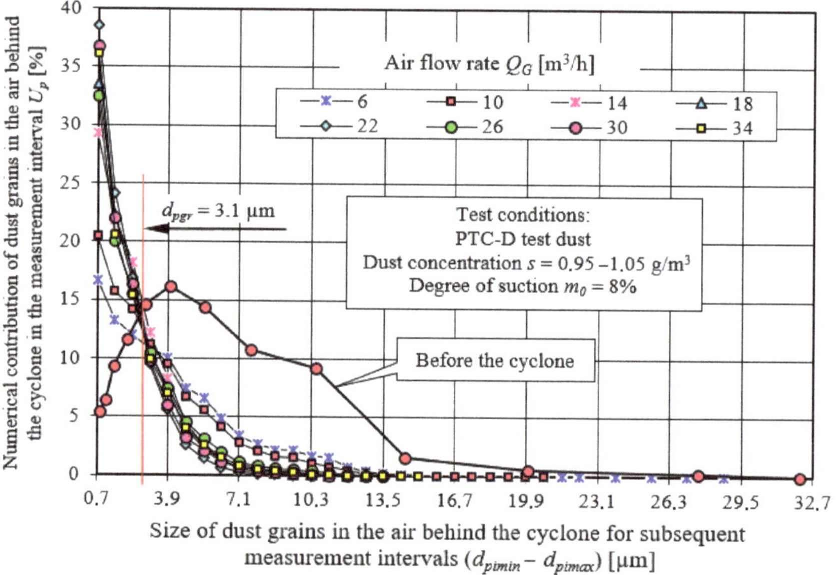

Figure 33. Numerical contributions of U_p of dust grains in the air upstream and downstream of the tested cyclone D-40 (No. 1—original version) for different values of air flow Q_G.

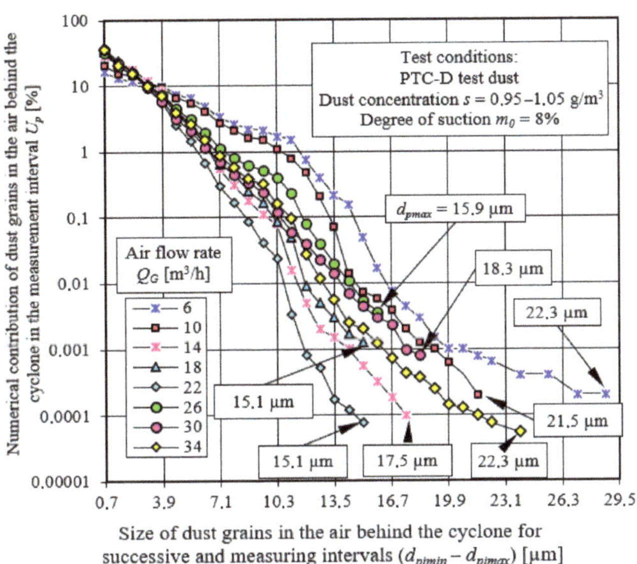

Figure 34. The numerical proportion of dust grains in the air downstream of the tested cyclone *D-40* (No. 1—original version) and the maximum dust grain size d_{pmax} for different values of airflow Q_G.

For the test dust used in this study, the grains do not exceed the size of $d_{pmax} = 80$ μm. The experimentally determined granulometric composition of the PTC-D test dust (Up number shares of dust grains) in the inlet air to cyclone No. 1 (original version) is shown in Figure 33. With the increase in the size of dust grains, the U_p number shares take on increasingly larger values to reach a maximum value of $U_{pmax} = 16\%$ for $d_p = 4$ μm, after which they decrease sharply again. For grain sizes $d_p = 15$ μm, the numerical proportion U_p does not exceed 1.5%. The dust in the outlet air from the cyclone has a significantly different granulometric composition than at the inlet. The maximum values of the U_p shares occur for dust grains with the smallest size registered by the counter, i.e., for $d_p = 0.7$ μm, and then, along with an increase in the size of grains, they decrease rapidly. In the last measurement interval, there is usually one dust grain of size $d_p = d_{pmax}$, and its value depends on the value of the air stream flowing through the cyclone (Figure 34). Similar results were obtained by the author of [100], testing a tangential inlet return cyclone with diameter $D = 31.1$ mm in the range of air flow rate $Q_G = 30.24–66.24$ m³/h using test dust with the density $\rho_{zp} = 0.98$ g/cm³.

The numerical shares of U_p of the dust grains and the size of the dust grain d_{pmax} in the air behind the cyclone decrease with the increasing air flow rate (flow velocity), but only up to a certain value of $Q_G \approx 22$ m³/h ($v_w = 14.7$ m/s). With a further increase in the flux Q_G, there is again an increase in the value of Up and the particle size of dust d_{pmax}.

In the case of the investigated cyclones, a characteristic dust grain size d_{pmax} of $d_p \approx 3.1$ μm can be observed, whose numerical shares of U_p in the air behind the cyclone do not depend on the value of the air flow Q_G. For the same measurement intervals, with an increase in the value of the air flow in the range $Q_G = 6–22$ m³/h, the numerical shares U_p of the dust grains with sizes above $d_p \approx 3.1$ μm decrease, and below this value they increase—Figure 35. An increase in the numerical shares U_p of the dust grains with sizes below $d_p \approx 3.1$ μm may indicate that these grains are not retained in the cyclone and are left together with the air.

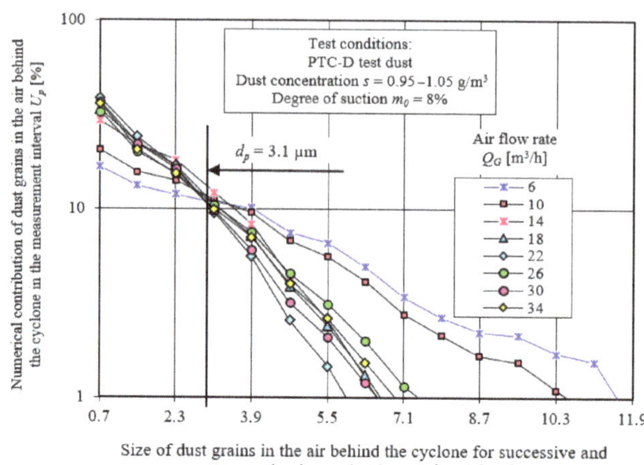

Figure 35. Numerical contribution of U_p of dust grains in the air behind the tested cyclone D-40 (cyclone No. 1—original version) for different values of air flow Q_G.

Theoretically, in the air behind the cyclone at fixed flow conditions (constant value of Q_G flux), there should be only dust grains below a certain limit size d_{pgr}, the value of which can be determined from the relation given in [34,38] and modified by the author:

$$d_{pgr} = \frac{3}{2}\sqrt{\frac{A_0}{A_w}\frac{D}{v_0}\left(\frac{D}{d_w}\right)^{-(2m+1)}\left(\frac{h_w}{d_w}\right)^{-1}\frac{\mu_g}{\rho_{zp}}} \qquad (15)$$

where m is the power exponent of the equation contained in the range 0.5–0.9, μ_g is the gas viscosity and ρ_{zp} is the density of dust grains.

In the case of the tested cyclone, calculated according to Equation (6) for the basic components of dust (SiO_2, Al_2O_3 and Fe_2O_3), the dimensions of the limit grains d_{pgr} take the values presented in Table 2.

Table 2. Results of calculations of the limit size d_{pgr} of the dust grains.

Q_G (m³/h)	v_0 (m/s)	d_{zgr} (μm)		
		SiO_2 $\rho_{zp} = 2.65$ g/cm³	Al_2O_3 $\rho_{zp} = 3.99$ (g/cm³)	Fe_2O_3 $\rho_{zp} = 5.24$ (g/cm³)
6	5.65	2.23	1.85	1.59
22	20.6	1.16	0.96	0.84
34	31.8	0.94	0.79	0.68

The calculated d_{zgr} sizes vary slightly and take very small values (0.68–2.23 μm), much smaller than the dust grain sizes d_{pmax} recorded in the air behind the cyclone under study. As the air flow rate through the cyclone Q_G and the density of dust grains increase, the d_{zgr} sizes take on smaller and smaller values, which is consistent with the general theory of cyclone operation.

3.3.3. Analysis of Test Results of Cyclones with Modifications

Results of tests of the influence of successive (A, B and C) elements of cyclones construction modifications on the characteristics of separation efficiency $\varphi_c = f(Q_G)$ and pressure drop $\Delta p_c = f(Q_G)$, are presented on the example of cyclone No. 1. Replacement of the symmetrical inlet with an asymmetrical inlet (version I of the cyclone—modification element A) causes an increase of approx. 4% of separation efficiency of cyclone c in the range of flux $Q_G = 6-22$ m³/h, with preservation of the previous character of the modification

Figure 36. This slight increase in separation efficiency could be caused by the fact that the asymmetrical inlet ensured obtaining such a distribution of dust in its cross-section, in which its basic mass was gathered at the external wall of the cylindrical part of the cyclone. Movement of dust grains towards the walls causes an increase in dust concentration in their vicinity. As a result of grains colliding with each other at the cyclone walls, dust coagulation may occur, which consequently increases the effect of gravitational force and accelerates dust release, especially of small grains. Changing the shape of the inlet port did not cause any noticeable changes in the pressure drop $\Delta p_c = f(Q_G)$ in relation to the course and value.

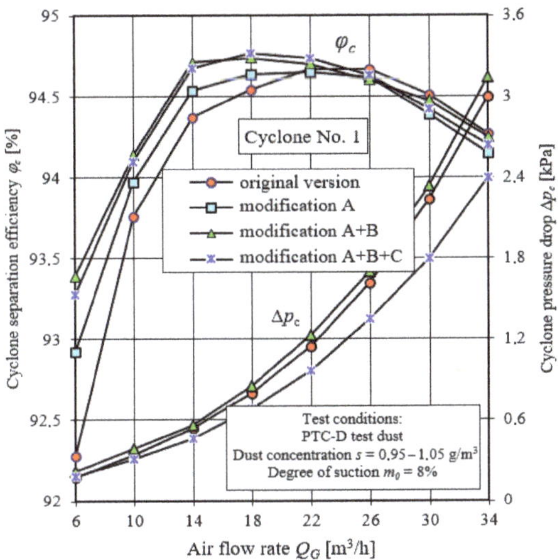

Figure 36. Change in separation efficiency $\varphi_c = f(Q_G)$ and pressure drop $\Delta p_c = f(Q_G)$ of cyclone D-40 (No. 1) for successive A, B and C cyclone design modification elements.

Application of the streamlined shape of the inlet hole of the outlet pipe (version II of the cyclone—elements of modifications A + B) caused a further increase (about 0.5%) of separation efficiency c of the cyclone in the range $Q_G = 6$–22 m^3/h. Above $Q_G = 22$ m^3/h, an opposite effect of A + B modification was observed, as a slight decrease of separation efficiency of the cyclone took place. At the same time, there was an approximately 6% increase in pressure drop pc caused most likely by the increase in aerosol velocity in the "z" constriction, which was created after making a streamlined ring on the outer surface of the outlet tube (Figure 28). The increase of aerosol velocity in the "z" constriction caused the dust grains of smaller size and mass to acquire the velocity and inertial force necessary to separate them from the air, which was reflected in the increase of cyclone separation efficiency in the range of smaller Q_G air flows. Thus, the maximum cyclone separation efficiency had a higher value of $\varphi_c = 94.77\%$ and moved its position from $Q_G = 22$ m^3/h to $Q_G = 18$ m^3/h.

Replacement of the cylindrical outlet tube with a conical one with preservation of the original inner diameter of the d_w hole and expanding towards the outlet together with the streamlined shape of the inlet hole (version III of the cyclone—elements of modification A + B + C) caused, first of all, a significant decrease. The decrease in pressure drop of the cyclone results from the decrease (from $v_w = 22.7$ m/s to $v_w = 14$ m/s for $Q_G = 34$ m^3/h) of the outlet velocity in the cyclone, which happens as a result of increasing the diameter of the cyclone outlet pipe. There was also a non-significant decrease in separation efficiency, compared to the cyclone version II, over the entire range of air flow values Q_G.

The change of separation efficiency c and the change of the size of the maximum d_{pmax} of the dust grains in the air behind the tested cyclone No. 1 in the original version and the cyclone version III (elements of the modifications A + B + C) are shown in Figure 37. There is a clear decrease in the size of the maximum d_{pmax} of the dust grains for the extreme values of the air flow ($Q_G = 6$ m^3/h and $Q_G = 34$ m^3/h) and an increase in this size for the middle values of Q_G.

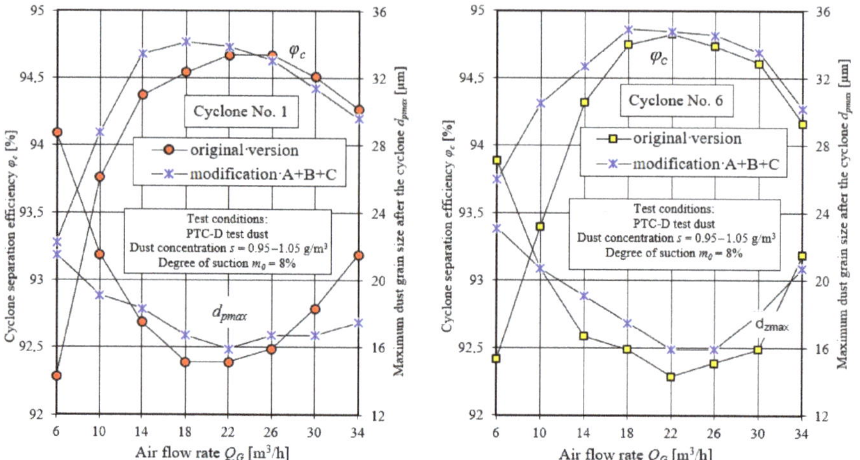

Figure 37. Variation of separation efficiency $\varphi_c = f(Q_G)$ and maximum dust grain size values $d_{pmax} = f(Q_G)$ in the Q_G exhaust air stream of cyclone No. 1 and No. 6 in the original version and after design modification—version III.

The design modification performed on the cyclones primarily resulted in a significant approximately 30% decrease in cyclone pressure drop compared to their original version. There was a slight (about 1%) increase in separation efficiency c of cyclones for lower values ($Q_G = 6$–22 m^3/h) of air flows.

For the remaining two specimens of cyclones D-40 (No. 6 and No. 9), the modification of the design into successive elements A, B and C caused similar changes as in cyclone No. 1 in the characteristics of separation efficiency $\varphi_c = f(Q_G)$ and filtration accuracy $d_{pmax} = f(Q_G)$ and pressure drop $\Delta p_c = f(Q_G)$ in relation to the course and values (Figure 37).

The modification of the cyclone resulted in an increase in the fractional efficiency of the cyclone, as evidenced by a decrease in the proportion of the maximum size d_{pmax} of dust grains in the air downstream of the cyclone in terms of the minimum and maximum values of the air flow rate Q_G passing through the cyclone.

From the test results presented in Figures 36 and 37, it can be seen that the D-40 cyclone should operate within a narrow range of changes in air flow $Q_G = 18$–26 m^3/h, resulting from maximum separation efficiency and minimum concentration of large-sized dust grains. Widening this range will cause the cyclone to operate at a lower separation efficiency and lower filtration accuracy, which will increase the mass of dust retained on the second filtration stage, and thus cause an accelerated increase in its pressure drop.

4. Conclusions

In this paper, an experimental evaluation of a tangential inlet reciprocating cyclone design modification of an air filter multicyclone was performed to increase the filtration efficiency of the inlet air to a tracked vehicle engine. Three modifications to the cyclone design were proposed and performed on three specimens. The symmetrical inlet of the cyclone inlet channel was replaced with an asymmetrical inlet. The cylindrical outlet tube was modified and replaced with a conical tube, and the edges of the inlet opening were additionally given a streamlined shape. After each modification performed, an experimental evaluation of the cyclone design changes was performed. The influence of the modifications on the cyclone's efficiency and flow resistance characteristics was examined. Next, modifications of the cyclone were performed on the same specimen without removing the previous modifications. Tests were performed in the air flow range $Q_G = 5\text{–}30\ \mathrm{m^3/h}$ resulting from the operation of the engine in the rotational speed range $n_{min}\text{–}n_N$. Polydisperse test dust with grain size range $d_{pmax} = 80\ \mu\mathrm{m}$ was used for testing. Improvements in cyclone performance were obtained in terms of filtration efficiency and accuracy as well as pressure drop. The following conclusions were drawn from the conducted research:

(1) Although there are many structural possibilities of increasing the effectiveness of cyclones, not all solutions, due to their too complicated construction, can be used in cyclones used in motor vehicle engine inlet air filters. The effect in the form of an increase in separation efficiency or a decrease in flow resistance may be disproportionally small in relation to the costs incurred.

(2) To modify the cyclones, which are the elements of the multicyclone of the motor vehicle air filter, the following solutions can be used practically: change of the shape of the inlet port—replacement of the symmetric inlet with an asymmetric one, introduction of a streamlined shape of the inlet opening of the outlet duct and change of the shape of the outlet duct (outlet tube) from cylindrical to conical. It may be technologically difficult to make a streamlined shape of the outlet tube inlet opening.

(3) The separation efficiency of the tested cyclones (reciprocating cyclones with tangential inlet) with respect to polydisperse dust with grain size up to $d_{pmax} = 80\ \mu\mathrm{m}$ increases with the increase of air flow Q_G (inlet velocity 0) up to the value of $\varphi_{cmax} = 94.8\%$, and then slightly decreases. Such a character of the course of $\varphi_c = f(Q_G)$ and the values of the obtained efficiencies are consistent with the information provided in the literature by the authors of research papers.

(4) For steady flow conditions ($Q_G = \mathrm{const.}$), the number of dust grains in the air behind the cyclone systematically decreases as their size increases until they disappear completely. In the last measurement interval, there is usually one dust grain with a maximum size of $d_p = d_{pmax}$, which indicates the filtration accuracy. As the air flow rate Q_G increases, the dimension of the dust grain d_{pmax} takes on smaller and smaller values, which corresponds to the maximum filtration efficiency, after which it increases again. There is no clear separation (at the boundary grain size d_{pgr}) between the dust grains retained by the cyclone and those leaving the cyclone with the air. The reason for this phenomenon is the polydispersity of the dust, the density, size and shape of the dust grains.

(5) Modification of the design of cyclones does not cause significant changes in the characteristics of separation efficiency φ_c and pressure drop Δp_c. A visible effect of the performed modifications of the cyclones' construction is, first of all, a significant 30% decrease in the pressure drop within the whole range of the examined flux Q_G, which is an effect of replacing the cylindrical cyclone outlet tube with a conical one. In addition, there is an approximately 1% increase in separation efficiency φ_c in the range of the smallest values of air flux Q_G and a shift in the maximum separation efficiency φ_{cmax} towards smaller values of air flux Q_G, which may result from the use of an asymmetrical inlet duct and the streamlined shape of the inlet opening of the outlet tube.

(6) Modification of the cyclones by elements A, B and C results in a clear decrease in the dimensions of the maximum d_{pmax} of the dust grains for the extreme ($Q_G = 6\ \mathrm{m^3/h}$ i $Q_G = 34\ \mathrm{m^3/h}$) values of the air stream (Figure 37) and an increase in the dimensions d_{pmax} for the air stream in the range $Q_G = (14\text{–}26)\ \mathrm{m^3/h}$. At the same time, in this range, the cyclone obtains the maximum values of cyclone filtration efficiency, which indicates the range of air stream in which the operation of this cyclone should take place.

In this study, the filtration efficiency of a single cyclone was improved by making three modifications to its design. The activities brought the expected results in terms of improving the filtration efficiency of the cyclone, so it is advisable to conduct further research on tangential inlet return cyclones in terms of increasing the separation efficiency and accuracy and reducing the pressure drop. Other modifications of this cyclone are possible, such as using a special cone on the outlet of separated dust, using more inlet channels or changing the inlet channel to a diagonal one that is directed upwards. This will be the subject of further research by the author. In the available literature there are results of positive research in this area. However, modifications should be performed and evaluated experimentally on a selected cyclone. Experimental research is expensive and labor intensive, but it is the most reliable research method.

Funding: The article was written as part of the implementation of the university research grant No. UGB 759/2022.

Institutional Review Board Statement: Not applicable.

Informed Consent Statement: Not applicable.

Data Availability Statement: Data are contained within the article.

Conflicts of Interest: The author declares no conflict of interest.

References

1. Barris, M.A. *Total FiltrationTM: The Influence of Filter Selection on Engine Wear, Emissions, and Performance*; SAE Technical Paper 952557; SAE: Warrendale, PA, USA, 1995.
2. Schaeffer, J.W.; Olson, L.M. Air Filtration Media for Transportation Applications. *Filtr. Sep.* **1998**, *35*, 124–129.
3. Jaroszczyk, T.; Pardue, B.A.; Heckel, S.P.; Kallsen, K.J. Engine air cleaner filtration performance—Theoretical and experimental background of testing. In Proceedings of the AFS Fourteenth Annual Technical Conference and Exposition, Tampa, FL, USA, 1 May 2001; Included in the Conference Proceedings (Session 16).
4. Dziubak, T. Zapylenie powietrza wokół pojazdu terenowego. *Wojsk. Przegląd Tech.* **1990**, *3*, 154–157. (In Polish)
5. Burda, S.; Chodnikiewicz, Z. Konstrukcja i badania pyłowe filtrów powietrza silnika czołgowego. *Bull. Mil. Univ. Technol.* **1962**, *3*, 12–34. (In Polish)
6. Juda, J.; Nowicki, M. *Urządzenia Odpylające*; Państwowe Wydawnictwo Naukowe: Warszawa, Poland, 1986. (In Polish)
7. Balicki, W.; Szczeciński, S.; Chachurski, R.; Kozakiewicz, A.; Głowacki, P.; Szczeciński, J. Problematyka filtracji powietrza wlotowego do turbinowych silników śmigłowcowych. *Trans. Inst. Aviat.* **2009**, *199*, 25–30.
8. Bojdo, N.; Filippone, A. Effect of Desert Particulate Composition on Helicopter Engine Degradation Rate. In Proceedings of the 40th European Rotorcraft Forum, Southampton, UK, 2–5 September 2014. [CrossRef]
9. Smialek, J.L.; Archer, F.A.; Garlick, R.G. Turbine airfoil degradation in the persian gulf war. *JOM* **1994**, *46*, 39–41. [CrossRef]
10. Vogel, A.; Durant, A.J.; Cassiani, M.; Clarkson, R.J.; Slaby, M.; Diplas, S.; Krüger, K.; Stohl, A. Simulation of Volcanic Ash Ingestion Into a Large Aero Engine: Particle–Fan Interactions. *ASME J. Turbomach.* **2019**, *141*, 011010. [CrossRef]
11. Summers, C.E. The physical characteristics of road and field dust. *SAE Trans.* **1925**, *20*, 178–197. [CrossRef]
12. Pinnick, R.G.; Fernández, G.; Hinds, B.D.; Bruce, C.W.; Schaefer, R.W.; Pendleton, J.D. Dust Generated by Vehicular Traffic on Unpaved Roadways: Sizes and Infrared Extinction Characteristics. *Aerosol Sci. Technol.* **1985**, *4*, 99–121. [CrossRef]
13. Long, J.; Tang, M.; Sun, Z.; Liang, Y.; Hu, J. Dust Loading Performance of a Novel Submicro-Fiber Composite Filter Medium for Engine. *Materials* **2018**, *11*, 2038. [CrossRef]
14. Jaroszczyk, T.; Wake, J.; Fallon, S.L.; Connor, M.J. *Development of Motor Vehicle Ventilation System Particulate Air Filters*; SAE Technical Paper Series 962241; International Truck & Bus Meeting & Exposition: Detroit, MI, USA, 1996.
15. Maddineni, A.K.; Das, D.; Damodaran, R.M. Numerical investigation of pressure and flow characteristics of pleated air filter system for automotive engine intake application. *Sep. Purif. Technol.* **2018**, *212*, 126–134. [CrossRef]
16. Jaroszczyk, T.; Petrik, S.; Donahue, K. Recent Development in Heavy Duty Engine Air Filtration and the Role of Nanofiber Filter Media. *J. KONES Powertrain Transp.* **2009**, *16*, 207–216.
17. Rieger, M.; Hettkamp, P.; Löhl, T.; Madeira, P.M.P. Effcient Engine Air Filter for Tight Installation Spaces. *ATZ Heavy Duty Worldw.* **2019**, *12*, 56–59. [CrossRef]
18. Tian, X.; Ou, Q.; Liu, J.; Liang, Y.; Pui, D.Y. Influence of pre-stage filter selection and face velocity on the loading characteristics of a two-stage filtration system. *Sep. Purif. Technol.* **2019**, *224*, 227–236. [CrossRef]
19. Muschelknautz, U. Design criteria for multicyclones in a limited space. *Powder Technol.* **2019**, *357*, 2–20. [CrossRef]
20. Muschelknautz, U. Comparing Efficiency per Volume of Uniflow Cyclones and Standard Cyclones. *Chem. Ing. Tech.* **2020**, *93*, 91–107. [CrossRef]
21. Otanocha, O.B.; Lin, L.; Zhou, Y.; Zhong, S.; Zhu, L. Picosecond laser surface micro-texturing for the modification of aerodynamic and dust distribution characteristics in a multi-cyclone system. *Cogent Eng.* **2016**, *3*, 1152746. [CrossRef]
22. Barbolini, M.; Di Pauli, F.; Traina, M. Simulation der Luftfiltration zur Auslegung von Filterelementen. *MTZ-Mot. Z.* **2014**, *75*, 52–57. [CrossRef]

23. Bojdo, N. Rotorcraft Engine Air Particle Separation. Doctoral Thesis, University of Manchester, Manchester, UK, 2012. Available online: https://www.escholar.manchester.ac.uk/uk-ac-man-scw:183545 (accessed on 2 April 2021).

24. Szczepankowski, A.; Szymczak, J.; Przysowa, R. The Effect of a Dusty Environment Upon Performance and Operating Parameters of Aircraft Gas Turbine Engines. In Proceedings of the Specialists' Meeting—Impact of Volcanic Ash Clouds on Military Operations NATO AVT-272-RSM-047, Vilnius, Lithuania, 17 May 2017.

25. Worek, J.; Badura, X.; Białas, A.; Chwiej, J.; Kawoń, K.; Styszko, K. Pollution from Transport: Detection of Tyre Particles in Environmental Samples. *Energies* **2022**, *15*, 2816. [CrossRef]

26. Cichowicz, R.; Dobrzański, M. Analysis of Air Pollution around a CHP Plant: Real Measurements vs. *Computer Simulations*. *Energies* **2022**, *15*, 553. [CrossRef]

27. Ciobanu, C.; Tudor, P.; Istrate, I.-A.; Voicu, G. Assessment of Environmental Pollution in Cement Plant Areas in Romania by Co-Processing Waste in Clinker Kilns. *Energies* **2022**, *15*, 2656. [CrossRef]

28. Kanageswari, S.V.; Tabil, L.G.; Sokhansanj, S. Dust and Particulate Matter Generated during Handling and Pelletization of Herbaceous Biomass: A Review. *Energies* **2022**, *15*, 2634. [CrossRef]

29. Luo, H.; Zhou, W.; Jiskani, I.M.; Wang, Z. Analyzing Characteristics of Particulate Matter Pollution in Open-Pit Coal Mines: Implications for Green Mining. *Energies* **2021**, *14*, 2680. [CrossRef]

30. Jaroszczyk, T.; Fallon, S.L.; Liu, Z.G.; Heckel, S.P. Development of a Method to Measure Engine Air Cleaner Fractional Effi-ciency. *SAE Trans.* **1999**, *1089*, 9–18.

31. Grafe, T.; Gogins, M.; Barris, M.; Schaefer, J.; Canepa, R. Nanofibers in Filtration Applications in Transportation. In Proceedings of the Filtration 2001 International Conference and Exposition of the INDA (Association of the Nowovens Fabric Industry), Chicago, IL, USA, 3–5 December 2001; Available online: https://www.yumpu.com/en/document/read/7555561/nanofiber-in-filtration-applications-in-transportation-donaldson- (accessed on 22 May 2022).

32. Schulze, M.; Taufkirch, G. Papierluftfilter Nutzfahrzeugen. *MTZ Mot. Z.* **1991**, *1991*, 52.

33. Duan, J.; Gao, S.; Lu, Y.; Wang, W.; Zhang, P.; Li, C. Study and optimization of flow field in a novel cyclone separator with inner cylinder. *Adv. Powder Technol.* **2020**, *31*, 4166–4179. [CrossRef]

34. Cortés, C.; Gil, A. Modeling the gas and particle flow inside cyclone separators. *Prog. Energy Combust. Sci.* **2007**, *33*, 409–452. [CrossRef]

35. Zhang, W.; Zhang, L.; Yang, J.; Hao, X.; Guan, G.; Gao, Z. An experimental modeling of cyclone separator efficiency with PCA-PSO-SVR algorithm. *Powder Technol.* **2019**, *347*, 114–124. [CrossRef]

36. Dziubak, T.; Bąkała, L.; Karczewski, M.; Tomaszewski, M. Numerical research on vortex tube separator for special vehicle engine inlet air filter. *Sep. Purif. Technol.* **2019**, *237*, 116463. [CrossRef]

37. Wang, J.; Kim, S.C.; Pui, D.Y.H. Figure of Merit of Composite Filters with Micrometer and Nanometer Fibers. *Aerosol Sci. Technol.* **2008**, *42*, 722–728. [CrossRef]

38. Dirgo, J.; Leith, D. Performance of Theoretically Optimised Cyclones. *Filtr. Sep.* **1985**, *22*, 119–125.

39. Juda, J. *Pomiary Zapylenia i Technika Odpylania*; WNT: Warszawa, Poland, 1968. (In Polish)

40. Kabsch, P. *Odpylanie i Odpylacze*; WNT: Warszawa, Poland, 1992. (In Polish)

41. Robinson, K.S.; Hamblin, C. Ivestigating Droplet Collection in Helices and a Comparison with Conventional Demisters. *Filtr. Sep.* **1986**, *24*, 166.

42. Swift, P. An Empirical Approach to Cyclone Design and Application. *Filtr. Sep.* **1986**, *23*, 24–27.

43. Warych, J. *Oczyszczanie Gazów-Procesy i Aparatura*; WNT: Warszawa, Poland, 1998. (In Polish)

44. Wheeldon, J.M.; Burnard, G.K. Performance of Cyclones in the Off-Gas Path of a Pressurised Fluidised Bed Combustor. *Filtr. Sep.* **1987**, *24*, 178–187.

45. Gimbun, J.; Chuah, T.G.; Choong, T.S.Y.; Fakhru'L-Razi, A. A CFD Study on the Prediction of Cyclone Collection Efficiency. *Int. J. Comput. Methods Eng. Sci. Mech.* **2005**, *6*, 161–168. [CrossRef]

46. Leith, D.; Mehta, D. Cyclone performance and design. *Atmosp. Environ.* **1973**, *7*, 527–549. [CrossRef]

47. Kim, J.C.; Lee, K.W. Experimental Study of Particle Collection by Small Cyclones. *Aerosol Sci. Technol.* **1990**, *12*, 1003–1015. [CrossRef]

48. Dirgo, J.; Leith, D. Cyclone Collection Efficiency: Comparison of Experimental Results with Theoretical Predictions. *Aerosol Sci. Technol.* **1985**, *4*, 401–415. [CrossRef]

49. Ramachandran, G.; Leith, D.; Dirgo, J.; Feldman, H. Cyclone Optimization Based on a New Empirical Model for Pressure Drop. *Aerosol Sci. Technol.* **1991**, *15*, 135–148. [CrossRef]

50. Iozia, D.L.; Leith, D. The Logistic Function and Cyclone Fractional Efficiency. *Aerosol Sci. Technol.* **1990**, *12*, 598–606. [CrossRef]

51. Karagoz, I.; Avci, A. Modelling of the Pressure Drop in Tangential Inlet Cyclone Separators. *Aerosol Sci. Technol.* **2005**, *39*, 857–865. [CrossRef]

52. Hsiao, T.-C.; Huang, S.-H.; Hsu, C.-W.; Chen, C.-C.; Chang, P.-K. Effects of the geometric configuration on cyclone performance. *J. Aerosol Sci.* **2015**, *86*, 1–12. [CrossRef]

53. Wu, J.-P.; Zhang, Y.-H.; Wang, H.-L. Numerical study on tangential velocity indicator of free vortex in the cyclone. *Sep. Purif. Technol.* **2014**, *132*, 541–551. [CrossRef]

54. Altmeyer, S.; Mathieu, V.; Jullemier, S.; Contal, P.; Midoux, N.; Rode, S.; Leclerc, J.-P. Comparison of different models of cyclone prediction performance for various operating conditions using a general software. *Chem. Eng. Process. Process Intensif.* **2004**, *43*, 511–522. [CrossRef]
55. Zhao, B. Development of a Dimensionless Logistic Model for Predicting Cyclone Separation Efficiency. *Aerosol Sci. Technol.* **2010**, *44*, 1105–1112. [CrossRef]
56. Zhou, F.; Sun, G.; Zhang, Y.; Ci, H.; Wei, Q. Experimental and CFD study on the effects of surface roughness on cyclone performance. *Sep. Purif. Technol.* **2018**, *193*, 175–183. [CrossRef]
57. Hoffmann, A.C.; Arends, H.; Sie, H. An Experimental Investigation Elucidating the Nature of the Effect of Solids Loading on Cyclone Performance. *Filtr. Sep.* **1991**, *28*, 188–193. [CrossRef]
58. Zhu, Y.; Lee, K. Experimental study on small cyclones operating at high flowrates. *J. Aerosol Sci.* **1999**, *30*, 1303–1315. [CrossRef]
59. Elsayed, K.; Lacor, C. The effect of cyclone inlet dimensions on the flow pattern and performance. *Appl. Math. Model.* **2011**, *35*, 1952–1968. [CrossRef]
60. Beaulac, P.; Issa, M.; Ilinca, A.; Brousseau, J. Parameters Affecting Dust Collector Efficiency for Pneumatic Conveying: A Review. *Energies* **2022**, *15*, 916. [CrossRef]
61. Le, D.K.; Yoon, J.Y. Numerical investigation on the performance and flow pattern of two novel innovative designs of four-inlet cyclone separator. *Chem. Eng. Process. Process Intensif.* **2020**, *150*, 107867. [CrossRef]
62. Safikhani, H.; Zamani, J.; Musa, M. Numerical study of flow field in new design cyclone separators with one, two and three tangential inlets. *Adv. Powder Technol.* **2018**, *29*, 611–622. [CrossRef]
63. Mazyan, W.; Ahmadi, A.; Brinkerhoff, J.; Ahmed, H.; Hoorfar, M. Enhancement of cyclone solid particle separation performance based on geometrical modification: Numerical analysis. *Sep. Purif. Technol.* **2018**, *191*, 276–285. [CrossRef]
64. Brar, L.S.; Elsayed, K. Analysis and optimization of multi-inlet gas cyclones using large eddy simulation and artificial neural network. *Powder Technol.* **2017**, *311*, 465–483. [CrossRef]
65. El-Emam, M.A.; Shi, W.; Zhou, L. CFD-DEM simulation and optimization of gas-cyclone performance with realistic macroscopic particulate matter. *Adv. Powder Technol.* **2019**, *30*, 2686–2702. [CrossRef]
66. Sun, Z.; Liang, L.; Liu, Q.; Yu, X. Effect of the particle injection position on the performance of a cyclonic gas solids classifier. *Adv. Powder Technol.* **2019**, *31*, 227–233. [CrossRef]
67. Lim, J.-H.; Park, S.-I.; Lee, H.-J.; Zahir, M.Z.; Yook, S.-J. Performance evaluation of a tangential cyclone separator with additional inlets on the cone section. *Powder Technol.* **2019**, *359*, 118–125. [CrossRef]
68. Wang, S.; Li, H.; Wang, R.; Wang, X.; Tian, R.; Sun, Q. Effect of the inlet angle on the performance of a cyclone separator using CFD-DEM. *Adv. Powder Technol.* **2018**, *30*, 227–239. [CrossRef]
69. Wasilewski, M.; Brar, L.S. Effect of the inlet duct angle on the performance of cyclone separators. *Sep. Purif. Technol.* **2018**, *213*, 19–33. [CrossRef]
70. Yoshida, H.; Ono, K.; Fukui, K. The effect of a new method of fluid flow control on submicron particle classification in gas-cyclones. *Powder Technol.* **2005**, *149*, 139–147. [CrossRef]
71. Mandsandulia, D.; Andersson, A.G.; Lundström, T.S. Effects of the inlet angle on the collection efficiency of a cyclone with helical-roof inlet. *Powder Technol.* **2017**, *305*, 48–55.
72. Misiulia, D.; Andersson, A.G.; Lundström, T.S. Effects of the inlet angle on the flow pattern and pressure drop of a cyclone with helical-roof inlet. *Chem. Eng. Res. Des.* **2015**, *102*, 307–321. [CrossRef]
73. Lim, K.; Kwon, S.-B.; Lee, K. Characteristics of the collection efficiency for a double inlet cyclone with clean air. *J. Aerosol Sci.* **2003**, *34*, 1085–1095. [CrossRef]
74. Zhao, B.; Shen, H.; Kang, Y. Development of a symmetrical spiral inlet to improve cyclone separator performance. *Powder Technol.* **2004**, *145*, 47–50. [CrossRef]
75. Bernardo, S.; Mori, M.; Peres, A.; Dionísio, R. 3-D computational fluid dynamics for gas and gas-particle flows in a cyclone with different inlet section angles. *Powder Technol.* **2006**, *162*, 190–200. [CrossRef]
76. Chen, L.; Ma, H.; Sun, Z.; Ma, G.; Li, P.; Li, C.; Cong, X. Effect of inlet periodic velocity on the performance of standard cyclone separators. *Powder Technol.* **2022**, *402*, 117347. [CrossRef]
77. Wei, Q.; Sun, G.; Gao, C. Numerical analysis of axial gas flow in cyclone separators with different vortex finder diameters and inlet dimensions. *Powder Technol.* **2020**, *369*, 321–333. [CrossRef]
78. Yao, Y.; Huang, W.; Wu, Y.; Zhang, Y.; Zhang, M.; Yang, H.; Lyu, J. Effects of the inlet duct length on the flow field and performance of a cyclone separator with a contracted inlet duct. *Powder Technol.* **2021**, *393*, 12–22. [CrossRef]
79. Dzierżanowski, P.; Kordziński, W.; Otyś, J.; Szczeciński, S.; Wiatrek, R. Napędy lotnicze. In *Turbinowe Silniki Śmigłowe i Śmigłowcowe*; WKŁ: Warszawa, Poland, 1985. (In Polish)
80. Lim, K.; Kim, H.; Lee, K. Characteristics of the collection efficiency for a cyclone with different vortex finder shapes. *J. Aerosol Sci.* **2004**, *35*, 743–754. [CrossRef]
81. Raoufi, A.; Shams, M.; Farzaneh, M.; Ebrahimi, R. Numerical simulation and optimization of fluid flow in cyclone vortex finder. *Chem. Eng. Process. Process Intensif.* **2008**, *47*, 128–137. [CrossRef]
82. Misiulia, D.; Elsayed, K.; Andersson, A.G. Geometry optimization of a deswirler for cyclone separator in terms of pressure drop using CFD and artificial neural network. *Sep. Purif. Technol.* **2017**, *185*, 10–23. [CrossRef]

83. Misyulya, D.I. New Designs of Devices for a Decrease in Power Consumption of Cyclone Dust Collectors. *Russ. J. Non Ferr. Met.* **2012**, *53*, 67–72. [CrossRef]
84. Qiang, L.; Qinggong, W.; Weiwei, X.; Zilin, Z.; Konghao, Z. Experimental and computational analysis of a cyclone separator with a novel vortex finder. *Powder Technol.* **2019**, *360*, 398–410. [CrossRef]
85. Wang, Z.; Sun, G.; Jiao, Y. Experimental study of large-scale single and double inlet cyclone separators with two types of vortex finder. *Chem. Eng. Process. Process Intensif.* **2020**, *158*, 108188. [CrossRef]
86. Yohana, E.; Tauviqirrahman, M.; Laksono, D.A.; Charles, H.; Choi, K.-H.; Yulianto, M.E. Innovation of vortex finder geometry (tapered in-cylinder out) and additional cooling of body cyclone on velocity flow field, performance, and heat transfer of cyclone separator. *Powder Technol.* **2022**, *399*, 117235. [CrossRef]
87. Elsayed, K.; Parvaz, F.; Hosseini, S.H.; Ahmadi, G. Influence of the dipleg and dustbin dimensions on performance of gas cyclones: An optimization study. *Sep. Purif. Technol.* **2020**, *239*, 116553. [CrossRef]
88. Parvaz, F.; Hosseini, S.H.; Elsayed, K.; Ahmadi, G. Influence of the dipleg shape on the performance of gas cyclones. *Sep. Purif. Technol.* **2019**, *233*, 116000. [CrossRef]
89. Kaya, F.; Karagoz, I. Numerical investigation of performance characteristics of a cyclone prolonged with a dipleg. *Chem. Eng. J.* **2009**, *151*, 39–45. [CrossRef]
90. Qian, F.; Zhang, J.; Zhang, M. Effects of the prolonged vertical tube on the separation performance of a cyclone. *J. Hazard. Mater.* **2006**, *136*, 822–829. [CrossRef]
91. Wasilewski, M. Analysis of the effect of counter-cone location on cyclone separator efficiency. *Sep. Purif. Technol.* **2017**, *179*, 236–247. [CrossRef]
92. Yoshida, H. Effect of apex cone shape and local fluid flow control method on fine particle classification of gas-cyclone. *Chem. Eng. Sci.* **2012**, *85*, 55–61. [CrossRef]
93. Yoshida, H.; Nishimura, Y.; Fukui, K.; Yamamoto, T. Effect of apex cone shape on fine particle classification of gas-cyclone. *Powder Technol.* **2010**, *204*, 54–62. [CrossRef]
94. Yoshida, H.; Kwan-Sik, Y.; Fukui, K.; Akiyama, S.; Taniguchi, S. Effect of apex cone height on particle classification per-formance of a cyclone separator. *Adv. Powder Technol.* **2003**, *14*, 263–266. [CrossRef]
95. Kępa, A. The efficiency improvement of a large-diameter cyclone—The CFD calculations. *Sep. Purif. Technol.* **2013**, *118*, 105–111. [CrossRef]
96. Obermair, S.; Woisetschläger, J.; Staudinger, G. Investigation of the flow pattern in different dust outlet geometries of a gas cyclone by laser Doppler anemometry. *Powder Technol.* **2003**, *138*, 239–251. [CrossRef]
97. Dziubak, T. Experimental Studies of Dust Suction Irregularity from Multi-Cyclone Dust Collector of Two-Stage Air Filter. *Energies* **2021**, *14*, 3577. [CrossRef]
98. Bukowski, J. *Mechanika Płynów*; PWN: Warszawa, Poland, 1968. (In Polish)
99. *PN-ISO 5011*; Filtry Powietrza do Silników Spalinowych i Sprężarek. PKNM: Warszawa, Poland, 1994. (In Polish)
100. Ma, L.; Ingham, D.; Wen, X. Numerical modelling of the fluid and particle penetration through small sampling cyclones. *J. Aerosol Sci.* **2000**, *31*, 1097–1119. [CrossRef]

Article

Experimental Study of a PowerCore Filter Bed Operating in a Two-Stage System for Cleaning the Inlet Air of Internal Combustion Engines

Tadeusz Dziubak

Faculty of Mechanical Engineering, Military University of Technology, Gen. Sylwestra Kaliskiego Street 2, 00-908 Warsaw, Poland; tadeusz.dziubak@wat.edu.pl

Abstract: Small dust grains cause a higher intensity of increase in the flow resistance of the fibrous filter bed, which, due to the established value of the permissible resistance, results in a shorter period of operation of the air filter and the vehicle. At the same time, the mass of dust per unit of filtration area takes on smaller values. Such a phenomenon occurs in the two-stage "multicyclone-baffle filter" engine inlet air filtration system. The main objective of this study was to experimentally determine the mass of dust retained per unit of filtration area (dust absorption coefficient k_m) of the PowerCore filter bed operating under two-stage filtration conditions, which cannot be found in the available literature. The original methodology and conditions for determining the dust absorption coefficient k_m of a PowerCore filter bed operating under two-stage filtration conditions are presented. Tests were carried out on the characteristics of filtration efficiency and accuracy, as well as on the flow resistance of a filtration unit consisting of a single cyclone and a PowerCore test filter with an appropriately selected surface area of filter material. During the tests, conditions corresponding to the actual conditions of vehicle use and air filter operation were maintained, including filtration speed and the dust concentration in the air. The experimentally determined dust absorption coefficient of the PowerCore research filters operating in a two-stage filtration system took on values in the range of k_m = 199–219 g/m^2. The dust absorption coefficient k_m of the PowerCore research filter operating under single-stage filtration conditions reached a value of k_m = 434 g/m^2, which is twice as high. Prediction of the mileage of a car equipped with a single-stage and two-stage "multi-cyclone-partition" filtration system was carried out, showing the usefulness of the experimentally determined dust absorption coefficients k_m.

Keywords: PowerCore filter bed; separation efficiency and accuracy; pressure drop; dust absorption coefficient; two-stage air filter; vehicle internal combustion engine

Citation: Dziubak, T. Experimental Study of a PowerCore Filter Bed Operating in a Two-Stage System for Cleaning the Inlet Air of Internal Combustion Engines. *Energies* **2023**, *16*, 3802. https://doi.org/10.3390/en16093802

Academic Editor: Anastassios M. Stamatelos

Received: 3 April 2023
Revised: 24 April 2023
Accepted: 26 April 2023
Published: 28 April 2023

1. Introduction

Along with the air drawn in from the atmosphere, engines suck in significant amounts of natural pollutants resulting from environmental activities (e.g., volcanic eruptions, forest fires, peat bogs, dust storms, flower pollen, mold particles, plant spores, bacteria, fungi, mites) and artificial (i.e., anthropogenic) pollutants, the source of which is industry and motorization. Industry is mainly a source of dust and toxic gases, while motorization is a source of toxic gases (e.g., CH, CO, NO$_x$), lead compounds, PM (particulate matter), dust from the wear of brake friction linings and clutch discs, and dust from the wear of wheel tires and road surfaces. Due to their mass, dirt particles fall to the ground at different speeds, the value of which depends on the particle's diameter and density. The reciprocal relationship between the drag force of the medium and the gravitational force is also important. The speed of descent increases significantly with increasing grain diameter. Particles smaller than 0.1 µm undergo random Brownian motion, resulting from collisions with gas particles and only slightly from collisions with other dust particles, whose motion, in turn, is mainly driven by moving gas particles. Particles between 0.1

and 1 μm have settling velocities in still air that are small compared to wind speeds. Particles larger than 1 μm have noticeable but small settling velocities. Particles above about 20 μm have high settling velocities and are removed from the air by gravity and other inertial processes. The approximate settling velocities for particles with a density of 1000 kg/m^3 are 0.1 μm—4×10^{-7} m/s, 1 μm—4×10^{-5} m/s, 10 μm—3×10^{-3} m/s, and 100 μm—3×10^{-1} m/s [1]. On the other hand, silica SiO_2 grains (density: 2650 kg/m^3) of 10, 50, and 100 μm fall with velocities of 0.08, 0.19, and 0.7 m/s, respectively. The mixture of many solid contaminants settled on the ground forms road dust, which—as a result of the movement of vehicles or by the wind—is lifted from the ground into the atmosphere, from where it is sucked up by engine intake systems.

The main component of engine inlet air pollution (road dust) is mineral dust, and its main components are silica (SiO_2) and alumina (Al_2O_3). The dust mass share of these two components is in the range of 65–95%, depending on the substrate. The presence of other components in the dust (e.g., Fe_2O_3, MgO, CaO, K_2O, Na_2O, and SO_3) is at the level of 1–5% [2–4].

Silica and alumina have hardness of 7 and 9 on the Mohs scale, respectively, where diamond—as the hardest mineral—has 10. The mineral grains have a very irregular poly-hedral shape with sharp edges. Therefore, they are the main cause of accelerated abrasive wear of two frictionally cooperating engine parts, including the T-P-C (piston–piston rings–cylinder liner) and crankshaft–pan pivot. Excessive wear of the T-P-C association causes an increase in combustion chamber leakage, which is the reason for the "escape" of fresh charge into the crankcase. As a result, there is a decrease in compression pressure and engine power, and an increase in specific fuel consumption and exhaust emissions [5,6].

In order to reduce friction losses and wear on the main components of a piston internal combustion engine, wear-resistant coatings applied to the sliding surfaces of piston rings and cylinder faces are appropriate [7,8]. A significant reduction in friction losses in an internal combustion engine can be achieved by ensuring the continuity of the oil film through appropriate selection of the shape of the sliding surfaces of both the upper and lower sealing rings [9].

Ensuring adequate cleanliness of the inlet air to the internal combustion engines of motor vehicles and work machinery, and thereby minimizing the wear of friction asso-ciations and achieving long life of the assemblies, is still an important operational and design problem, especially when tracked vehicles in a column are driven over sandy terrain. Air dust concentrations then take on values of more than 1 g/m^3. The condition of the ground due to precipitation, the direction of the wind, the conditions in which the vehicles are moving, and the type of running gear cause the dust concentration in the air to reach different values (Table 1).

Table 1. Dust concentrations in the air for different vehicle traffic conditions.

Author	Ambient Conditions	Value (g/m^3)
[10,11]	Moving column of tracked vehicles in dry desert conditions	20
[12]	Dusty environments	0.001–10
[13]	Cars moving on highways	0.0004–0.1
[13]	Moving columns of vehicles on dry, sandy terrain	do 0.03–8
[14]	An all-terrain vehicle moving a distance of several meters behind the column driving on sandy terrain at 30 km/h: A column of tanks; A column of armored personnel carriers; A column of trucks.	 1.17 0.62 0.18

According to the authors of [15], at dust concentrations in the range of 0.05–0.7 g/m^3, visibility is reduced, and when the value of 1.5 g/m^3 is exceeded, visibility becomes zero.

For filtration of the intake air of modern passenger car engines, air filters are used with a filter cartridge, most often made of pleated paper, with a low and limited absorbency (in the range of 150–250 g/m^2), but characterized by high accuracy above d_p = 2–5 μm and high separation efficiency above φ = 99.5%. A mid-range passenger car requires about 200–300 m^3 of air per hour for the fuel–air mixture. At a dust concentration in the vehicle's environment of s = 5 mg/m^3, the engine sucks in less than 400 g of dust along with the air over the course of 20,000 km, requiring 2 m^2 of filter paper area. Trucks, working and agricultural machinery, and special vehicles are equipped with large-displacement and high-power CI engines, which require an air flow of 2000–4000 m^3/h or more. More than 170 kg of dust enters the engine of a special vehicle (V_{ss} = 38.8 dm^3) operated at an average speed of V = 20 km/h on dirt roads (s = 1 g/m^3) along with the air during a 1000 km run. Stopping such a large mass of dust is possible only through two-stage filters, where the first stage of filtration is an inertia filter (multicyclone, i.e., dozens of cyclones arranged in parallel), and the second is a filter paper cartridge (Figure 1). The essence of the operation of the two-stage filter is that inertia filters are able to separate significant masses of dust from large streams of polluted air (with an efficiency of 86–97%), without an increase in pressure drop, but with a low accuracy below 15–35 μm. This is evident from the literature [16–20], as well as from the single-cyclone tests performed by the author [21,22].

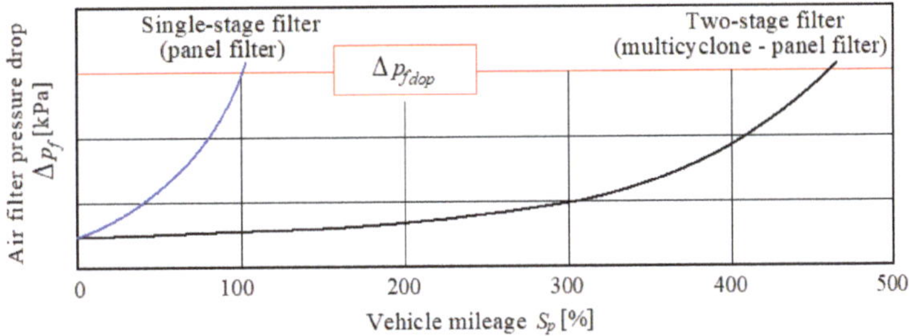

Figure 1. Variation in the pressure drop of a single-stage air filter (paper cartridge) and a two-stage air filter operating in a "multicyclone-paper cartridge" system.

Only a small fraction (about 10%) of the dust mass sucked in with the inlet air reaches the second stage of filtration (paper cartridge). Since the porous baffle has limited absorption capacity, the operating time of the two-stage engine air filtration system to reach the permissible pressure drop of the filter Δp_{fdop} (5–8 kPa) is much longer than that of the baffle filter alone under the same dusty air conditions, increasing the vehicle's service interval (Figure 1).

A Multicyclone is a device built of individual cyclones; the diameter of the cylindrical part of the cyclone D does not exceed 40 mm. Several cyclones are set in parallel (side by side) and connected by common plates, which guarantee a common air inlet and outlet. The dust retained in the cyclones is collected in a dust-settling tank, which is a common element for all cyclones.

Multicyclones can be built with return or through-feed cyclones. Due to their simple and robust design, lack of moving parts, and because of their low pressure drops and small changes in pressure during operation, cyclones have become a device that is popularly used for the pre-filtration of air sucked in by vehicles' internal combustion engines. Cyclones have the advantage of being able to operate under conditions of high dust concentrations in the air and high temperatures. A two-stage (multicyclone–filter cartridge) system for the filtration of air drawn in by an internal combustion engine is shown in Figure 2. The second stage of filtration is a cylindrical filter cartridge with an appropriately sized surface of filter material, arranged in series behind the multicyclone.

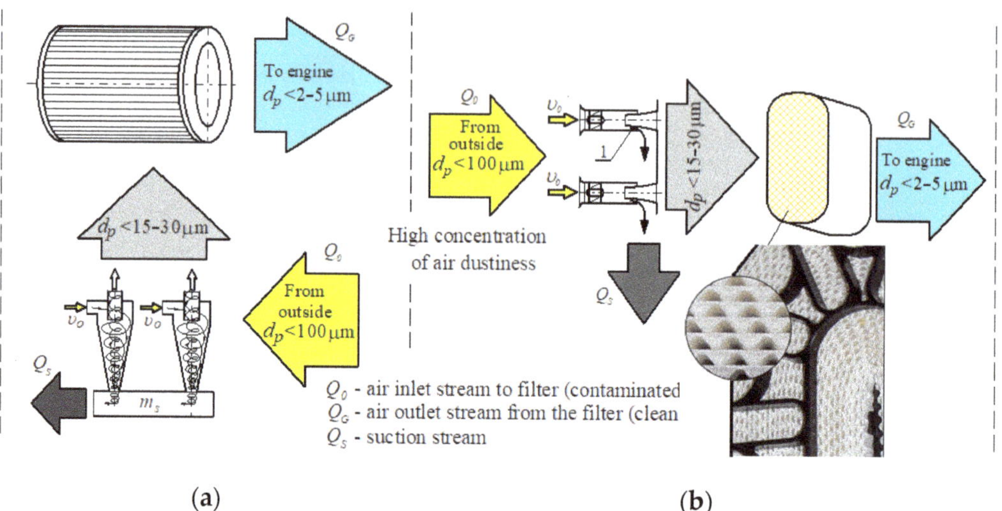

Figure 2. Two-stage (multicyclone–filter cartridge) inlet air filtration system for an off-road vehicle engine: (**a**) tangential inlet return cyclones–paper cylindrical cartridge, (**b**) through-feed cyclones with axial inlet–PowerCore cartridge.

The large values of pressure drop (2–2.5 kPa) that tangential inlet return cyclones generate in the engine intake system have recently caused researchers to focus on the development of axial inlet pass cyclones. The overall structure of a through-feed cyclone differs from that of a tangential inlet return cyclone. The basic element of a through-feed cyclone is a cylindrical hull, inside which a swirler is located in its front part, which usually has four blades (vanes) with a helicoidal (helical) surface. When using a multicyclone with through-feed cyclones, it is more advantageous to use a filter cartridge of the PowerCore type (Figure 2b). Air flow then takes place in the same direction, reducing the pressure drop of the entire system.

The PowerCore filter cartridge has a completely different design from previous cylindrical cartridges. It is a monolith constructed of layers of smooth and pleated paper arranged alternately. The resulting parallel channels have either a blinded inlet or a blinded outlet. A channel with a free inlet has a blinded outlet, and vice versa. The resulting design forces air to flow through the channel wall—which is a filter material—into an adjacent channel that has a free outlet. This avoids excessive turbulence and allows the aerosol to flow directly into the filter outlet, reducing the flow resistance.

With the same flow rate, filter cartridges made with PowerCore technology have several advantages; firstly, they are dimensionally 2–3 times smaller than conventional pleated filter paper cartridges made in the form of a cylinder [23,24], and they achieve efficiency $\varphi_f = 99.85\%$ [25–27]. PowerCore cartridges have a higher dust capacity and, thus, a longer life, which means less frequent replacement and lower operating costs.

The air filtration process in the filter cartridge of a single-stage filter proceeds differently from that on the filter cartridge that is the second filtration stage set in series behind the multicyclone. In the case of a single-stage filter, dust directly from the environment flows into the filter cartridge along with the air. These are large dust grains, but not exceeding the value of $d_p = 100$ μm. In a two-stage air filtration system (multicyclone–filter cartridge), the dust at the exit of the cyclone has a different fractional composition than at the inlet to the cyclone. In the multicyclone, dust grains with sizes above $d_p = 15–35$ μm are retained, and then small dust grains flow onto the second filtration stage (i.e., paper filter). This is evident from the information presented in the literature [28–31].

Retained in the porous bed (filter paper, non-woven fabric), small dust grains form a more compact—and, thus, less permeable to air—structure. This results in a more rapid increase in flow resistance. As a result, the filter obtains the established value of the permissible resistance Δp_{fdop} much earlier and for a smaller mass of retained dust. At the same time, the mass of dust retained per unit area of filtration takes smaller values. Such a phenomenon was observed during numerical studies of fibrous materials [32–35] and experimental studies of the "cyclone-filter paper" assembly [22,36].

Limitations of the filter's operation due to the achievement of the permissible value of pressure drop Δp_{fdop} result in a reduction in its service life. This property is important when analyzing the required mileage of the vehicle/air filter operating time.

This property is characterized by the dust absorption coefficient k_m of the filter paper, which can be defined—assuming a uniform distribution of dust over the entire active surface of the test cartridge filter paper—by the following relationship [37–40]:

$$k_m = \frac{m_w}{A_w} \ \left[\text{g/m}^2\right],\tag{1}$$

where m_w is the total mass of dust retained by the filter medium around the time of the adopted value of the permissible resistance Δp_{fdop}.

Cellulose-based filter materials loaded with standard dust, the grain size of which usually does not exceed 100 μm, are in the range k_m = 220–240 g/m^2 [41–43]. Any other filter material will have a different value of this coefficient. Therefore, for the design of the filter element of a two-stage filter, it is necessary that the operating properties of the filter material are suitable for the operating conditions of the second stage of the air filter, set in series behind the multicyclone.

During the design (i.e., selection of filter paper) of air filters for heavy-duty vehicles' CI engines, the permissible filtration velocity should not exceed the value v_{Fdop} = 0.03–0.06 m/s [42–44].

For the nominal air demand of the engine Q_{Ns}, and assuming the permissible value of the filtration speed from the above range, the active filter area is then determined from the following relation:

$$A_w = \frac{Q_{Ns}}{3600 \cdot v_{Fdop}} \ \left[\text{m}^2\right].\tag{2}$$

The filter operating time can be determined during in-service testing of complete air filters on a vehicle or during laboratory testing. However, such tests are very expensive, labor-intensive, and complicated. In the available literature, a theoretical relationship is given to determine the operation time τ_p of a two-stage air filter [45]:

$$\tau_p = \frac{A_w \cdot k_m \cdot k_c}{Q_{Ns} \cdot s \cdot (1 - \varphi_m) \cdot \varphi_p} h,\tag{3}$$

where A_w is the active surface area of the filter material of the second stage of filtration (m^2), k_m is the dust absorption coefficient of the filter material at Δp_{fdop} (g/m^2), k_c is the coefficient that takes into account the difference between the parameters of the actual pollutants contained in the engine inlet air and the test pollutants during laboratory testing, Q_{Ns} is the nominal air demand of the engine (m^3/h), s is the average concentration of dust in the air sucked from the environment into the filter (g/m^3), φ_m is the separation efficiency of the first stage of air filtration (multicyclone), and φ_p is the separation efficiency of the filter material of which the filter cartridge is made.

For a constant assumed driving speed V_p (km/h), the distance traveled by the vehicle S_p during τ_p is expressed by the following relation:

$$S_p = \tau_p \cdot V_p \ [\text{km}].\tag{4}$$

Relation (5) then determines the distance traveled by the vehicle until the filter reaches an acceptable resistance Δp_{fdop}:

$$S_p = \frac{A_c \cdot k_m \cdot k_c \cdot V_p}{Q_{Ns} \cdot s \cdot (1 - \varphi_m) \cdot \varphi_p} \, [km]. \tag{5}$$

It follows from the presented relationship that its practical use requires knowledge of many data that characterize a particular filter material, including the absorption coefficient k_m, the separation efficiency φ_p, and the coefficient k_c for a specific type of air pollutant. Manufacturers of filter materials only provide data describing their structure with the following parameters: bed thickness, pore dimensions, grammage, mechanical strength, pressure drop (permeability), and density. The efficiency of modern filter materials used in automotive technology is known and takes values in the range of $\varphi_p = 99.5$–99.9%. The coefficient $k_c = 1$ when tested with test dust. It follows that the only unknown parameter in Relation (5) is the dust absorption coefficient k_m.

For pleated cellulose-based filter materials operating in a two-stage filtration system, values of the dust absorption coefficient k_m can be found [34,35,46]. Lacking in the available literature are values of the dust absorption coefficient k_m for filter beds of the PowerCore type, which have a different design than pleated beds. Therefore, in the present work, it was decided to partially fill this gap by conducting experimental studies on PowerCore beds operating in two-stage and single-stage systems. To this end, an original and uncomplicated methodology for determining the dust absorption coefficient k_m of filter materials for the second stage of filtration in a two-stage "multicyclone-partition" filter is presented. The methodology involves testing a set in the form of a single cyclone and a test filter medium set in series behind it, which is an appropriately sized section of the actual filter medium. If during the testing of the "filter set" the operating conditions of the test filter cartridge and the cyclone are maintained as they occur during the operation of a complete air filter of real dimensions, the results obtained can be treated as the characteristics of the actual-size filter cartridge.

2. Materials and Methods

2.1. Materials

The subject of experimental and analytical analyses was the filter material from the original PowerCore (Figure 3a) cartridge, from which test filters with a filter area of Aw = 0.153 m² were made and shaped into cylindrical cores (Figure 3b).

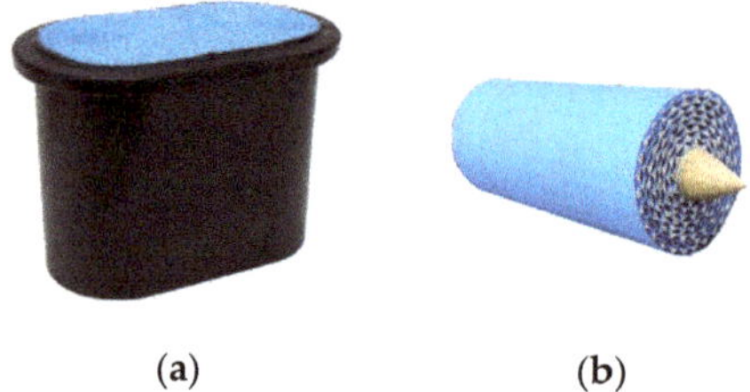

(a)   (b)

Figure 3. (**a**) Fabricated research filter; (**b**) original PowerCore G2 filter cartridge (author's photos).

The purpose of this study was to determine and compare the filtration properties (separation efficiency and accuracy, pressure drop and dust absorption coefficient k_m) of research filters made from a PowerCore filter bed operating in single-stage and two-stage (in series behind the cyclone) filtration systems.

A pass-through cyclone, which is a component of off-road vehicles' intake air filters, was used as the first filtration stage of the "cyclone-infiltrator" filtration set (Figure 4).

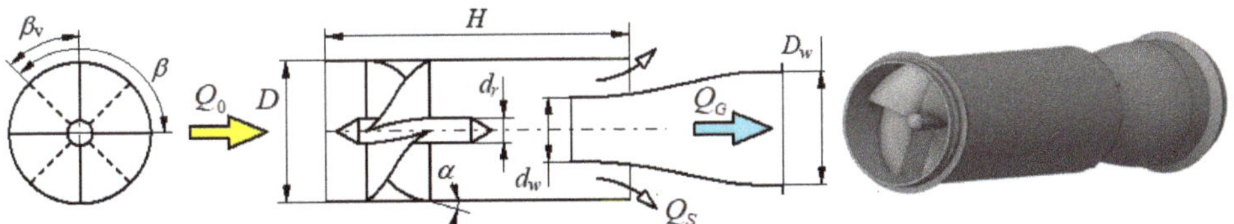

Figure 4. Through-feed cyclone: Q_0—inlet stream, Q_G—outlet stream, Q_S—dust extraction stream, swirl diameter $D = 36$ mm, height of the cylindrical part of the cyclone $H = 100$ mm, inlet diameter of the outlet tube $d_w = 18$ mm, angle of turn of the guide vanes $\beta = 135°$, angle of overshoot of the guide vanes $\beta v = 45°$, outlet diameter of the outlet tube $D_w = 36$ mm, diameter of the core $d_r = 6$ mm, outlet angle of the guide vanes $\alpha = 30°$.

The scope of testing of test filters operating with and without a cyclone–filter system included the determination, at a fixed value of the flux Q_{Gmax}, of the following characteristics involving test dust:

- Separation efficiency $\varphi_w = f(k_m)$;
- Filtration accuracy $d_{pmax} = f(k_m)$;
- Pressure drop $\Delta p_w = f(k_m)$.

where k_m is the dust absorption coefficient of the filter material, defined as the total mass of dust m_z retained on 1 m^2 of the active surface of the filter bed until a fixed value of flow resistance is reached.

2.2. Experimental Method

In this study, an experimental research method was applied using a universal stand that was built to determine the performance of single cyclones, research filters, and the characteristics of research filters that are the second stage of filtration (set in series behind a single cyclone)–Figure 5. A single cyclone and a research filter cartridge set directly behind it form a "filter set".

Immediately upstream and downstream of the test filter was a measuring line, to which a U-tube water-pressure gauge line was connected, which was used to measure the current pressure drop downstream of the test filter and determine the pressure drop Δp_w. The measuring line was terminated with a gauge filter, which was used to collect the mass of m_{AG} dust passed through the test filter. The mass of m_{AG} dust was used to determine the separation efficiency φ_w of the test filter. At the same time, the gauge filter protected the rotameter from dust entering it. The delivery of test dust to the test system was carried out using a vibrating metering device and a stream of compressed air. The dust supplied to the dust chamber was mixed with the inlet air stream Q_0, and then the contaminated air was sucked into the cyclone or test cartridge.

The dust retained by the cyclone fell by the force of gravity into a sealed settling tank, where it was not stored but was removed on an ongoing basis by means of an additional Q_S flux (suction flux) produced by a special pump. The value of the Q_S flux was determined from the following relation for the assumed value of the suction stage $m_0 = 15\%$:

$$Q_S = Q_G \cdot m_0. \tag{6}$$

The air flow rate Q_G was determined according to Relation (7) for the assumed maximum filtration velocity $v_F = 0.06$ m/s in the filter bed, which for the filters of trucks and special vehicles is recommended in the range $v_F = 0.03$–0.06 m/s.

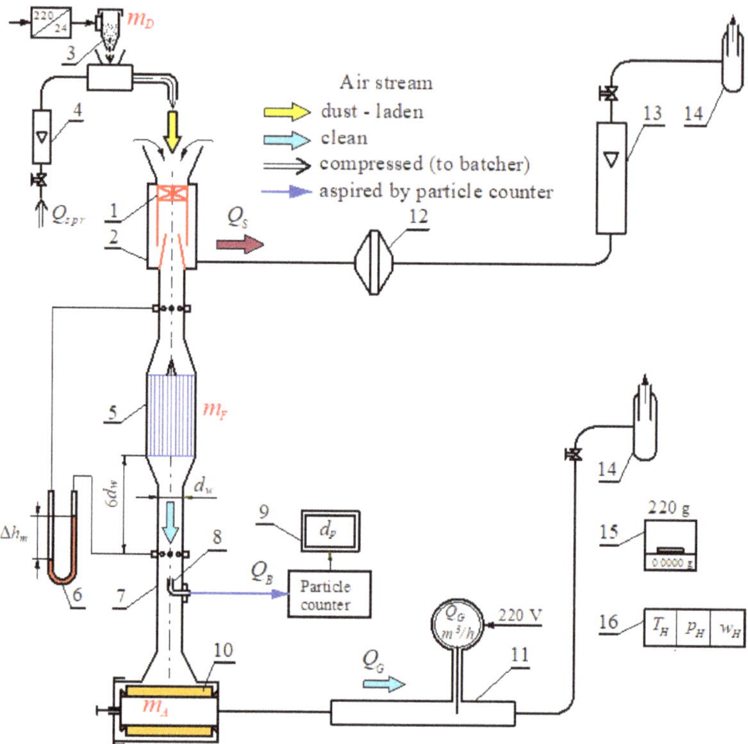

Figure 5. Functional diagram of the test stand: 1—cyclone, 2—dust settler, 3—dust dispenser, 4—rotameter, 5—test cartridge, 6—U-tube water-pressure gauge, 7—measuring tube, 8—dust probe, 9—particle counter, 10—measuring (absolute) filter, 11—air flow meter, 12—suction absolute filter, 13—rotameter, 14—suction fans, 15—analytical balance, 16—instrument for recording air temperature, pressure, and humidity.

For the filtration velocity in the PowerCore filter bed adopted for the test (v_{Fw} = 0.06 m/s), the maximum value of the test flux calculated using the following relationship was Q_{wmax} = 34 m^3/h, while the suction flux for the value of the suction degree m_0 = 15% took the value Q_S = 5.1 m^3/h.

$$Q_{wmax} = A_w \cdot v_{Fw} \cdot 3600 \left[\text{m}^3/\text{h} \right].$$ (7)

At a certain distance behind the test filter, the tip of the dust probe was placed centrally in the axis of the measurement tube, and it was used to draw in the air with dust to the sensor of the particle counter. The counter recorded the number and size of dust grains in the air stream Q_G behind the test filter element in the range of 0.7–100 μm, in i = 32 measurement intervals, which were limited by the adopted diameters (d_{pimin}–d_{pimax}).

The separation efficiency characteristics $\varphi_w = f(k_m)$ of the test filters were determined using the mass method, which consisted of measuring the dust masses retained on the test filter m_{Fz} and on the absolute filter m_A, as well as the dust mass metered m_D onto the filter, after each measurement cycle. The tests were carried out in j successive measurement cycles with a specified duration τ_p (the time of uniform dust dosing into the system). The dust mass was determined with an analytical balance. After each measurement cycle, the mass of dust retained on the tested filter m_{Fz}, on the absolute filter m_A, and dosed to the filter system m_D was determined with an analytical balance. The tests were conducted at an assumed dust concentration s = 1 g/m^3 in the air at the cyclone inlet. When testing test filters without a cyclone, a dust concentration of $s = 0.5$ g/m^3 was used. PTC-D test dust was used, which is a substitute for AC fine test dust in Poland, the chemical and fractional composition of which is shown in Figure 6. The measurement duration, which is the time for uniform dust dosing into the system, was set at $\tau_p = 120$ s during the initial period of filtration (first I) and $\tau_p = 360$ s during the main period (second II) of test filter operation. During the measurement cycle (60 s before the scheduled end of dust dosing), the procedure for measuring the number and size of dust grains in the air behind the research filter was started in the particle counter.

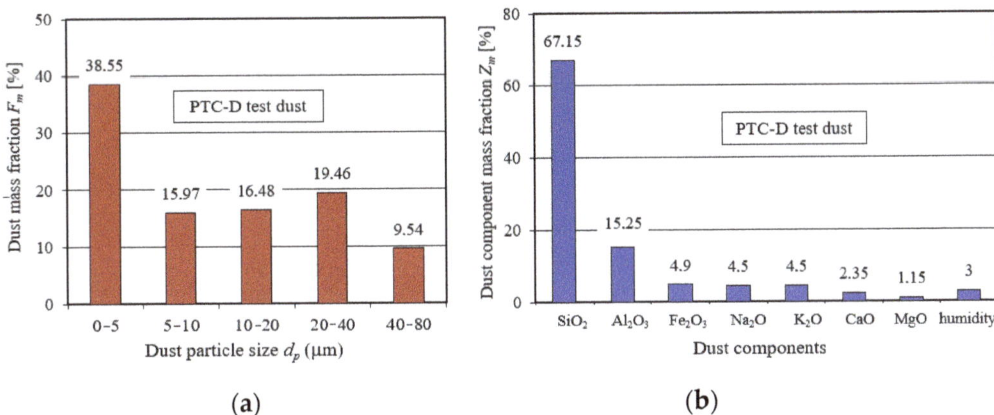

(a)   **(b)**

Figure 6. PTC-D test dust: (**a**) mass proportion of individual fractions in the dust; (**b**) mass proportions of components in the dust.

After each measurement cycle j, we determined the parameters necessary to determine the efficiency, filtration accuracy, pressure drop, and dust absorption coefficient of the filter cartridge.

1. The pressure drop Δp_{fj} of the filter cartridge was determined as the static pressure drop upstream and downstream of the filter based on the measured (i.e., after the dust dosing was completed) height Δh_{mj} on a U-tube water-pressure gauge.

2. The separation efficiency of the test filter was determined by an indirect method, as the quotient of the mass of dust m_{Fzj} retained by the filter cartridge and the mass of dust m_{Fdj} fed to the filter cartridge during the subsequent measurement cycle, based on the following relation:

$$\varphi_j = \frac{m_{Fzj}}{m_{Dj}} = \frac{m_{Fzj}}{m_{Fzj} + m_{Aj}} 100\%. \tag{8}$$

3. The dust absorption coefficient k_{mj} of the tested filter material was determined from the following relationship:

$$k_{mj} = \frac{\sum_{j=1}^{n} m_{Fzj}}{A_w} \left[\text{g/m}^2 \right]. \tag{9}$$

4. The N_{pi} number of dust grains passing through the test filter (located in the air stream downstream of the filter) was determined in measurement intervals bounded by fixed diameters (d_{pimin}–d_{pimax}).

5. The filtration accuracy was determined as the largest dust grain size $d_{pj} = d_{pmax}$ in the air stream downstream of the filter.

6. The percentage of each dust grain fraction in the air stream downstream of the filter for a given measurement cycle was calculated from the following relation:

$$U_{pi} = \frac{N_{pi}}{N_p} = \frac{N_{pi}}{\sum_{i=1}^{32} N_{pi}} 100\%, \tag{10}$$

where $N_p = \sum_{i=1}^{32} N_{pi}$, which is the total number of dust grains that passed through the test filter (from all measurement intervals) during the measurement cycle.

According to the above methodology, we determined the filtration characteristics (separation efficiency $\varphi_w = f(k_m)$ and filtration accuracy $d_{pmax} = f(k_m)$, as well as pressure drop $\Delta p_w = f(k_m)$) of research filters made from a PowerCore bed working in one-stage and two-stage systems. The tests were carried out until the filter medium reached the assumed value of acceptable resistance $\Delta p_{fdop} = 3$ kPa.

2.3. Analytical Method

An analytical method was used to explain the phenomenon of changes in the flow resistance of the filter bed depending on grain size, which involved analyzing the arrangement of dust grains in a compact structure. Mineral dust drawn in with air has an irregular grain shape and is characterized by high polydispersity in the range up to 100 μm. Therefore, a proxy diameter is adopted for theoretical considerations. Most often, the diameter of dust grains is assumed to be spherical. Figure 7 shows a slice of the deposit model in the form of a cube with side d formed by regularly arranged spheres of equal diameter d.

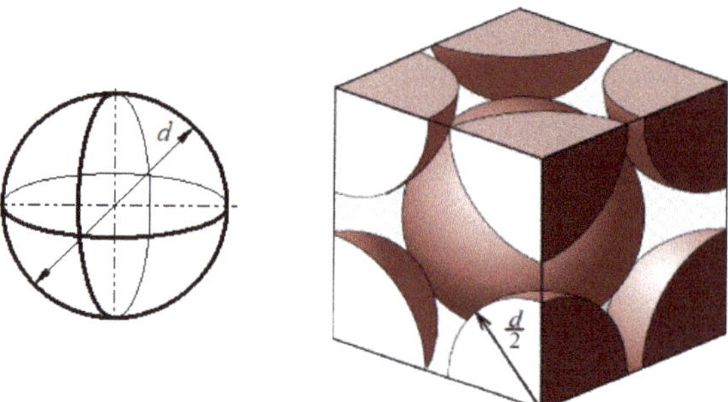

Figure 7. Bed model of similar spherical particles.

The volume V_1 of the space between the particles is equal to the volume V of the cube with side d minus the volume V_2 occupied by the particles, according to the following equation:

$$V_1 = V - V_2 \tag{11}$$

$$V_1 = d^3 - \left(\frac{4}{3} \pi \left(\frac{d}{2} \right)^3 \right) \tag{12}$$

$$V_1 = d^3 \left(1 \frac{\pi}{6} \right). \tag{13}$$

From the above equation, it follows that the volume V_1 of the space between the particles depends only on the diameter d of the particles in the third power. The volumes V_1 between successive particles form spaces that, when connected together, form a channel d_ε for air flow.

At the same time, as the diameter d of the particles decreases, the number of microchannels d_ε distributed over the filter area also increases, in proportion to $1/d^2$, since the total dimensions of the filter area do not change. Thus, the total cross-section through which the air flows also does not change, from which it follows that the average air velocity in the microchannels does not change and is independent of d.

The pressure drop Δp of a filter bed laden with dust particles can be described by a modified Darcy–Weisbach formula:

$$\Delta p = \lambda (Re) \frac{g_w}{d_\varepsilon} \frac{\rho}{2} v_\varepsilon^2, \tag{14}$$

where ρ is the density of the fluid, λ is the coefficient of friction, g_w is the thickness of the filter bed, d_ε is the hydraulic diameter of the channels formed by the particles, and v_ε is the actual velocity in the channels between the particles.

It follows from the above relationship that the flow resistance of the filter bed is directly proportional to its thickness g_w, fluid density ρ, and actual velocity v_ε in the channels between the particles, and inversely proportional to the conventional diameter d_ε of the channels formed by the particles. Assuming an invariant value of the thickness of the filter bed g_w and the velocity in the channels between the particles d_ε, the flow resistance then depends on the conventional diameter d_ε of the channels formed by the particles.

3. Results and Discussion

The characteristics of separation efficiency $\varphi_w = f(k_m)$, pressure drop $\Delta p_w = f(k_m)$, and filtration accuracy $d_{pmax} = f(k_m)$ as a function of the dust absorption coefficient k_m of two specimens of PowerCore bed test filters (No. 1 and No. 2), which operated in the "filter set" as the second filtration stage (after the through-feed cyclone), are shown in Figure 7. The obtained characteristics had a similar course and values. Due to the achieved values of separation efficiency, the operating time of the studied filters can be conventionally divided into two periods: The first period (i.e., initial filtration) is characterized by small values of separation efficiency, which systematically increase (sometimes very sharply) with the amount of dust mass retained by the filter bed and, thus, the increase in the dust absorption coefficient k_m. The initial filtration period lasts from the start of the filtration process until the filter material reaches the maximum set value of separation efficiency. In the case of the conducted research, the zone of separation of the two periods was assumed at the moment when the filters reached the separation efficiency $\varphi = 99.5\%$. This value was achieved by the filters at similar values of the dust absorption coefficient i.e., filter No. 1: $k_{m1} = 8.87$ g/m^2; filter No. 2: $k_{m2} = 12$ g/m^2. After the first measurement cycle, the separation efficiency of both filters reached similar values: research filter No. 1 reached $\varphi_{wc1} = 96.6\%$, and filter No. 2 reached $\varphi_{wc2} = 95.8\%$ (Figure 8), showing the homogeneity of the filter bed and the high accuracy of the research filters. During the tests, the cyclone efficiency was at 85.2%. During the tests, the cyclone's efficiency was 85.2%, and its pressure drop did not exceed $\Delta p_c = 0.6$ kPa.

Following the initial filtration period (I), the second period of operation of the PowerCore filter—called the main period (II)—was characterized by a course of efficiency ($\varphi_w = 99.7$–99.99%) and filtration accuracy ($d_{pmax} = 3$–6 µm) stabilized at a certain level, and a slow increase in the pressure drop. When the assumed flow resistance $\Delta p_w = 3$ kPa was reached, the filters obtained the dust absorption coefficient km at similar levels, i.e., filter No. 1: $k_m = 219$ g/m^2, and filter No. 2: $k_m = 199$ g/m^2. The data from the literature show that standard pleated cellulose-based filter materials operating in a two-stage system achieve an absorption coefficient of $k_m = 90$ g/m^2, which is half that of PowerCore filter beds operating under the same conditions, which may be due to the possibility of accumulation of additional dust mass inside the channels of the PowerCore bed, as shown in Figure 9. Once the filter bed (walls of the channel) has been saturated with dust, as the operation of the filter continues, the dust accumulates inside the channel, which acts as a kind of reservoir. Figure 9b shows the dust accumulated inside the channels of the PowerCore filter under test. Thus, the absorption capacity of the filter bed is enlarged. The dust settles at the bottom of the channel, and its level rises. Thus, the free filter area decreases, causing an increase in the flow velocity through the channel walls, which results in an increase in the filter's flow resistance (Figure 8).

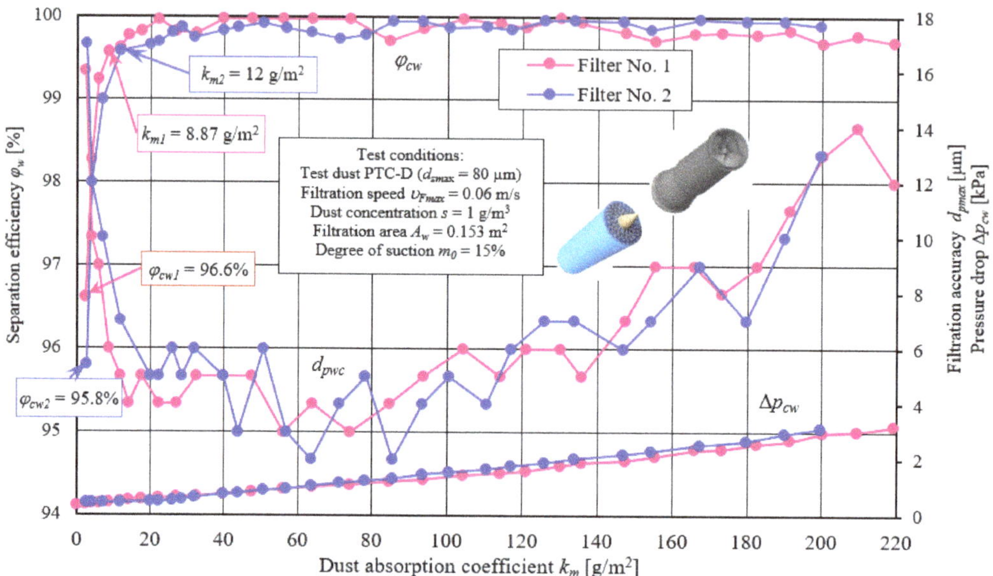

Figure 8. Characteristics: separation efficiency $\varphi_w = f(k_m)$, pressure drop $\Delta p_w = f(k_m)$, and filtration accuracy $d_{pmax} = f(k_m)$ as a function of the dust absorption coefficient k_m of filter beds working in the "filter set" as the second filtration stage (after the through-feed cyclone).

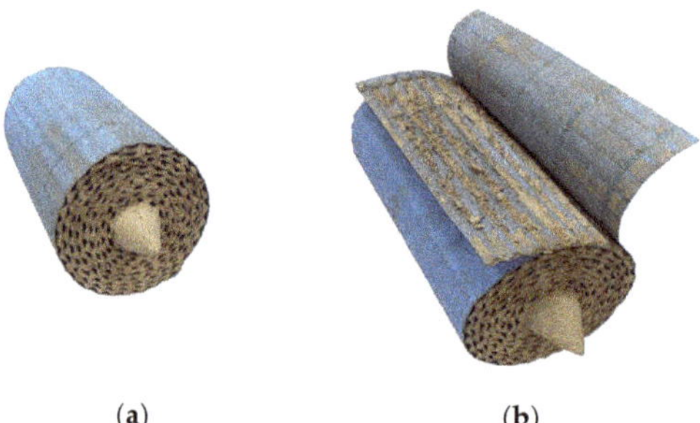

(a) (b)

Figure 9. PowerCore filter after testing with test dust: (**a**) view of the filter from the side of the air and dust inlet; (**b**) dust accumulated in the bed channels.

Associated with changes in efficiency were the maximum sizes of the dust grains d_{pmax} in the air behind the filter under test. There was a sharp decline, from $d_{pmax} = 17$ μm after the first measurement to $d_{pmax} = 4$–5 μm when the filter reached an efficiency of 99.5%, establishing the end of the first stage of filtration. In the second stage of the filter's operation, the separation efficiency was maintained at 99.5–99.9%, and the maximum grain size took on values in the range of $d_{pmax} = 3$–6 μm. After the filters achieved a dust absorption coefficient of $k_m = 150$ g/m² and a pressure drop of $\Delta p_{wc} = 2.5$ kPa, increasingly larger dust grains dpmax and a slight decrease in separation efficiency were registered in the exhaust air. As soon as the filters achieved a pressure drop of $\Delta p_{cw} = 3$ kPa, dust grains of $d_{pmax} = 14$ μm were found in the exhaust air.

For the process of aerosol filtration in fiber baffles, the phenomenon of continuous influx and retention of dust grains on the surface of individual fibers is characteristic. This is mainly the effect of inertial, direct trapping, and diffusion mechanisms. Settling dust grains form characteristic dendrites, which grow to a considerable size, filling the free spaces (i.e., pores) between fibers. Thus, the free field of air flow decreases, which causes an increase in velocity between fibers. As a result, there is an increase in the pressure drop, which is a function of velocity in the second power. The increase in the mass of dust accumulated in the filter bed causes the pressure drop to increase (Figure 8). Due to the small thickness of the filter papers (g_p = 0.4–0.8 mm), they have a limited dust absorption capacity. When this is exhausted, the dust settles on the surface of the bed in the form of a "filter cake", which manifests itself in an additional increase in flow resistance. The large pressure drop downstream of the filter and the high velocity of air flow in the narrow channels between the fibers can cause dust grains to be locally detached from the formed dendrites and transported toward the outlet. These can be dust grains of considerable size. In the case of the studied filters, the beginning of this process was noticed when the value of the pressure drop behind the filter reached about Δp_f = 2.5 kPa.

Figure 10 shows a comparative analysis of test filter No. 1 operating in a two-stage system (after the cyclone), and of test filter No. 3 with the same parameters but operating in a single-stage system. The characteristics of the efficiency and accuracy of filtration of both filters in the initial period were similar with respect to the course and values. It can be seen that there was a more intense increase in the pressure drop of the filter operating in a two-stage system (downstream of the cyclone). The value Δp_f = 3 kPa was reached by this filter after loading the bed with dust k_m = 219.2 g/m^2. For the research filter operating in a single-stage system, the value of Δp_f = 3 kPa was reached after obtaining a dust load of k_m = 434.2 g/m^2, which is twice as much. This is because smaller dust particles more quickly form dendrites on the fibers, which fill the free spaces (i.e., pores) in the fiber bed, and the flow of air through the layer of dense dust grains is hindered. Hence, the greater pressure drop, which grows more intensively and limits the filter's operating time and the car's mileage.

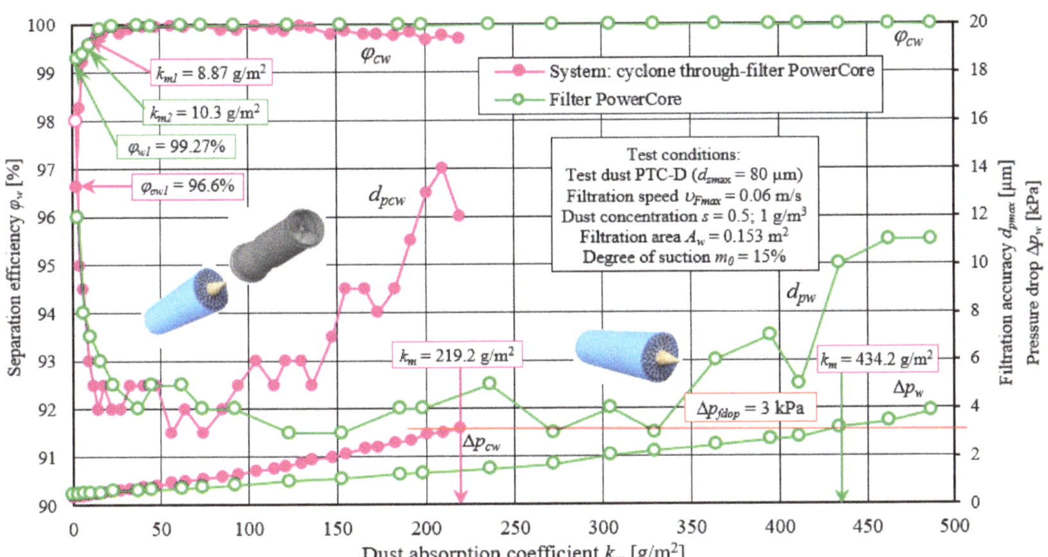

Figure 10. Characteristics of efficiency $\varphi_w = f(k_m)$ and accuracy $d_{pmax} = f(k_m)$ of filtration and flow resistance $\Delta p_w = f(k_m)$, depending on the k_m factor of the PowerCore filter bed tested in the "cyclone-test cartridge" filter set, and without cyclone.

Figure 11 shows the predicted mileage (for different values of dust concentration in the air) of a diesel truck equipped with filters with a single-stage filtration system (PowerCore filter) and with a two-stage system (cyclone–PowerCore filter). Calculations were carried out using Relation (2), along with the dust absorption coefficient k_m of the PowerCore filter and the separation efficiency values of the cyclones φ_m and the PowerCore filter φ_w, determined experimentally during the tests. Regardless of the air filtration system present in the vehicle, as the dust concentration in the air increased, the mileage of the vehicle to reach the permissible resistance decreased, with higher mileage values obtained by the vehicle with a two-stage filtration system.

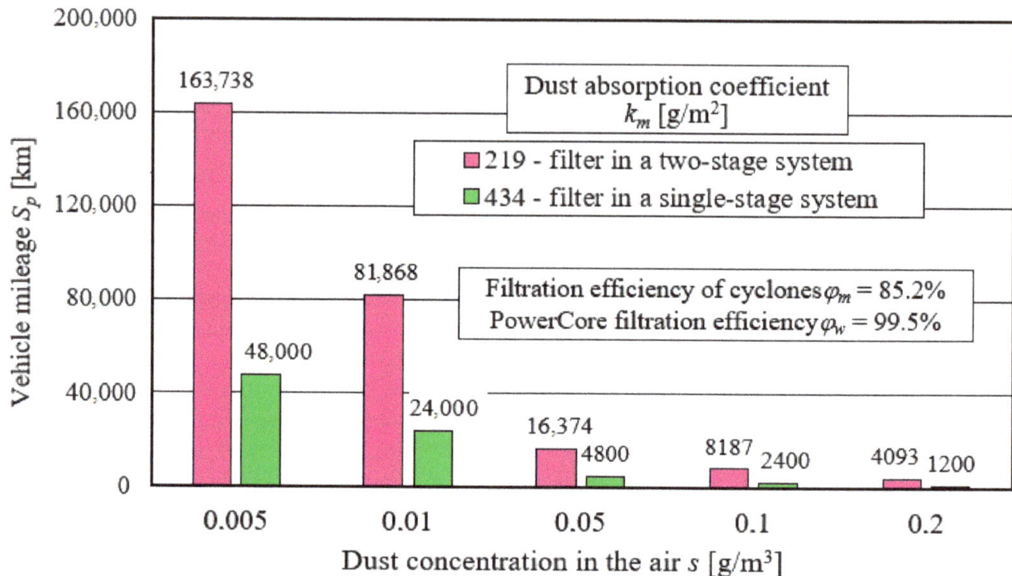

Figure 11. Projected mileage of the car to reach Δp_{fdop} = 3 kPa under different operating conditions (dust concentration in the air) with an air filter operating in single-stage and two-stage filtration systems.

A vehicle equipped with a PowerCore single-stage filtration system, for a dust concentration of s = 5 mg/m^3 and a dust absorption coefficient of k_m = 434 g/m^2, achieves a projected mileage of about 48,000 km. The same vehicle, equipped with a two-stage "PowerCore multicyclone-filter" filtration system, for a dust concentration of s = 5 mg/m^3 and a much lower ratio of k_m = 219 g/m^2, achieves three times the mileage (more than 160,000 km) when the filter reaches the permissible resistance value Δp_{fdop} = 3 kPa.

To analyze the mileage of vehicles with two-stage and single-stage systems, the pressure drop (3 kPa) on the filter cartridge was assumed to be the same. In the two-stage system, there is an additional pressure drop across the cyclones. In the case of the tested two-stage system, the pressure drop on the through-feed cyclone reached Δp_c = 0.6 kPa at a nominal air flow rate of Q_{wmax} = 34 m^3/h. In order to maintain the same operating conditions of filter materials in both systems (single-stage and two-stage), the operation of the filter operating in the two-stage system should be limited by the permissible resistance of Δp_{fdop} = 3.6 kPa. Obviously, a higher pressure drop on the filter means additional energy losses from the engine. From the available literature, it is known that the values of the permissible resistance of air filters of trucks and special vehicles are in the range Δp_{fdop} = 6–8 kPa.

Projecting the mileage of a vehicle equipped with a filter with a two-stage filtration system (cyclones–baffle), and using a dust absorption coefficient with the same value as for a single-stage filter (k_m = 434 g/m^2), results in an artificial extension of this mileage. For a dust concentration of s = 0.005 g/m^3, the vehicle mileage is extended to 324,328 km, i.e., by 100%. Such calculations are inappropriate and misleading. Therefore, it is expedient to conduct research to determine the actual dust absorption coefficient k_m of the filter material proposed for operation in a two-stage system, which can then be used to determine the real mileage of the vehicle.

The number of dust grains found in the cyclone's inlet air stream, and before and after the test filter operating in the "cyclone-inlet" set, in successive measurement cycles for different values of the dust absorption coefficient k_m is shown in Figure 12. It was found that with the increase in separation efficiency in each successive measurement cycle, there was a decrease in the total number N_p of dust grains in the air after the test filter. A clear relationship can be seen between the size of the grains d_p and their number N_p. As the size of the dust grains increases, their number decreases until they disappear completely. Located in the last size interval, grains (or a single grain) have the largest size $d_p = d_{pmax}$ (Figure 12). It was assumed that the dust grain with the largest size d_{pmax} located in the exhaust air stream from the test filter expresses the filtration accuracy of the filter material in measurement cycle j. In the exhaust air stream from the cyclone, there were dust grains whose number had decreased significantly compared to the number of grains in the air before the cyclone. This applies to dust grains above 3 µm. The larger the dust grains, the more effectively they are retained by the cyclone. If we denote cyclone efficiency as the quotient of the number of dust grains retained (N_{pu}–N_{pw}) and delivered to the N_{pu}, system, dust grains of 5 µm were retained by the cyclone with 53.6% efficiency, 10 µm grains achieved 86.8% efficiency, and 20 µm grains had 97.2% efficiency. The cyclone's increasing separation efficiency was due to the increasing inertial force of grains with increasing volume and, thus, mass. Grains larger than 35 µm were retained by the cyclone with 100% efficiency. Proceeding in a similar manner, it was determined that after the first measurement cycle, dust grains with a size of 10 µm were retained by the test filter material with 99.5% efficiency.

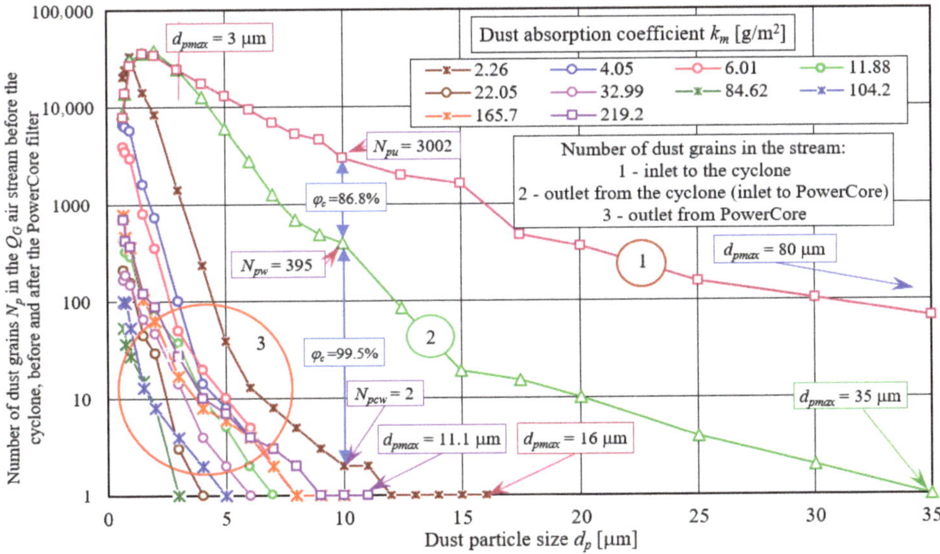

Figure 12. The number of dust grains present in the cyclone's inlet air stream, upstream and downstream of the PowerCore test filter operating in the "cyclone-test filter" set, in successive measurement cycles for different values of the dust absorption coefficient k_m.

The granulometric composition of dust in the air stream shows the change in the numerical shares of U_p determined by Relation (5) as the size of the dust grains d_p increased. The granulometric composition of dust in the air stream flowing into and out of the cyclone was similar in terms of the course, but there were significant differences as to the values (Figure 13). As the size of the dust grains d_p increased, there was a rather sharp increase in their U_p numerical shares, reaching a maximum at $d_p = 1.5$ µm, followed by a gentle decrease. The U_{pc} number shares of dust grains in the air behind the cyclone were found to be larger in the range of 0.7–3.8 μm and smaller above 3.8 μm than the U_{pc} number shares of dust in the air entering the cyclone. As a result of the filtration process in the cyclone, larger grains (above 3.8 μm) were separated, resulting in a decrease in their shares of the total number of dust grains and an increase in the shares of dust grains with sizes below 3.8 μm, with this value being a conventional limit for the purposes of this research. For example, the share of dust grains with a size of 2 μm in the air behind the cyclone increased from $U_{p0} = 16.5\%$ to $U_{pc} = 20.7\%$, while the share of dust grains with a size of 6 μm in the air behind the cyclone decreased from $U_{pc} = 4.56\%$ to $U_{p0} = 1.59\%$. Another filtration stage (PowerCore filter) significantly changed the granulometric composition of the dust. In the air downstream of the filter there were mainly smaller dust grains—less than 3.8 mm; hence, their shares were high. The proportion of dust grains with a size of 1 μm was $U_{p1} = 32.6\%$ after the first test cycle and $U_{p5} = 37.4\%$ after cycle No. 5.

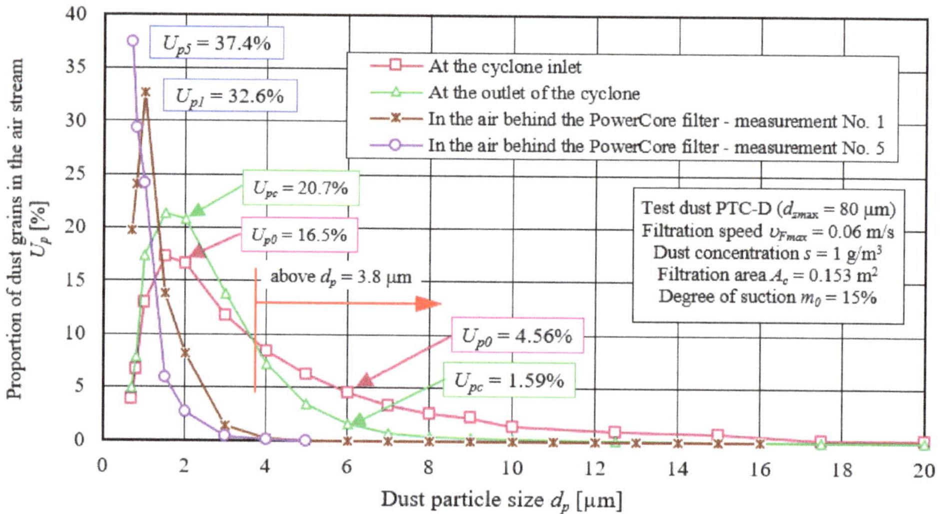

Figure 13. Granulometric composition of dust in the air stream before and after the cyclone and the PowerCore filter.

The low efficiency and accuracy of air filtration by fiber filters is characteristic of their initial stage of operation. This phenomenon occurs in the engine intake system after replacing a contaminated filter element with a new one. This is an unfavorable phenomenon, as the engine's inlet air stream may then contain dust grains of considerable size, which can cause accelerated wear of the internal combustion engine's components. According to the authors of [9–11,47], dust grains with diameters of more than 1 μm are the cause of surface wear, and dust grains with diameters of more than 5 μm are the cause of accelerated wear of two frictionally cooperating engine components. Consequently, such filtration accuracy is required of intake air filters for internal combustion engines. From the above, it follows that the engine's air filter cartridge should not be replaced too often, except for when there is a loss of its filtering properties. The criterion for replacing the air filter cartridge is the achievement of the manufacturer's set value of acceptable resistance. Therefore, the test results obtained by the author are essential for making the right decision as to when to service the air filter. At the final stage of operation of the tested filters (above $\Delta p_f = 2.5$ kPa), dust grains were noted, the size of which ($d_{pcw} = 14$ μm) could cause accelerated wear. From the above analysis, it follows that the air filters should not be used above the specified pressure drop value. Based on the test results obtained, it is reasonable to believe that in the case of the PowerCore filters tested, this value is above $\Delta p_f = 2.5$ kPa.

Although the dust absorption capacity of the filter material operating in a two-stage "cyclone-filter test" filtration system is lower than that of the same material operating in a single-stage system, the operating time of the two-stage system until the permissible resistance $\Delta p_f = 3$ kPa is reached is longer. This was confirmed by the results of the characteristics separation efficiency $\varphi_w = f(m_D)$, pressure drop $\Delta p_w = f(m_D)$, and filtration accuracy $d_{pmax} = f(m_D)$ as a function of the mass of dust m_D delivered to the "cyclone-test filter" filtration system and directly to the PowerCore test filter (Figure 14).

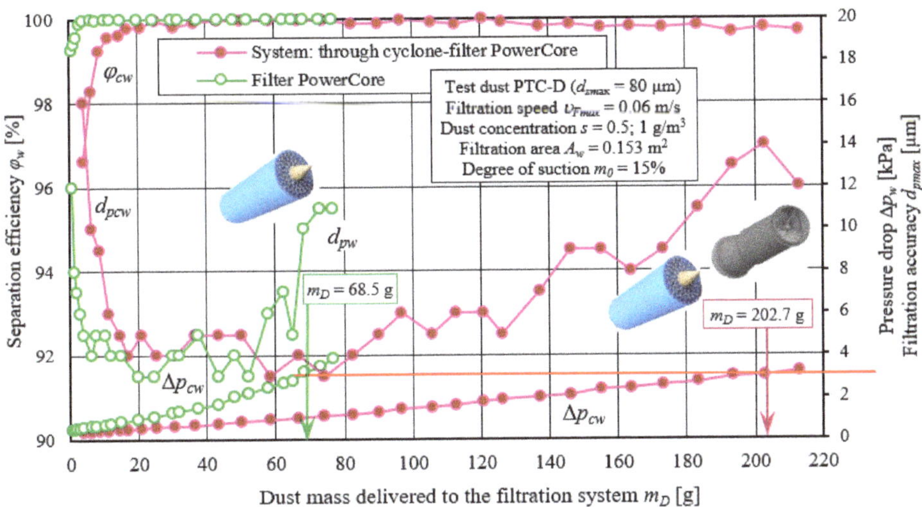

Figure 14. Characteristics: separation efficiency $\varphi_w = f(m_D)$, pressure drop $\Delta p_w = f(m_D)$, and filtration accuracy $d_{pmax} = f(m_D)$ as a function of the mass of dust m_D delivered to the two-stage system "cyclone-filter research PowerCore" and the single-stage system (directly to the research filter PowerCore).

From the presented characteristics, it can be seen that the achievement of the permissible resistance of 3 kPa by the test filter operating in the "cyclone-test filter" system occurs after 202.7 g of dust is supplied to the system. If the test filter operates individually, it achieves a pressure drop of 3 kPa after delivering only 68.5 g of dust. This has a definite impact on the vehicle's mileage, especially when it is operated under conditions of high dust concentration in the air. The filter's achievement of the permissible resistance is a prerequisite for its service, i.e., filter element replacement. Although the filter material working in series behind the multicyclone has a lower absorbency, the operating time of the "multicyclone-pore-partition" filter system to reach the permissible resistance is much longer than that of the same cartridge working individually, which is the primary advantage of two-stage filtration. This explains the need for two-stage filters in vehicles operated under conditions of high dust concentration in the air.

If the first stage of filtration of the inlet air to the engine is an inert filter (multicyclone) and the second is a filter cartridge made of fibrous material, then most (i.e., more than 90%) of the dust supplied to the system is retained by the multicyclone. In the case of the tested "cyclone-filter PowerCore" set, the cyclone efficiency was at the level of about 84.6%. Thus, only 15% of the dust mass that was introduced into the "multicyclone-filter" filtration system reached the filter cartridge. The retained dust accumulates in the settling tank or is removed from it on an ongoing basis. During the tests, ejective (ongoing) removal of dust from the cyclone settling tank was used, with a suction rate of $m_0 = 15\%$. Therefore, with a constant flow of air to the engine, the flow resistance of the multicyclone is unchanged—in contrast to baffle filters, where the retained mass causes an increase in flow resistance at the same time.

The increase in flow resistance of a cartridge operating in a two-stage system (cyclone-test filter) is more intense, despite the same mass of dust being retained by the bed per unit area (Figure 10). In the case of two-stage filtration, small dust particles accumulate on the filter bed, which form dendrites on the fibers that grow faster than when the bed is loaded with larger particles. Small dust grains, which make up a much larger percentage of the total number of particles entering the filter when the bed is operated with an upstream cyclone, penetrate the filter paper structure much more easily and fill it more tightly compared to larger-diameter grains. Free spaces are created between the charged dust particles, through which air flows. However, the spaces between small grains are much smaller than those between large dust grains, increasing the velocity of aerosol flow through them and, thus, increasing the flow resistance.

The volume V_1 of space between particles of equivalent diameter d = 80 μm, according to Relation (13), takes the following value:

$$V_1 = d^3 \left(1\frac{\pi}{6}\right)$$

$$V_{1(80)} = 80^3 \left(1\frac{\pi}{6}\right)$$

$$V_{1(80)} = 267{,}946.7 \text{ μm}^3.$$

For smaller and smaller particle diameters d, the volume V_1 of space between particles takes on smaller and smaller values. The volume V_1 of space between particles with an equivalent diameter of d = 20 μm, which is four times smaller, has a value of $V_{1(20)}$ = 4186.7 μm^3, which is 64 times smaller than for d = 80 μm. For d = 8 μm, $V_{1(8)}$ is 1000 times smaller, and for d = 2 μm, $V_{1(2)}$ is 64,000 times smaller.

From the Darcy–Weisbach relation (Relation (14)) describing the pressure drop Δp of a filter bed loaded with dust particles, it follows that the flow resistance is directly proportional to g_w, ρ, and velocity v_ε, and inversely proportional to the conventional diameter d of the channels formed by the particles. For a filter bed loaded with dust particles of small diameters, the conventional diameter d_ε of the channels formed by the particles will take smaller values, resulting in an increase in flow resistance. This is the main reason for the increase in pressure drop with decreasing particle size.

In addition, particles with smaller and smaller diameters have a higher and higher ratio of surface area A_c to volume V_c. Consequently, they have a higher resistance and pressure drop per unit volume (i.e., mass of particles). The ratio of the surface area A_c of a particle with the shape of a sphere to its volume V_c. is expressed by the following relation:

$$a_g = \frac{A_c}{V_c} = \frac{6}{d_p}. \tag{15}$$

If we take particles with different equivalent diameters for analysis (i.e., d_p = 80, 50, 20, 10, 2.0, 1.0, and 0.7 μm), the ratio of particle area to volume is then a_g = 0.075, 0.12, 0.3, 0.6, 3, 6, and 8.57, respectively.

4. Conclusions

In this paper, an experimental study of PowerCore deposits operating in two-stage and single-stage systems was carried out to determine the characteristics of separation efficiency, accuracy, and pressure drop depending on the dust absorption coefficient. To this end, an original and uncomplicated methodology for determining the dust absorption coefficient of k_m filter materials for the second stage of filtration in a two-stage "multicyclone-partition" filter is presented. The methodology consists of testing a set in the form of a single-pass cyclone and a PowerCore test filter medium set in series behind it, which is an appropriately sized section of the actual filter medium. The following conclusions were obtained from the testing and analysis:

1. PowerCore filter beds operating in a "cyclone-pore barrier" system achieve an absorption coefficient of k_m = 219.2 g/m^2, which is twice as low as when operating in a single-stage filtration system—undoubtedly influenced by the small dust grains resulting from the cyclone filtration process.

2. PowerCore filter beds operating in a two-stage system achieve more than twice the absorption coefficient of standard pleated cellulose-based filter materials, whose k_m = 90 g/m^2, which may be due to the possibility of dust accumulating inside the channels of the PowerCore bed.

3. Filter materials operating in the "multicyclone-porous baffle" system are characterized by a more intensive increase in flow resistance, which has a direct impact on reducing the operating time of the air filter to reach the permissible resistance, thereby reducing the mileage of the vehicle. This is the result of the formation of a more compact and tight structure from small-diameter dust grains, which have a higher ratio of surface area A_c to volume V_c than larger-diameter grains and, therefore, have a higher resistance and pressure drop per unit volume (mass of dust particles).

4. Forecasting the mileage of a vehicle equipped with a filter with a two-stage filtration system (cyclones-baffle), and using a dust absorption coefficient with the same value as for a single-stage filter (k_m = 434 g/m^2), results in an artificial extension of mileage by more than 100%, which is inappropriate and misleading. Therefore, it is expedient to conduct research to determine the actual dust absorption coefficient k_m of the filter material proposed for two-stage operation, which can then be used to determine the real mileage of the vehicle.

5. The low (φ = 95.8–96.6%) efficiency of the filter material and the presence of large (d_{pmax} = 12–16 μm) dust grains in the purified air during the initial but short period of operation of the baffle filter may affect the accelerated wear of the engine associations working together in a tray—especially the piston–piston ring–cylinder (P-PR-C) association. Dust grains of similar size can also be located behind the air filter when it exceeds a certain pressure drop value. In the case of the tested bed, this value is about 2.5 kPa. Therefore, the use of the air filter after exceeding the specified value of the permissible resistance is not advisable.

6. Although the filter material working in series behind the multicyclone has a lower level of absorption, the operating time of the two-stage "multicyclone-fiber baffle" system is much longer to reach the permissible resistance than for the same fiber material working in a single-stage system. This is due to the essence of the operation of the two-stage "inertia filter-baffle filter" filtration system. This explains the need for two-stage filters in vehicles operating under conditions of high dust concentrations in the air.

7. The originality of the developed test methodology lies in the fact that dust is applied to the filter material working behind the cyclone, and the granulometric composition of this dust is changed as a result of the actual air filtration process in the cyclone, characterized by the cyclone inlet speed, dust concentration, and suction rate. The methodology allows for the experimental determination of the characteristics of any fibrous filter materials predicted for the second stage of air filtration. These can be materials with any structural parameters, and the tests can be carried out over a wide range of changes in filtration conditions that correspond to the operation of an air filter under conditions of high dust concentration in the air.

Funding: This research received no external funding.

Data Availability Statement: Data is contained within the article.

References

1. Su, W.H. Dust and atmospheric aerosol. *Resour. Conserv. Recycl.* **1996**, *16*, 1–14. [CrossRef]
2. Bojdo, N.; Filippone, A. Effect of desert particulate composition on helicopter engine degradation rate. In Proceedings of the 40th European Rotorcraft Forum, Southampton, UK, 2–5 September 2014.
3. Smialek, J.L.; Archer, F.A.; Garlick, R.G. Turbine airfoil degradation in the persian gulf war. *JOM* **1994**, *46*, 39–41. [CrossRef]
4. Haig, C.; Hursthouse, A.; McIlwain, S.; Sykes, D. The effect of particle agglomeration and attrition on the separation efficiency of a Stairmand cyclone. *Powder Technol.* **2014**, *258*, 110–124. [CrossRef]
5. Dziubak, T.; Dziubak, S.D. A Study on the Effect of Inlet Air Pollution on the Engine Component Wear and Operation. *Energies* **2022**, *15*, 1182. [CrossRef]
6. Dziubak, T.; Karczewski, M. Experimental Studies of the Effect of Air Filter Pressure Drop on the Composition and Emission Changes of a Compression Ignition Internal Combustion Engine. *Energies* **2022**, *15*, 4815. [CrossRef]
7. Wróblewski, P. Reduction of friction energy in a piston combustion engine for hydrophobic and hydrophilic multilayer nanocoatings surrounded by soot. *Energy* **2023**, *271*, 126974. [CrossRef]
8. Wróblewski, P.; Rogólski, R. Experimental Analysis of the Influence of the Application of TiN, TiAlN, CrN and DLC1 Coatings on the Friction Losses in an Aviation Internal Combustion Engine Intended for the Propulsion of Ultralight Aircraft. *Materials* **2021**, *14*, 6839. [CrossRef] [PubMed]
9. Wróblewski, P.; Koszalka, G. An Experimental Study on Frictional Losses of Coated Piston Rings with Symmetric and Asymmetric Geometry. *SAE Int. J. Engines* **2021**, *14*, 853–866. [CrossRef]
10. Barris, M.A. *Total Filtration TM: The Influence of Filter Selection on Engine Wear, Emissions, and Performance*; SAE Technical Paper 952557; SAE: Warrendale, PA, USA, 1995.
11. Schaeffer, J.W.; Olson, L.M. Air Filtration Media for Transportation Applications. *Filtr. Sep.* **1998**, *35*, 124–129. [CrossRef]
12. Jaroszczyk, T.; Pardue, B.A.; Heckel, S.P.; Kallsen, K.J. Engine air cleaner filtration performance—Theoretical and experimental background of testing. In Proceedings of the AFS Fourteenth Annual Technical Conference and Exposition, Tampa, FL, USA, 1 May 2001. Included in the Conference Proceedings (Session 16).
13. Barbolini, M.; Di Pauli, F.; Traina, M. Simulation der luftfiltration zur auslegung von filterelementen. *MTZ* **2014**, *75*, 52–57. [CrossRef]
14. Dziubak, T. Zapylenie powietrza wokół pojazdu terenowego. *Wojsk. Przegląd Tech.* **1990**, *3*, 154–157. (In Polish)
15. Szczepankowski, A.; Szymczak, J.; Przysowa, R. The Effect of a Dusty Environment Upon Performance and Operating Parameters of Aircraft Gas Turbine Engines. In Proceedings of the Specialists' Meeting—Impact of Volcanic Ash Clouds on Military Operations NATO AVT-272-RSM-047, Vilnius, Lithuania, 17 May 2017.
16. Wasilewski, M.; Brar, L.S.; Ligus, G. Effect of the central rod dimensions on the performance of cyclone separators—Optimization study. *Sep. Purif. Technol.* **2021**, *274*, 119020. [CrossRef]
17. Nejad, J.V.N.; Kheradmand, S. The effect of arrangement in multi-cyclone filters on performance and the uniformity of fluid and particle flow distribution. *Powder Technol.* **2022**, *399*, 117191. [CrossRef]
18. Acharya, A.S.; Kurian, J.; Lowe, K.T.; Ng, W.F. Spectral Proper Orthogonal Decomposition Downstream of a Vortex Tube Separator. In Proceedings of the AIAA SciTech Forum, National Harbor, MD, USA, Online, 23–27 January 2023; Available online: https://www.researchgate.net/publication/367311519 (accessed on 27 March 2023).
19. Fushimi, C.; Yato, K.; Sakai, M.; Kawano, T.; Kita, T. Recent Progress in Efficient Gas–Solid Cyclone Separators with a High Solids Loading for Large-scale Fluidized Beds. *KONA Powder Part. J.* **2021**, *38*, 94–109. [CrossRef]
20. Muschelknautz, U. Comparing Efficiency per Volume of Uniflow Cyclones and Standard Cyclones. *Chem. Ing. Tech.* **2020**, *93*, 91–107. [CrossRef]
21. Dziubak, T. Experimental Investigation of Possibilities to Improve Filtration Efficiency of Tangential Inlet Return Cyclones by Modification of Their Design. *Energies* **2022**, *15*, 3871. [CrossRef]
22. Dziubak, T.; Szwedkowicz, S. Operating properties of non-woven fabric panel filters for internal combustion engine inlet air in single and two-stage filtration systems. *Eksploat. I Niezawodn. Maint. Reliab.* **2015**, *17*, 519–527. [CrossRef]
23. Dziubak, T. Experimental Studies of PowerCore Filters and Pleated Filter Baffles. *Materials* **2022**, *15*, 7292. [CrossRef] [PubMed]
24. Wei, S.; Qian, F.; Cheng, J.; Xiao, P.; Tang, L.; Jiang, R. Flow field analysis and structure optimization of honeycomb air filter. *Chin. J. Process Eng.* **2019**, *19*, 271–278. [CrossRef]
25. PowerCore®Air Intake Filters. Available online: https://www.donaldson.com/en-us/engine/filters/products/air-intake/replacement-filters/powercore-filters/ (accessed on 27 March 2023).
26. Engine Air Cleaners, Service Parts and Accessories. Available online: https://www.donaldson.com/content/dam/donaldson/engine-hydraulics-bulk/catalogs/air-intake/emea/f116005/Air-Intake-Product-Guide.pdf (accessed on 27 March 2023).
27. Engine Air Filtration for Light, Medium, & Heavy Dust Conditions. Available online: https://www.donaldson.com/content/dam/donaldson/engine-hydraulics-bulk/catalogs/air-intake/north-america/F110027-ENG/Air-Intake-Systems-Product-Guide.pdf (accessed on 27 March 2023).
28. Yang, J.; Sun, G.; Gao, C. Effect of the inlet dimensions on the maximum-efficiency cyclone height. *Sep. Purif. Technol.* **2013**, *105*, 15–23. [CrossRef]
29. Duan, J.; Gao, S.; Lu, Y.; Wang, W.; Zhang, P.; Li, C. Study and optimization of flow field in a novel cyclone separator with inner cylinder. *Adv. Powder Technol.* **2020**, *31*, 4166–4179. [CrossRef]

30. Baltrėnas, P.; Chlebnikovas, A. Experimental research on the dynamics of airflow parameters in a six-channel cyclone-separator. *Powder Technol.* **2015**, *283*, 328–333. [CrossRef]
31. Jaroszczyk, T.; Ptak, T. Experimental Study of Aerosol Separation Using a Minicyclone. In Proceedings of the 10th Annual Powder and Bulk Solids Conference, O'Hare Expo Center, Rosemont, IL, USA, 7–9 May 1985.
32. Teng, G.; Shi, G.; Zhu, J. Influence of pleated geometry on the pressure drop of filters during dust loading process: Experimental and modelling study. *Sci. Rep.* **2022**, *12*, 20331. [CrossRef] [PubMed]
33. Allam, S.; Mimi Elsaid, A. Parametric Study on Vehicle Fuel Economy and Optimization Criteria of the Pleated Air Filter Designs to Improve the Performance of an I.C Diesel Engine: Experimental and CFD Approaches. *Sep. Purif. Technol.* **2020**, *241*, 116680. [CrossRef]
34. Saleh, A.M.; Tafreshi, H.V.; Pourdeyhimi, B. An analytical approach to predict pressure drop and collection efficiency of dust-load pleated filters. *Sep. Purif. Technol.* **2016**, *161*, 80–87. [CrossRef]
35. Saleh, A.M.; Vahedi Tafreshi, H. A simple semi-numerical model for designing pleated air filters under dust loading. *Sep. Purif. Technol.* **2014**, *137*, 94–108. [CrossRef]
36. Dziubak, T. Performance characteristics of air intake pleated panel filters for internal combustion engines in a two-stage configuration. *Aerosol Sci. Technol.* **2018**, *52*, 1293–1307. [CrossRef]
37. Seok, J.; Chun, K.M.; Song, S.; Lee, S. Study on the filtration behavior of a metal fiber filter as a function of filter pore size and fiber diameter. *J. Aerosol Sci.* **2015**, *81*, 47–61. [CrossRef]
38. Sun, Z.; Liang, Y.; He, W.; Jiang, F.; Song, Q.; Tang, M.; Wang, J. Filtration performance and loading capacity of nano-structured composite filter media for applications with high soot concentrations. *Sep. Purif. Technol.* **2019**, *221*, 175–182. [CrossRef]
39. Thomas, D.; Penicot, P.; Contala, P.; Leclerc, D.; Vendel, J. Clogging of fibrous filters by solid aerosol particles Experimental and modelling study. *Chem. Eng. Sci.* **2001**, *56*, 3549–3561. [CrossRef]
40. Walsh, D.C.; Stenhouse, J.I.T. The effect of particle size, charge, and composition on the loading characteristics of an electrically active fibrous filter material. *J. Aerosol Sci.* **1997**, *28*, 307–321. [CrossRef]
41. Bugli, N.J. *Service Life Expectations and Filtration Performance of Engine Air Cleaners*; No. 2000-01-3317; SAE Technical Paper; SAE: Warrendale, PA, USA, 2000.
42. Bugli, N. *Automotive Engine Air Cleaners—Performance Trends*; No. 2001-01-1356; SAE Technical Paper; SAE: Warrendale, PA, USA, 2001.
43. Durst, M.; Klein, G.; Moser, N. Filtration in fahrzeugen. In *Mannṭhummel GMBH*; Mann ṭ Hummel GMBH: Ludwigsburg, Germany, 2005.
44. Taufkirch, G.; Mayr, G. Papierluftfilter für Motoren in Nutzfahrzeugen. *MTZ* **1984**, *45*, 95–105.
45. Melzer, H.H.; Brox, W. Ansauggerauschdampfer und Luftfilter für BMW 524 td. *MTZ* **1984**, *45*, 223–227.
46. Dziubak, T.; Bąkała, L.; Dziubak, S.D.; Sybilski, K.; Tomaszewski, M. Experimental Research of Fibrous Materials for Two-Stage Filtration of the Intake Air of Internal Combustion Engines. *Materials* **2021**, *14*, 7166. [CrossRef] [PubMed]
47. Needelman, W.; Madhavan, P. *Review of Lubricant Contamination and Diesel Engine Wear*; SAE Technical Paper; SAE: Warrendale, PA, USA, 1988. [CrossRef]

Article

Computational Fluid Dynamics Simulation Approach for Scrubber Wash Water pH Modelling

Marian Ristea [1], Adrian Popa [2,*] and Ionut Cristian Scurtu [2]

[1]   Jalmare Oy, 20500 Turku, Finland; marian.ristea@jalmare.com
[2]   Mircea cel Batran Naval Academy, 900218 Constanta, Romania; ionut.scurtu@anmb.ro
*   Correspondence: adrian.popa@anmb.ro

Abstract: In the current article, we will use a CFD approach for the scrubber wash water dilution simulation, by considering the current MEPC (Marine Environment Protection Committee, a subsidiary of IMO—International Maritime Organization) regulations that are in force. The necessity for scrubber wash water pH modelling and its importance in the current environmental framework is emphasized. The presented 3D model is considered as a 400 mm hydraulic diameter fluid domain with two outlets and a discharge water flow rate of 3050 m^3/h for the considered pH value of 3, obtained within a state-of-the-art exhaust gas scrubber solution developed by a major EGCS (Exhaust Gas Cleaning Systems) supplier. The CFD study was developed by considering a k-ε turbulence model. In order to achieve accurate results, a structured mesh with two levels of refinement volumes was realized. Based on the obtained data and the various parameters discussed, the paper presents a way to investigate the optimal results for further analytical research of the scrubber washwater dilution process within the exhaust gas cleaning system.

Keywords: CFD; scrubber; wash water; pH modelling

Citation: Ristea, M.; Popa, A.; Scurtu, I.C. Computational Fluid Dynamics Simulation Approach for Scrubber Wash Water pH Modelling. *Energies* **2022**, *15*, 5140. https://doi.org/10.3390/en15145140

Academic Editors: Tadeusz Dziubak, Karczewski Mirosław and Piotr Wróblewski

Received: 29 May 2022
Accepted: 13 July 2022
Published: 15 July 2022

Publisher's Note: MDPI stays neutral with regard to jurisdictional claims in published maps and institutional affiliations.

1. Introduction

Within the shipping industry, large quantities of fossil fuels are burned by the ship's diesel engines, with the exhaust gases having carbon oxides (CO_x) and water (H_2O) as the main components. Together with the main fractions [1–3], the combustion process also generates sulfur oxides (SO_x), nitrogen oxides (NO_x), and carbon-based matter (soot, smoke), all of them with huge environmental impact, such as acid rain and carbon-based airborne particles, which are detrimental to human health.

Based on the real global concern about environmental issues, determined by the exhaust gas emissions and their impact, there is a huge interest in developing technical solutions for reducing the level of pollution [4–7].

Therefore, for both new builds and existing ships, a fitting/retrofitting race is ongoing—increasingly ships are using various solutions for cleaning the exhaust gases. A scrubber technology was developed with unique features to enable a more sustainable operating environment for the shipping industry [8].

The main objective of this study is to evaluate the open-loop solution, with an emphasis on the aspects regulated within MEPC 259(68).

The open-loop cleaning process is based on exhaust gases "washing" with seawater, thus resulting in large quantities of residuals—sulfuric acid (H_2SO_4) or sulfurous acid (H_2SO_3) diluted in the wash water. The obtained product is seawater with increased acidity, which is to be discharged overboard (either treated in a second stage or diluted).

2. Study Aim, Materials and Methods

2.1. Marine Environment and Seawater Alkalinity

The wash water with low pH values reacts with the salt in the seawater, forming carbonic acid, which is considered to be unharmful for the marine environment. The main source of hydrogen ions (H$^+$) absorbents in the seawater is formed by the carbonate and bicarbonate salts naturally present [8–10]. The natural presence of the salts generates an alkaline seawater. The general values of seawater alkalinity around the globe range from 2200–2400 µmol/kg.

Nevertheless, diluted sulfuric acid from the wash water is a threat to the marine environment and, therefore, additional rules are required in order to create the lowest possible environmental impact due to ship wash water discharge.

Guidelines for exhaust gas cleaning systems, applicable for this study, are published in the Resolution MEPC.259(68) in Annex 1 and are stated as follows: 10.1.2.1. "The pH discharge limit, at the overboard monitoring position, is the value that will achieve as a minimum pH 6.5 at 4 m from the overboard discharge point with the ship stationary".

2.2. Main Parameters and Initial Setup Geometry

The study assesses the compliance with MEPC 259(68) paragraph 10.1.2.1 and relevant DNV-GL (Det Norske Veritas) and BV (Bureau Veritas) rules and regulations for the ship wash water discharge.

Based on existing references, a computational model was created that takes into account the existing rules and the specific functional status of the ship's propulsion and power generation systems [11,12].

The starting point for the initial setup geometry is presented in Figure 1. For the simulation in CFD, all elements designed for enhancing the flow turbulence were removed and the scrubber wash water inlet surface was assigned to be at 1 m distance from the outlet piping elbow.

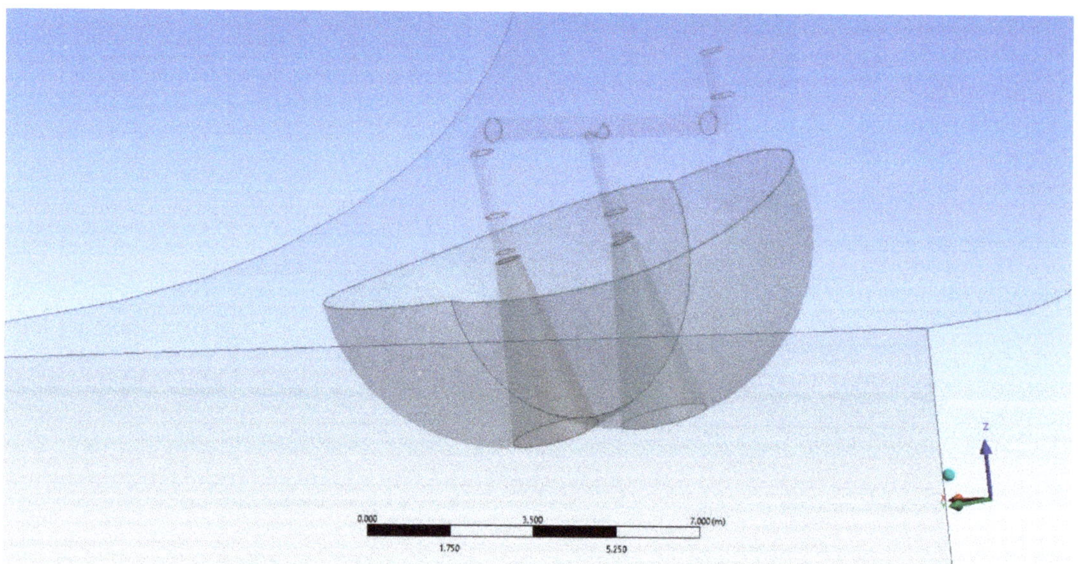

Figure 1. Spherical volumes in CFD simulation.

In this way, the simulation is developed with a conservative approach, starting with the geometry setup, by decreasing the turbulence values affected by the existing valves or piping elements. This objective, of achieving the most conservative approach, is accomplished by decreasing the dilution quality obtained when using such turbulence-enhancing attachments.

The simulation setup is developed using the following initial operational setup:

- Seawater temperature: 25 °C;
- Discharge water temperature: 35 °C;

- Discharge water flow rate: 3050 m³/h—considered to be the flow at full load while the ship is stationary;
- Number of outlets: 2;
- Discharge water pH: 3;
- Considered ship speed: 0 kn (stationary).

To assess the dilution pattern in the initial setup, the hull geometry of a ship that actually has the system onboard is used [13,14]. To reduce the influence of turbulence in the numerical analysis, all diffusers and turbulence-increasing elements were removed (for example, all valves on the terminal line of the scrubber wash water outlet were disregarded).

For a proper assessment of the dilution pattern, at the required 4 m distance from the outlets, there were used two spherical volumes, with a radius of 4 m and the center in the outlet center (Figure 1). In the same figure, we can observe the intersection of the two spherical volumes with the ship hull and the position of the scrubber outlet shell penetrations [15,16].

The outlet location is identified based on the penetration plan, as presented in Figure 2.

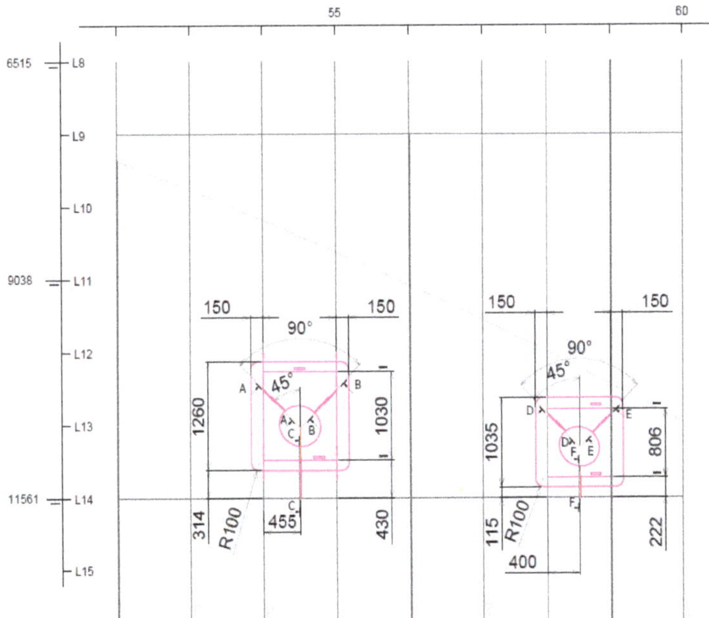

Figure 2. Shell penetration plan for scrubber outlet.

The generated geometry (Figure 1) follows the position and dimensions stated in the construction drawings, available during the retrofit project of the ship which was the subject of certification (Figure 2).

The bounding box of the computational domain for scrubber wash water pH modelling has the following dimensions: Height = 20 m, Width = 35 m, Length = 60 m.

The origin of the coordinate system is located on the ship hull surface, the Ox axis positive toward fore, the Oy axis positive toward portside, and the Oz axis positive upwards [17]. For boundaries definition, the details identified in the initial setup geometry are as shown in Figure 3.

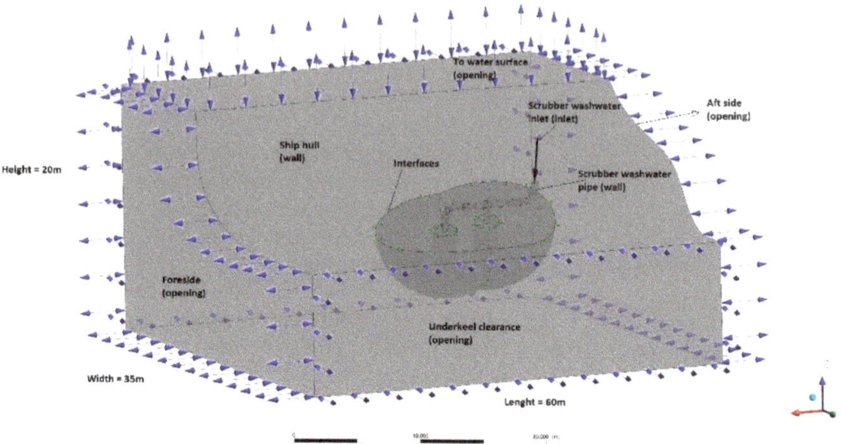

Figure 3. Boundaries of the computational domain.

The computational domain was simplified as shown in Figure 3, by taking into account the following aspects:

1. The need to decrease the computational effort leads to the model being generated to consider only the hull section in the area of interest.
2. Since the main requirement is to have no flow around the hull, without any current alongside [13–17], it is reasonable to assume that there will also be no wash water flow in the hull proximity.

2.3. Domain Boundaries

Based on Figure 3, the boundaries used are [18,19]:

— opening, for the water volume situated on the fore and aft side of the ship, for the outer water domain facing from ship shell and to the water surface, and also under the keel area; this boundary condition allows the fluid to cross the boundary surface in either direction. For example, all of the fluid might flow into the domain at the opening, or all of the fluid might flow out of the domain, or a mixture of the two might occur.
— inlet, only for the pipe surface situated before the two openings in the side shell; for the subsonic inlet, the magnitude of the inlet velocity is specified, and the direction is taken to be normal to the boundary. The direction constraint requires that the flow direction, D_i, is parallel to the boundary surface normal, which is calculated at each element face on the inlet boundary.

The boundary velocity components are specified, with a non-zero resultant into the domain.

$$U_{inlet} = U_{spec}i + V_{spec}j + W_{spec}k \tag{1}$$

The mass flow rate, defined within the simulation, is considered to be uniform and normal to the boundary. The mass influx [20] will be calculated by using:

$$\rho U = \frac{\dot{m}}{\iint dA} \tag{2}$$

where $\iint dA$ is the integrated boundary surface area at a given mesh resolution.

— wall, for the ship shell and pipe walls; there was considered a No Slip wall, with the velocity of the fluid at the wall boundary of 0.

2.4. Mesh Details

In order to achieve appropriate results, a structured mesh was used based on the hex-dominant method. Within the ship shell volume of influence, the maximum length for elements was considered to be 0.3 m for all points at a 4 m distance from the center of the spherical volumes. In the center of the spherical volumes, we designed a mesh based on refinement cones using a maximum element length of 0.04 m. This is considered to be the main area of interest and the decision for these local refinements is made for simulation accuracy. The described mesh with refinement cones is presented in Figure 4. The metrics for the current mesh are presented in Table 1. The selected mesh for 3D scrubber wash water discharge pH modelling has near 1 million elements, contributing to a precise result and a faster computational process for the identified initial conditions.

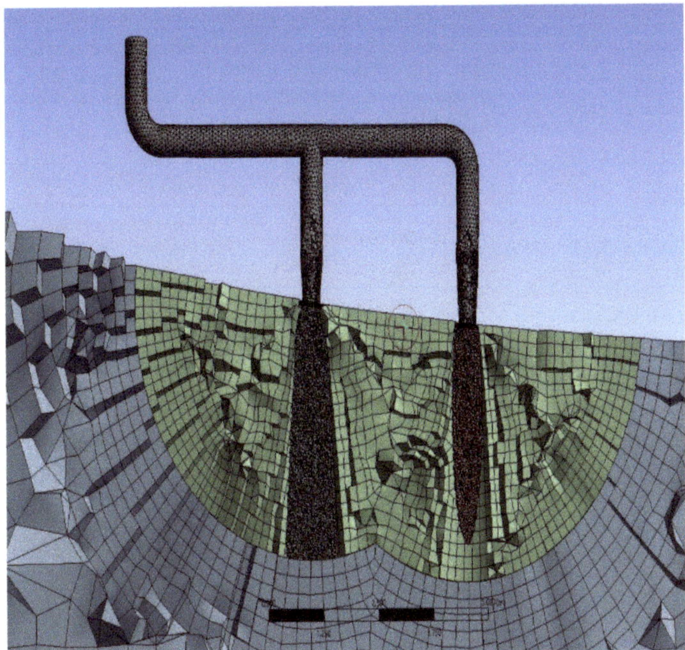

Figure 4. Computational domain mesh with refinement cones.

Table 1. Data for the considered mesh with refinement cones.

Domain	Nodes	Elements	Tetrahedra	Hexaedra
D 400 LC	1684594	925743	774419	129569

3. Simulation Model Description

3.1. Turbulence Model

The flow within the computational domain is a fully turbulent flow and, given the simulation objectives, the near-wall flow prediction is not necessary. Based on these assumptions, the standard k–ε; turbulence model was used. The model is a semi-empirical one [20,21]; the rate of dissipation (ε) and turbulence kinetic energy (k) are based on the below transport equations [20]:

$$\frac{\partial}{\partial t}(\rho k) + \frac{\partial}{\partial x_i}(\rho k u_i) = \frac{\partial}{\partial x_j}\left[\left(\mu + \frac{\mu}{\sigma_k}\right)\frac{\partial k}{\partial x_j}\right] + G_k + G_b - \rho\varepsilon - Y_M + S_k \tag{3}$$

$$\frac{\partial}{\partial t}(\rho\varepsilon) + \frac{\partial}{\partial x_i}(\rho\varepsilon u_i) = \frac{\partial}{\partial x_j}\left[\left(\mu + \frac{\mu}{\sigma_\varepsilon}\right)\frac{\partial\varepsilon}{\partial x_j}\right] + C_{1\varepsilon}\frac{\varepsilon}{k}(G_k + C_{3\varepsilon}G_b) + C_{2\varepsilon}\frac{\varepsilon^2}{k} + S_\varepsilon \qquad (4)$$

where G_k represents the generation of turbulence kinetic energy due to the mean velocity gradients, G_b is the generation of turbulence kinetic energy due to buoyancy, Y_M represents the contribution of the fluctuating dilatation in compressible turbulence to the overall dissipation rate, $C_{1\varepsilon}$, $C_{2\varepsilon}$, and $C_{3\varepsilon}$ are constants, σ_k and σ_ε are the turbulent Prandtl numbers for k and ε, while S_k and S_ε are user-defined terms.

3.2. Simulation Sensitivity and Stability

In order to identify a proper way to investigate the scrubber wash water dilution process, we will consider two factors that are affecting the simulation in a 3D computational domain: the mesh sensitivity and the solution stability. To assess the mesh influence in the current simulation, three different cases are considered: the initial mesh, the rougher mesh, and the refined mesh. For each case, the mesh metrics are presented in Table 2. The variation of maximum element length in the area of interest (0.08 m, 0.04 m, and 0.03 m). The results are presented in Figures 5–7.

Table 2. Mesh metrics for the initial mesh, the rougher mesh, and the refined mesh.

Mesh Case	Max Element Length throughout Domain	Probe Volume	Max Element Length in the Interest Area	Nodes	Elements
Initial mesh	0.7 m	0.3 m	0.04 m	1684594	925743
Rougher mesh	0.7 m	0.3 m	0.06 m	1223629	620681
Refined mesh	0.7 m	0.18 m	0.03 m	2072666	1188839

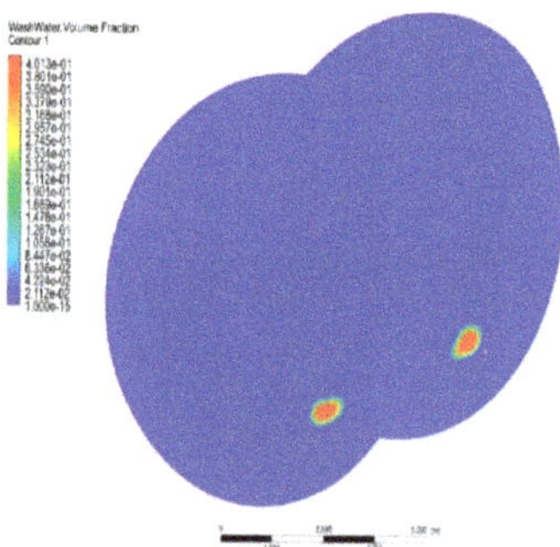

Figure 5. Maximum Volume Fraction values for the Scrubber Wash Water (initial mesh).

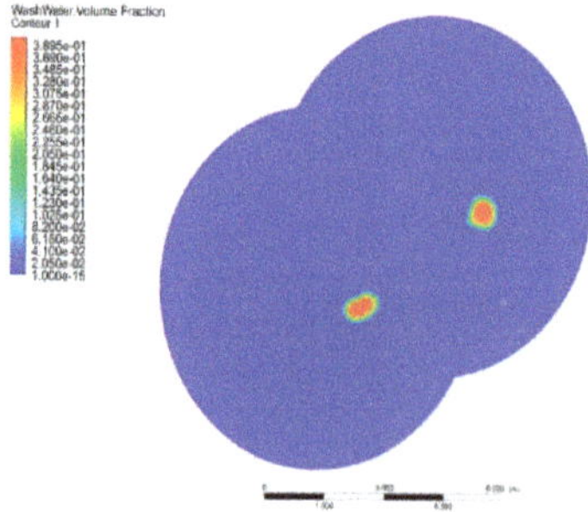

Figure 6. Maximum Volume Fraction values for the Scrubber Wash Water (rough mesh).

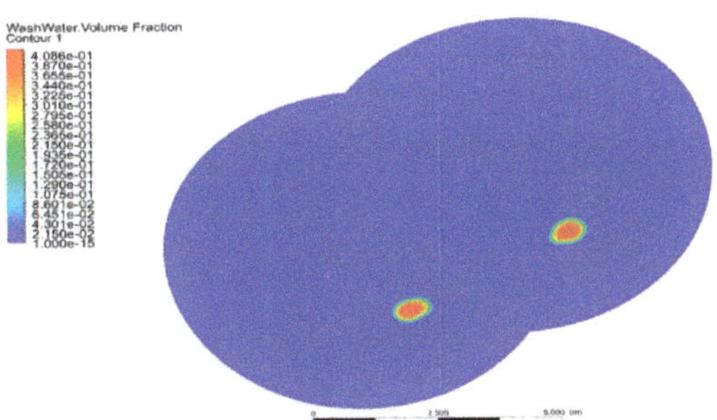

Figure 7. Maximum Volume Fraction values for the Scrubber Wash Water (refined mesh).

For all considered mesh developments, the solution was considered to be converging after the first 1000 iterations, and the residual values development within the carried iterations was clearly stable. The results for scrubber wash water pH modelling are stable and present similar shape; the obtained results values are not highly dependent on the mesh refinement level.

For a better comparison and understanding of the simulation results, we used the peak volume fraction results for the area of interest (4 m radius probe sphere). The values obtained in the presented cases are shown in Table 3. The achieved differences are small, established in a range of max 3% variation, related to the initial mesh setting, which leads to the conclusion that the solution is stable and can also be considered acceptable.

Table 3. Peak volume fraction for each case.

Mesh Case	Peak Volume Fraction
Initial mesh	0.4013
Rougher mesh	0.3895
Refined mesh	0.4086

The results from the variation of maximum element length in the area of interest (0.08 m, 0.04 m, and 0.03 m) bring us to the conclusion that the initial used mesh meets the required conditions for consistency and stability of the results based on the stability of the results and on the convergence trend and achievement. The results are less sensitive to mesh metrics in the range presented, and each case can be a way to investigate the optimal results for further analytical research. Due to this conclusion and the lower number of elements, the initial setup will be used in the 3D scrubber wash water dilution study.

3.3. Scrubber Wash Water pH Development Study

The usage of the initial mesh, with a 400 mm hydraulic diameter for the two outlets, as stated in the initial setup, represents a difficult computational task. Results in the simulation run should meet the stability and convergence criteria. The convergence criteria for the residuals value were set at 10^{-9} and required 6500 iterations to obtain a solution.

The considered load case was developed using a 400 mm hydraulic diameter outlet and uses the highest operational value of the wash water flow (3050 m^3/h), according to MEPC.259(68) and IMO's PPR 2/2/3. This evaluates the dilution ratio at a distance of 4 m from the hull in any direction during stationary operation. Data for this load case are presented in Table 4.

Table 4. Load case parameters.

Ship Speed [kn]	ME Load [%]	Power Plant Load [%]	Wash Water Flow [m^3/h]	Wash Water Mass Flow [kg/s]	Wash Water Velocity [m/s]	Hydraulic Diameter [mm]
0	85	75	2600 + 450	868.4	4.31	400

4. Results

The simulation results for velocity, pressure, and wash water volume fraction are presented in Figures 8–20. During the post-processing process, we use isosurfaces to show the results at the MEPC's required distance from the outlet.

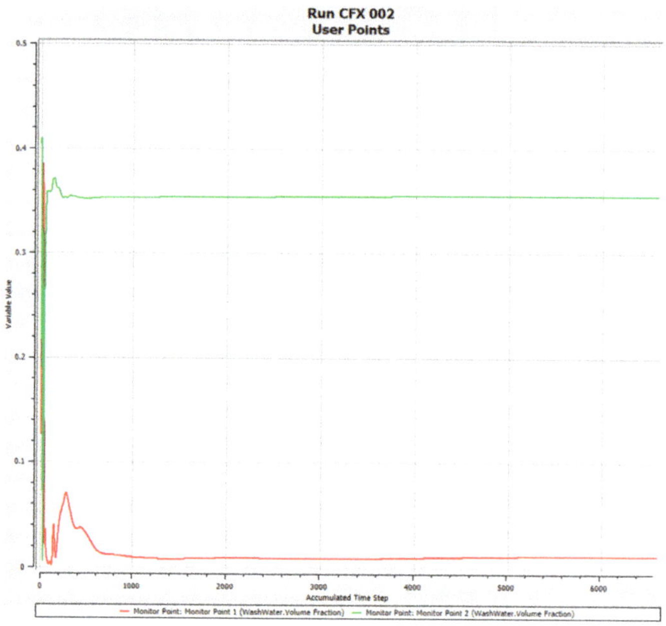

Figure 8. Volume fraction plot for convergence check (initial mesh).

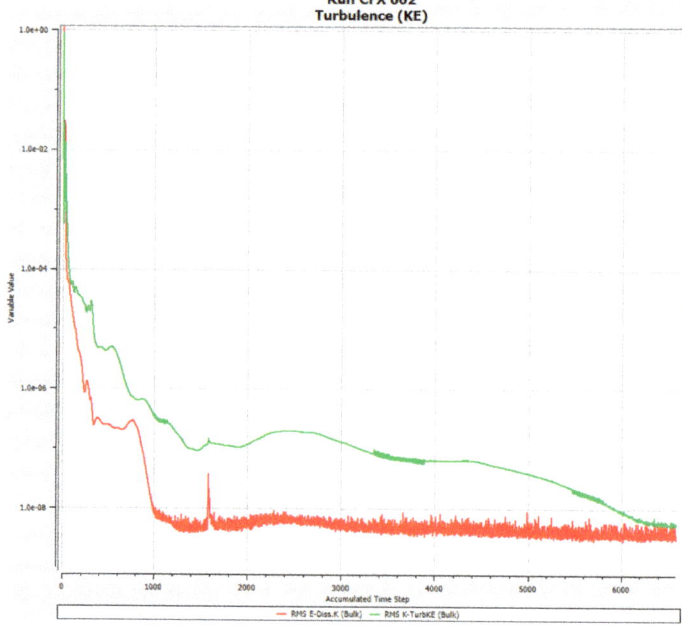

Figure 9. RMS residuals for k and ε values (initial mesh).

Figure 10. Velocity variation—pipe longitudinal section.

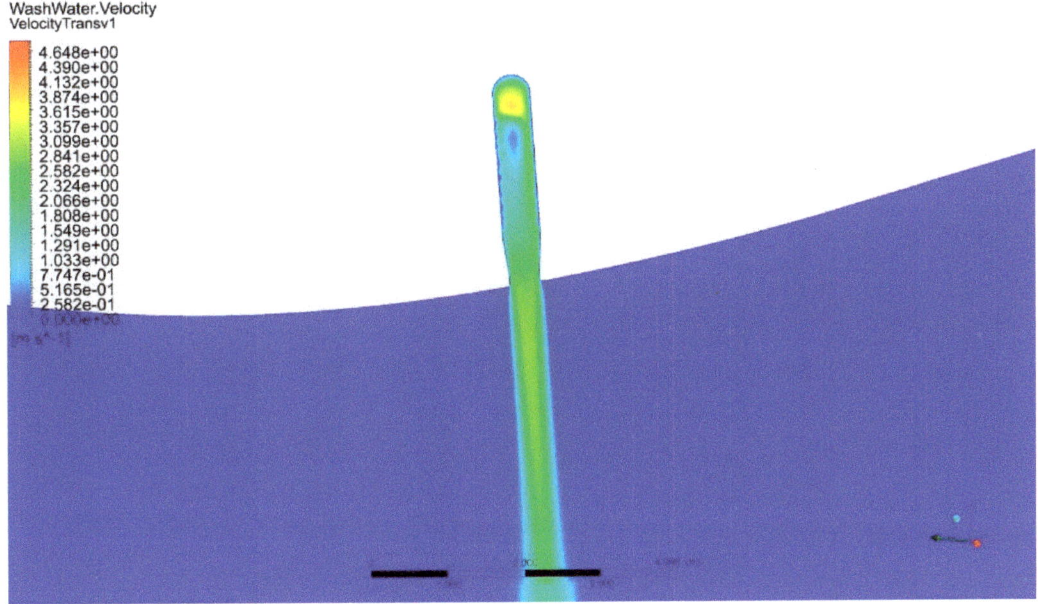

Figure 11. Velocity variation—pipe transversal section.

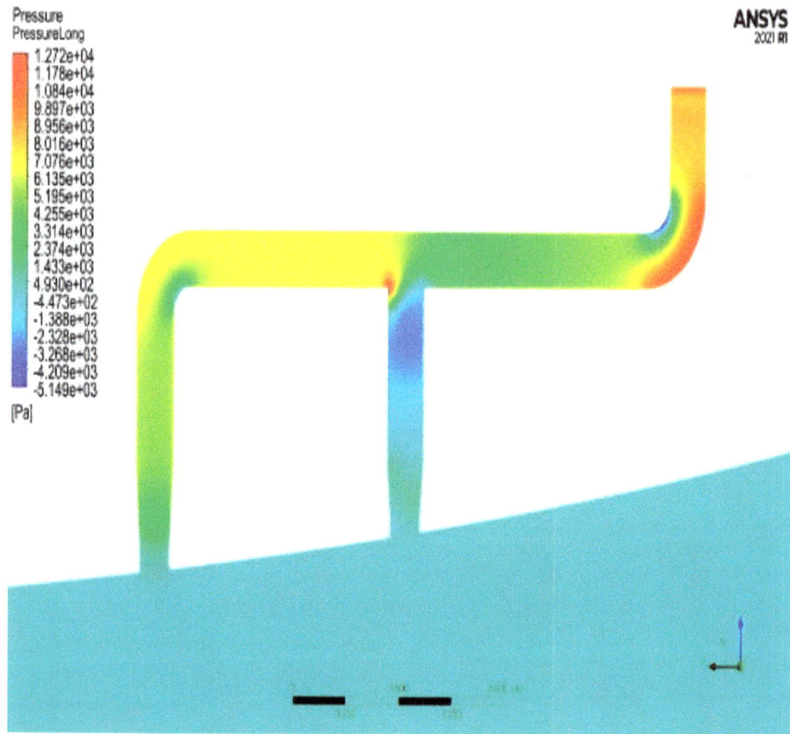

Figure 12. Pressure variation—pipe longitudinal section.

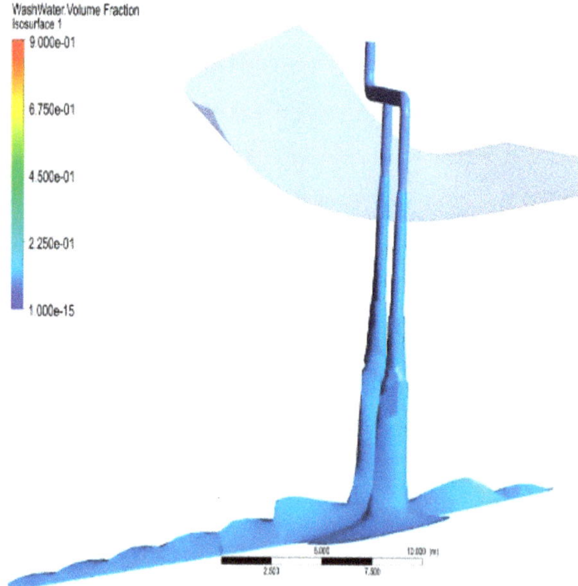

Figure 13. Scrubber Wash Water Volume Fraction of 0.1 isosurface development.

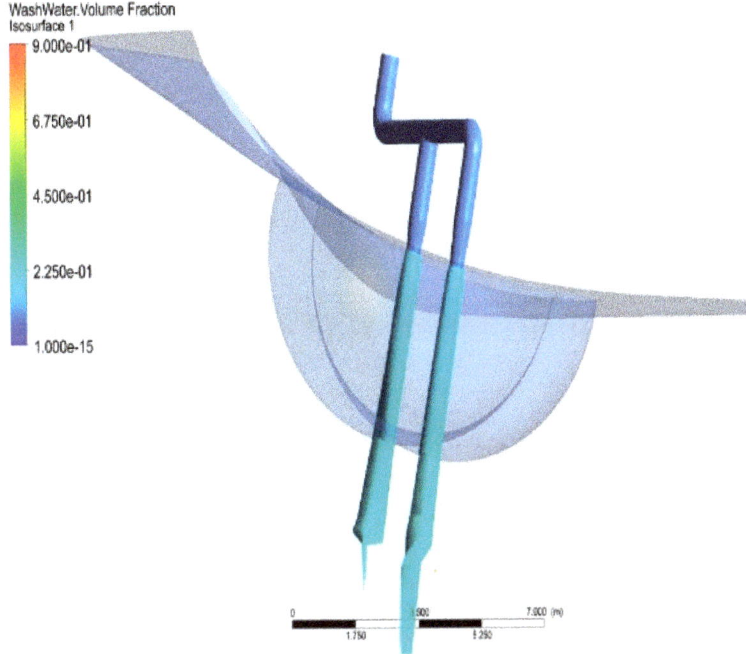

Figure 14. Scrubber Wash Water Volume Fraction of 0.2 isosurface development.

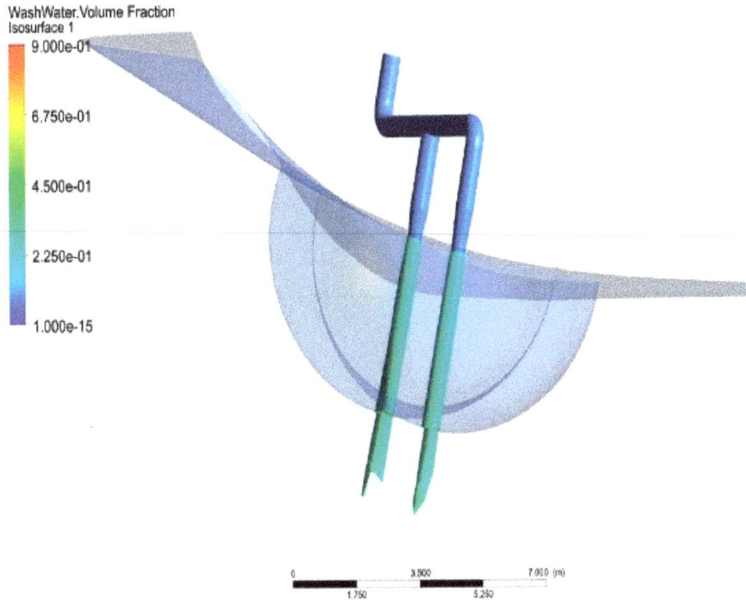

Figure 15. Scrubber Wash Water Volume Fraction of 0.3 isosurface development.

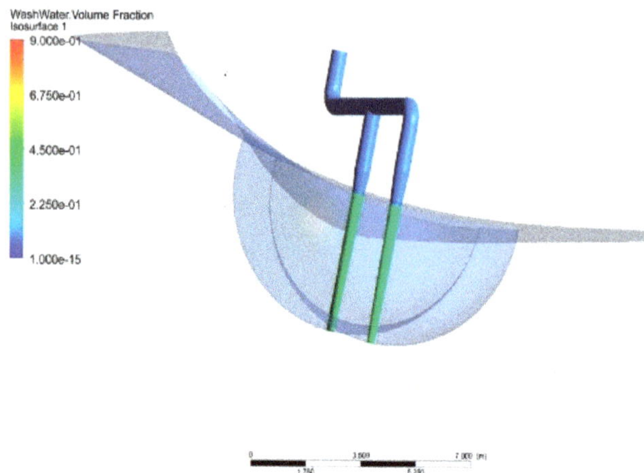

Figure 16. Scrubber Wash Water Volume Fraction of 0.4 isosurface development.

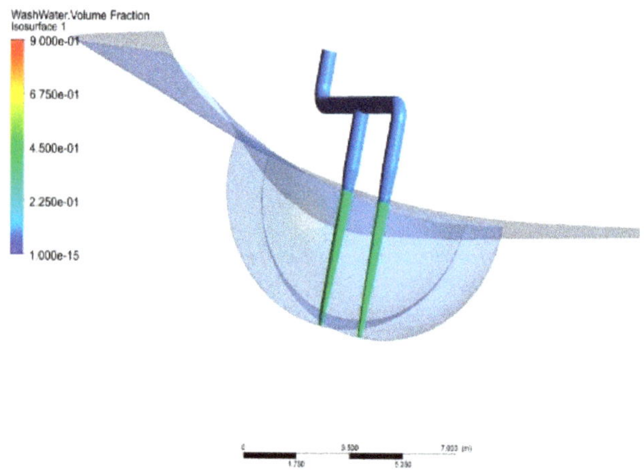

Figure 17. Scrubber Wash Water Volume Fraction of 0.41 isosurface development.

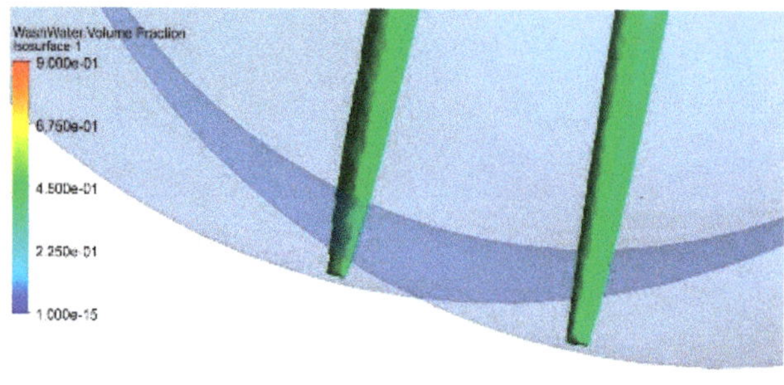

Figure 18. Scrubber Wash Water Volume Fraction of 0.41 isosurface development—probe sphere detail.

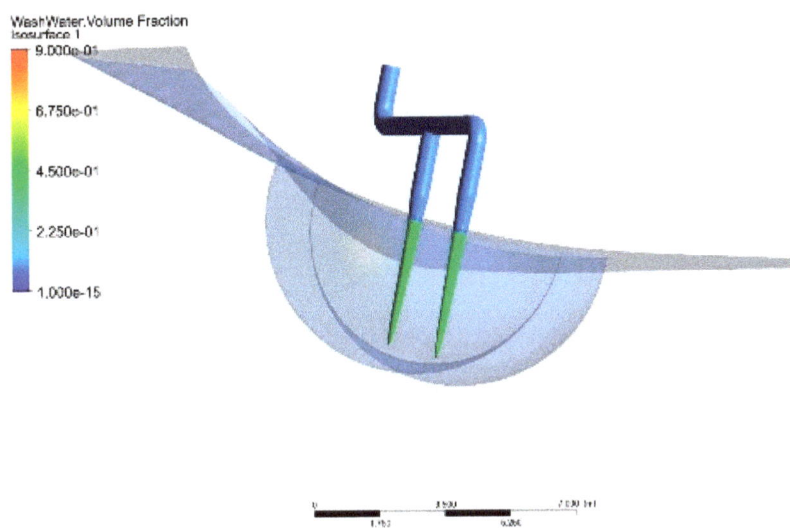

Figure 19. Scrubber Wash Water Volume Fraction of 0.42 isosurface development.

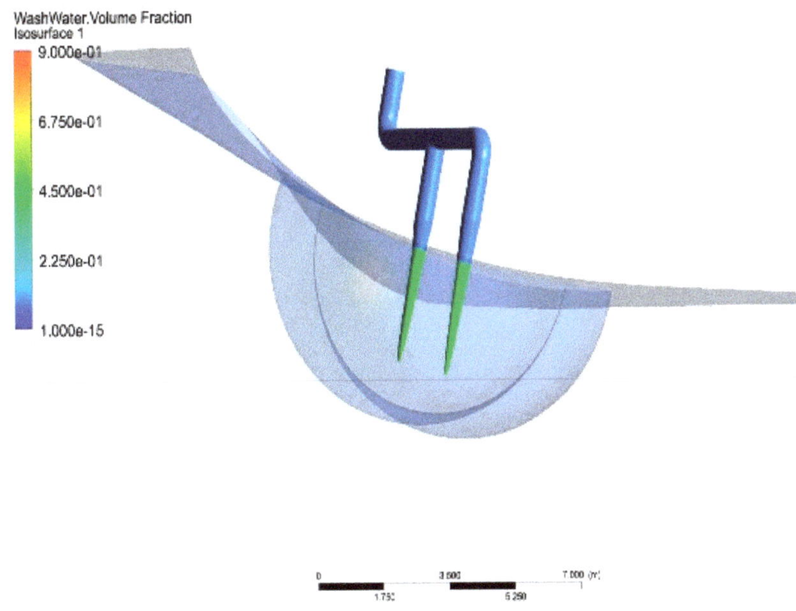

Figure 20. Scrubber Wash Water Volume Fraction of 0.43 isosurface development.

In Figures 10–12, the pressure and velocity values in the longitudinal and transversal sections of the pipe are shown, and they can be considered as an additional check regarding the validity of the computational setup, taking into account the discharge water flow rate of 3050 m^3/h and the pipe outlet hydraulic diameter. Figures 13–22 show the isosurface representations for relevant scrubber wash water volume fractions.

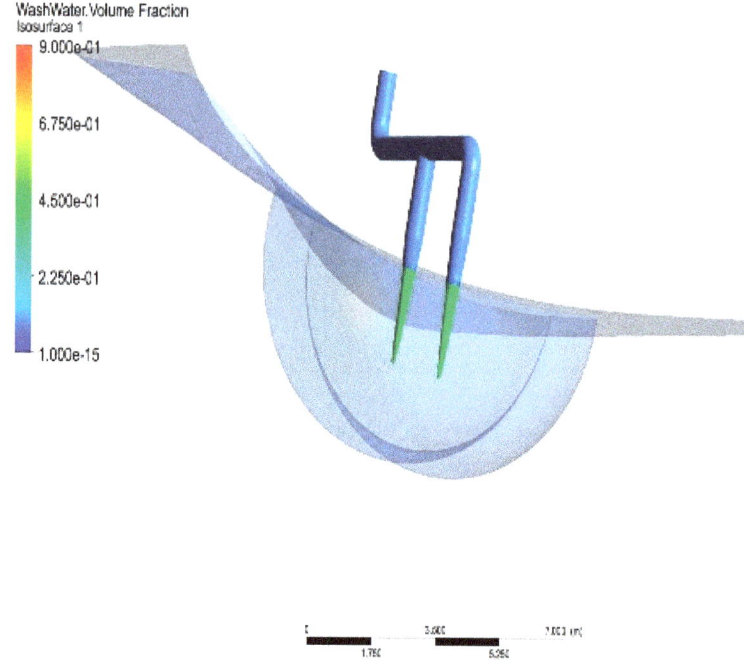

Figure 21. Scrubber Wash Water Volume Fraction of 0.44 isosurface development.

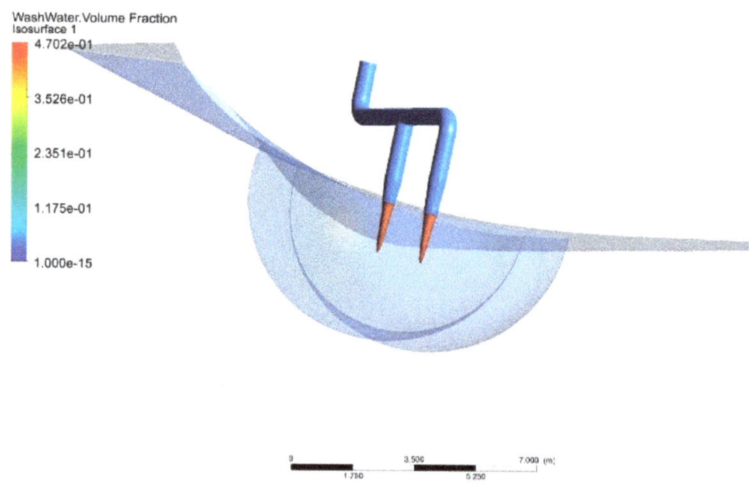

Figure 22. Scrubber Wash Water Volume Fraction of 0.45 isosurface development.

5. Discussion

As stated within the introductory section of this article, based on the existing rules and regulations [13,16,17] and taking into account the available findings in the theoretical and practical studies and results [4,12], the maximum acceptable volume fraction of the scrubber wash water flow, at 4 m distance from the outlets, is deemed to be 0.47, for which the pH value of the diluted wash water is a minimum of 6.5.

Based on the initial mesh for the geometry used, specific results were obtained and can be considered as a starting point for further discussions and studies regarding the optimal values for the scrubber wash water dilution process and related requirements. A comparative analysis based on the obtained results is made in terms of dilution for the considered load case.

The numerical results presented for the scrubber wash water pipes in terms of velocity (Figures 10 and 11) and pressure (Figure 12) highlight the flow parameters, taking into account the possible vibrations that can appear at both pipe level and hull opening connection. The higher the velocity, the higher the vibration level, especially for strainers, valves, and elbows.

The flow modelling results contains basic information for ship wash water investigated according to applicable international standards, in compliance with MEPC 259(68).

During operation of the scrubber, wash water is fed from two hull openings (corresponding with the wash water outlet) and the result is a turbulent jet behavior that can be visualized and processed via the simulation results as a scalar scale or isosurface. The most relevant isosurface shown in the results is the one based on Volume Fraction (VF) because the scrubber wash water and seawater are modelled as Volume Fraction models.

As a measure of quality check, we use unity check, and the resulting value is 0.853, which confirms the results within the acceptable results parameter range. The wash water concentrations obtained were checked against the acceptable values and centralized in Table 5.

Table 5. The centralized values for wash water dilution.

Model Particulars	Maximum Volume Fraction	Acceptable Volume Fraction	Unity Check
400 mm hydraulic diameter	0.4013	0.47	0.853

Based on the results obtained, it can be observed (Figures 17–19) that the maximum Volume Fraction value for the scrubber wash water is 0.4013 on the 4 m radius control sphere. A unity check value of 0.85 can be observed, which is deemed to be within acceptable limits (by considering a conservative approach). The main interest for the flow development is the dilution ratio achieved against the required value in order to meet the approval and certification criteria.

The present simulation can be considered as developed based on an ultraconservative approach, obtained in this case by the applied simplifications, e.g., the removal of the turbulence enhancing appendages and using a No Slip wall, without any material roughness, as boundaries, but also taking into consideration the real operational situation, which is further developed in the paragraph below.

The MEPC rule, as interpreted both in the subsequent guidelines and also in the simulation setup, requires two totally opposite operational conditions to be met: on the one hand, the flow around the ship hull is null, therefore, the ship is considered to be static, without any current in the area; on the other hand, the load of the scrubber system must be considered as the nominal load, which corresponds to an ME load of a minimum of 85%, a power plant load of a minimum of 75% (see Table 4), and the nominal wash water flow of 3500 m^3/h. This particular situation will never be found in real life; therefore, we can easily consider that the operational situation for a static ship is closer with a 75% load of the power plant and 0% load for ME, resulting in a substantially decreased exhaust gas flow, "washed" with the nominal scrubber wash water flow, of 3500 m^3/h. The natural assumption, based on these facts, is that the real initial value of the scrubber wash water pH at the ship's openings will be higher than the value already considered within the simulation setup (of 3), and in such a situation, the minimum acceptable wash water Volume Fraction values are, in a real operational situation, more relaxed than the ones used in the present approach.

6. Conclusions

The 3D model and the computational domain used for developing the scrubber wash water dilution modelling are adopted based on a real ship hull design, taking into consideration the existing IMO rules which are regulating the process.

The increased number of retrofitting projects aiming to decrease the exhaust gas emissions SO_x content in the shipbuilding industry through the use of exhaust gas scrubbers comes with various side challenges, such as preserving the overall water quality during the scrubber operation and not affecting the water pH (or achieving the smallest possible impact). These requirements are enforced by international bodies such as IMO (International Maritime Organization) via MEPC resolutions, and they are monitored or certified by Class Registers. Based on the aforementioned situation, the requirements for carrying out specific dilution studies through the use of CFD analysis are, sometimes, generating a need for determining good practice in carrying out the simulations, and also for creating a benchmarking database for the simulation approach, technical solution pursued, and the results obtained.

The existing results and simulation setup, corroborated with the real operational situation, leads to a possible conclusion that the actual rules and their interpretation are forcing an ultraconservative way of developing the related studies and the calculation of the allowable values in the area of interest. Nevertheless, a database with experimental results in the 4 m distance range, together with the functional parameters of the system in various cases are required to defend this affirmation and can represent future research related to this topic.

Future scrubber wash water research will be directed both to the simulation using various flows and velocities for the washwater, and also regarding the possible influence of acidic water on other inlets from other systems which are placed in the vicinity of the scrubber wash water outlet.

Author Contributions: Conceptualization, M.R.; Formal analysis, M.R. and A.P.; Methodology, M.R. and A.P.; Project administration, M.R.; Resources, M.R., I.C.S. and A.P.; Software, M.R. and A.P.; Supervision, M.R. and A.P. All authors have read and agreed to the published version of the manuscript.

Funding: This research was conducted by Jalmare Oy, during several new build and retrofitting projects, where dilution process modelling was required for certification purposes.

Institutional Review Board Statement: Not applicable.

Informed Consent Statement: Not applicable.

Conflicts of Interest: The authors declare no conflict of interest.

References

1. Atkins, P.W.; De Paula, J. *Physical Chemistry*; Oxford University Press: Oxford, UK, 2006.
2. Batchelor, G.K. *An Introduction to FLUID Dynamics*; Cambridge University Press: Cambridge, UK, 2001.
3. Bronsted, J.N. Some remarks on the concept of acids and bases. *Recl. Trav. Chim. Des Pays-Bas* **1923**, *42*, 718–728. Available online: https://www.chemteam.info/Chem-History/Bronsted-Article.html (accessed on 6 January 2022).
4. Choi, Y.S.; Lim, T.W. Numerical Simulation and Validation in Scrubber Wash Water Discharge from Ships. *J. Mar. Sci. Eng.* **2020**, *8*, 272. [CrossRef]
5. Renforth, P.; Henderson, G. Assessing Ocean alkalinity for carbon sequestration. *Rev. Geophys.* **2017**, *55*, 636–674. [CrossRef]
6. Flagiello, D.; Esposito, M.; Di Natale, F.; Salo, K. A Novel Approach to Reduce the Environmental Footprint of Maritime Shipping. *J. Marine. Sci. Appl.* **2021**, *20*, 229–247. [CrossRef]
7. Qureshi, M.A.A.; Tahir, R.M. The Implementation of Water Salinity Tester as a quantitative tool to determine the Total Dissolved Solids in water. *Tech. Rom. J. Appl. Sci. Technol.* **2021**, *3*, 1–8.
8. Available online: www.valmet.com (accessed on 28 March 2021).
9. Teuchies, J.; Cox, T.J.S.; Van Itterbeeck, K.; Meysman, F.J.; Blust, R. The impact of scrubber discharge on the water quality in estuaries and ports. *Environ. Sci. Eur.* **2020**, *32*, 103. [CrossRef]
10. Khan, Z.H. Relation Among Temperature, Salinity, pH and DO of Seawater Quality. *Tech. Rom. J. Appl. Sci. Techn.* **2020**, *2*, 39–45. [CrossRef]
11. Brown, K.; Kalata, W.; Schick, R. Optimization of SO2 Scrubber using CFD Modelling. In Proceedings of the "SYMPHOS 2013", 2nd International Symposium on Innovation and Technology in the Phosphate Industry, Agadir, Morocco, 6–10 May 2013.
12. Ulpre, H. Turbulent Acidic Discharges into Seawater. Ph.D. Thesis, University College London, London, UK, 2015.
13. Marine Environment Protection Committee. *Resolution MEPC.259 (68)—Guidelines for Exhaust Gas Cleaning Systems*; Marine Environment Protection Committee: London, UK, 2015.

14. Bureau Veritas—Rules for the Classification of Steel Ships—PART C and PART F—Machinery, Electricity, Automation and Fire Protection. Available online: https://marine-offshore.bureauveritas.com/nr467-rules-classification-steel-ships (accessed on 6 January 2022).
15. DNV-GL. *Rules for Classification, Part 4—Systems and Components*; DNV-GL: Høvik, Norway, 2018.
16. DNV-GL. *pH Discharge Criteria for Exhaust Gas Cleaning Systems according to MEPC.259(68)*; DNV-GL: Høvik, Norway, 2015.
17. International Maritime Organization—Sub-committee on Pollution Prevention and Response (PPR 2/2/3). Application of Calculation-Based Methodology for Verification of the Washwater Discharge Criteria for pH in Section 10.1.2.1(ii) of the 2009 Guidelines for Exhaust Gas Cleaning Systems. Available online: https://edocs.imo.org/Processing/English/PPR (accessed on 13 November 2014).
18. Ansys Inc. Ansys CFX Solver Theory Guide. *Ansys CFX Release* **2009**, *15317*, 724–746.
19. Stachnik, M.; Jakubowski, M. Multiphase model of flow and separation phases in a whirlpool: Advanced simulation and phenomena visualization approach. *J. Food Eng.* **2020**, *274*, 109846. [CrossRef]
20. Le, T.C.; Hong, G.H.; Lin, G.Y.; Li, Z.; Pui, D.Y.; Liou, Y.L.; Wang, B.-S.; Tsai, C.-J. Honeycomb wet scrubber for acidic gas control: Modelling and long-term test results. *Sustain. Environ. Res.* **2021**, *31*, 21. [CrossRef]
21. Dendy Adanta, I.M.; Rizwanul, F.; Nura Muaz, M. Comparison of standard k-ε and SST k-ω turbulence model for breastshot waterwheel simulation. *J. Mech. Sci. Eng.* **2020**, *7*, 39–44. [CrossRef]

MDPI

St. Alban-Anlage 66

4052 Basel

Switzerland

www.mdpi.com

Energies Editorial Office

E-mail: energies@mdpi.com

www.mdpi.com/journal/energies